The Aqueous Chemistry of the Elements

The Aqueous Chemistry of the Elements

George K. Schweitzer

Lester L. Pesterfield

OXFORD
UNIVERSITY PRESS

2010

OXFORD
UNIVERSITY PRESS

Oxford University Press, Inc., publishes works that further
Oxford University's objective of excellence
in research, scholarship, and education.

Oxford   New York
Auckland   Cape Town   Dar es Salaam   Hong Kong   Karachi
Kuala Lumpur   Madrid   Melbourne   Mexico City   Nairobi
New Delhi   Shanghai   Taipei   Toronto

With offices in
Argentina   Austria   Brazil   Chile   Czech Republic   France   Greece
Guatemala   Hungary   Italy   Japan   Poland   Portugal   Singapore
South Korea   Switzerland   Thailand   Turkey   Ukraine   Vietnam

Published by Oxford University Press, Inc.
198 Madison Avenue, New York, New York 10016

www.oup.com

Oxford is a registered trademark of Oxford University Press

Library of Congress Cataloging-in-Publication Data

Schweitzer, George Keene, 1924–
The aqueous chemistry of the elements / George K. Schweitzer,
Lester L. Pesterfield.
    p. cm.
Includes bibliographical references and index.
ISBN 978-0-19-539335-4
1. Solution (Chemistry) 2. Inorganic compounds—Solubility. 3. Chemical
elements—Solubility. I. Pesterfield, Lester L. II. Title.
QD544.3.S39 2010
541'.3422—dc22   2009010326

9   8   7   6   5   4   3   2   1

Printed in the United States of America
on acid-free paper

Preface

The chemistry of the elements in aqueous solution is a very important aspect of the chemistry discipline because of its numerous applications in many sciences, applied sciences, engineering fields, and technologies. Years ago, much of the subject matter was taught in courses under the title of qualitative analysis. But those courses disappeared and in many universities, nothing replaced the loss of subject matter. Courses in descriptive inorganic chemistry came to be offered in many schools, but only a portion of the subject matter actually dealt with the detailed chemical behavior of the elements in aqueous solution.

One of the difficulties in the teaching of the descriptive aspects of inorganic solution chemistry is that there are too few overarching frameworks which can be used to avoid the chronicling of empirical fact after empirical fact. In this volume, we have addressed ourselves to using the theoretical construct of E–pH diagrams to provide a platform for the correlation of the data. Such diagrams permit the systematic treatment of the empirical data with reference to the influences of pH, redox phenomena, free energy changes, insolubilities, and complexation on the solution properties of the elements and their compounds.

Our aim has been to provide a textbook for courses in descriptive inorganic chemistry and a reference book for the numerous other fields in which inorganic solution chemistry is of importance.

The authors would like to thank Ms. Alicia McDaniel of Western Kentucky University for pertinent consideration of some recent advances.

George K. Schweitzer
Lester L. Pesterfield
October 2009

Contents

The Aqueous Chemistry of the Elements

1

E–pH Diagrams

1. Introduction

This volume is intended to employ E–pH diagrams to describe the inorganic solution chemistry of the chemical elements. Such diagrams are very useful in numerous fields of investigation, including electrochemistry, analytical chemistry, inorganic chemistry, geochemistry, environmental chemistry, corrosion chemistry, hydrometallurgy, water chemistry, agricultural chemistry, toxicology, biochemistry, chemical engineering, materials science, health physics, and nutrition. It is assumed that the reader is acquainted with the following major topics which are treated in elementary chemistry: stoichiometry, equilibrium, acid–base phenomena, solubility, complexation, elementary thermodynamics, and electrochemistry.

In 1923, W. M. Clark and B. Cohen published a paper in which they introduced the idea of plotting the electromotive force as referred to the hydrogen electrode E against the pH for several chemical systems.[1] In 1928, Clark continued to develop this graphical presentation in his text on the determination of pH.[2] The utility of the method was further extended by numerous other investigators such as M. Pourbaix, G. Valensi, G. Charlot, T. P. Hoar, R. M. Garrels, N. de Zoubov, J. Van Muylder, E. Deltombe, C. Vanleugenhaghe, J. Schmets, M. Maraghini, P. Van Rysselberghe, A. Moussard, J. Brenet, F. Jolas, K. Schwabe, J. Besson, W. Kunz, A. L. Pitman, J. N. Butler, P. Delahay, H. Freiser, H. A. Laitinen, L. G. Sillen, P. L. Cloke, and others. In 1963, M. Pourbaix in collaboration with

N. de Zoubov published Atlas d'equilibres electrochimiques, a collection of E–pH diagrams for 90 chemical elements. This volume was translated into English in 1966 by J. A. Franklin and published as Atlas of Electrochemical Equilibria in Aqueous Solutions.[3] Subsequently other investigators published computer programs for constructing the diagrams: L. Santoma; B. G. Williams, and W. H. Patrick; P. B. Linkson, B. D. Phillips, and C. D. Rowles; K. Osseo-Asare, A. W. Asihene, T. Xue, and V. S. T. Ciminellie; D. R. Drewes; M. Mao and E. Peters; H-H. Huang and C. A. Young; J. P. Birk and Laura L. Tayer; G. P. Glasby and H. D. Schulz; and Q. Feng, Y. Ma, and Y. Lu.[4]

2. The Na E–pH Diagram

Figures 1.1 through 1.17 exemplify the general nature of E–pH diagrams. Each E–pH diagram is a plot of E against pH for aqueous solutions.

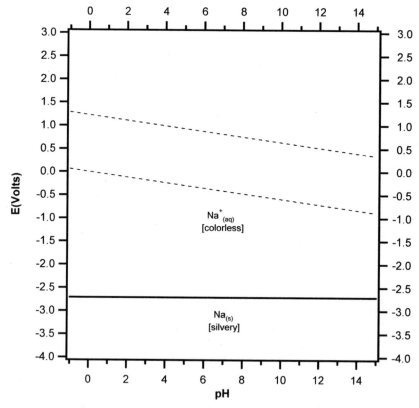

Figure 1.1 E–pH diagram for Na species. Soluble species concentrations (except H^+) = $10^{0.0}$ M. Soluble species and most solids are hydrated. No agents producing complexes or insoluble compounds are present other than HOH and OH^-.

The E (vertical) axis is a reflection of the potential values in volts (v) of reduction half-reactions describing the conditions under which changes in the aqueous oxidation state of the element occur. These E values range from +3.00 v to −4.00 v. The pH (horizontal) axis gives pH values ranging from a pH of −1.0 ($10^{1.0}$ molar hydrogen ion) to a pH of 15.0 ($10^{-15.0}$ molar hydrogen ion). The sloped dashed lines have to do with the behavior of the solvent water. This will be discussed in detail later.

The Na E–pH diagram will now be examined. Figure 1.1 shows the diagram for standard conditions, namely, a temperature T at 298 K (25.0°C), all dissolved species at 1.00 molal activity (the Na^+ ion), and all gases at 1.00 fugacity. In treatments of this system and all systems hereafter, the molar concentration M will be substituted for the molal activity, and the pressure in atmospheres will be substituted for the fugacity. These substitutions will usually introduce only small errors for the concentrations and pressures that will be employed.[5] The labels for the species (Na^+ and Na) indicate the predominant species under various E and pH conditions. This may be seen by examining the three vertical lines (constant pH values) in Figure 1.2 and the three horizontal lines (constant E values) in Figure 1.3. Take a look at the vertical line at a constant pH of 0.0. Start at the top of the line where E is equal to 3.0 volts (v). As one scans down, the Na^+ species is the predominant one until the line at about −2.7 v is reached. Below this value the predominant species is metallic Na. The same analysis applies to the other two vertical lines at pH values of 7.0 and 14.0.

Now observe the horizontal line in Figure 1.3 at a constant voltage of 2.0 v. Starting at the right and scanning to the left, it can be seen that Na^+ is the predominant species at all pH values. The same is the case for the horizontal line at a constant voltage of 0.0 v. Scanning from right to left across the third horizontal line at a constant voltage of −3.0 v shows that the predominant species is metallic Na at all pH values. Similar analyses may be made for vertical lines at any given constant pH and for horizontal lines at any constant E.

Now, the question arises as to the origin of the horizontal line at about −2.7 v. This is obtained from the reduction half-reaction which relates the two species (Na^+ and Na) on the two sides of the line. A reduction half-reaction is one in which the electron or electrons appear on the left side of the equation. For example, in Figure 1.1 (as well as Figures 1.2 and 1.3), the voltage of the horizontal line comes from the following reduction half-reaction at all pH values:

$$e^- + Na^+_{(aq)} \rightarrow Na_{(s)} \quad E° = -2.71 \text{ v.}$$

In this equation, e^- stands for the electron, Na^+ for the sodium ion, (aq) for the aqueous state, Na for elemental sodium, (s) for the solid state, and E° represents the standard electrode potential given in volts (v). The superscript ° on E indicates that the reaction is taking place under standard conditions (T = 298 K, Na^+ concentration of 1.00 M). Values of E° may be readily obtained from tables in reference works such as A. J. Bard, R. Parsons, and

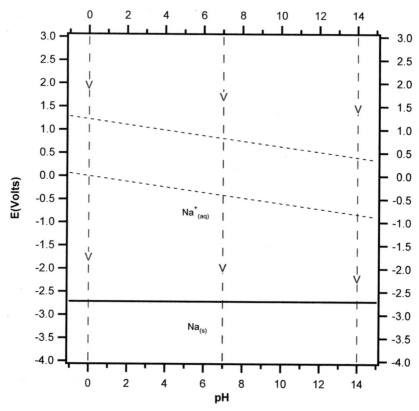

Figure 1.2 E–pH diagram for Na species. Soluble species concentrations (except H^+) = $10^{0.0}$ M. Soluble species and most solids are hydrated. No agents producing complexes or insoluble compounds are present other than HOH and OH^-. Species transformations take place at constant pH.

J. Jordan; D. R. Lide; and J. A. Dean.[6] The potential for a half-reaction, E° or E, may be thought of as the driving force for the electron or electrons in the reaction. The horizontal line at −2.71 v represents this reduction half-reaction. In the region below this line, at E° values more negative than −2.71 v, the half-reaction proceeds to the right such that the predominant species is $Na_{(s)}$. In the region above this line, at E° values more positive than −2.71 v, the reaction proceeds to the left. As a result, in the region above this line the predominant species is Na^+(aq).

For Figure 1.4, the Na^+(aq) concentration has been changed to 0.10 M, and it is to be noted that the E value for the horizontal line between predominant species Na^+ and Na has changed to about −2.8 v. The same sort of remarks as before regarding the three vertical and the three horizontal lines apply to this figure. To ascertain why the E value has changed, it is important to note that the half-reaction of interest is now conducted under non-standard state concentration conditions. The concentration of Na^+ has been altered from

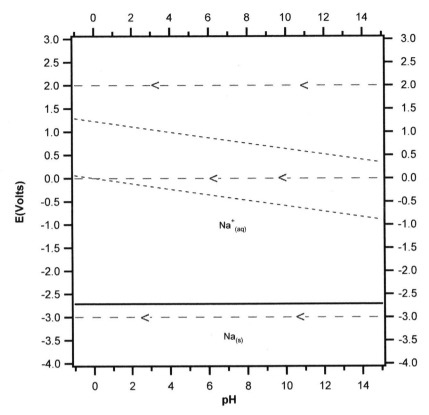

Figure 1.3 E–pH diagram for Na species. Soluble species concentrations (except H^+) = $10^{0.0}$ M. Soluble species and most solids are hydrated. No agents producing complexes or insoluble compounds are present other than HOH and OH^-. Species transformations take place at constant E.

1.00 M to 0.10 M. The E° value is therefore not applicable and must be changed to an E value. The half-reaction now reads

$$e^- + Na^+_{(aq)} \rightarrow Na_{(s)} \quad E = -2.8 \text{ v (estimated from diagram)}.$$

The horizontal line at about -2.8 v represents this reduction half-reaction. At E values below (more negative than) -2.8 v, the half-reaction proceeds to the right, and the predominant species in the region below the line is $Na_{(s)}$. At E values above (more positive than) -2.8 v, the reaction proceeds to the left, and the predominant species in the region above the line is $Na^+_{(aq)}$. Calculation of the change from E° to E can be made from the Nernst equation, which takes the following form:

$$E = E° - (2.303RT/nF)\log\left(\Pi\, [\text{products}]^x / \Pi[\text{reactants}]^y\right)$$

where R = 8.314 J/mol K (joules per mole per degree absolute), T = 298 K, F = 96,490 C/mole (coulombs per mole), n = moles of electrons involved

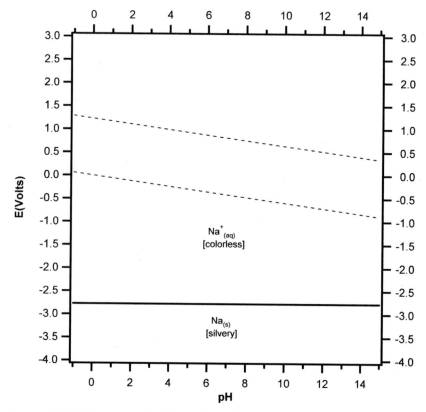

Figure 1.4 E–pH diagram for Na species. Soluble species concentrations (except H^+) = $10^{-1.0}$ M. Soluble species and most solids are hydrated. No agents producing complexes or insoluble compounds are present other than HOH and OH^-.

in the reaction, Π [products]^x signifies the product of the concentrations of the resulting species with each concentration raised to the power x which appears as its prefix in the half-reaction equation, and Π [reactants]^y signifies the product of the concentrations of the reacting species with each concentration raised to the power y which appears as its prefix in the half-reaction equation. Substituting the constants into the above equation yields the following simplified form:

$$E = E^\circ - (0.0591/n)\log(\Pi\,[\text{products}]^x/\Pi[\text{reactants}]^y) \qquad (1)$$

Further substituting into Equation (1) for the specific $Na^+_{(aq)} \rightarrow Na_{(s)}$ half-reaction at 0.10 M sodium ion concentration gives:

$$E = E^\circ - (0.0591/n)\log([\text{Na}]/[e^-][\text{Na}^+])$$
$$= -2.71 - (0.0591/1)\log([1]/[1][0.10]) = -2.77 \text{ v.}$$

The reader should recall that the concentrations of the electron, pure solids, and the solvent (water), are defined as 1. The calculated value of -2.77 v matches the value of -2.8 v which was estimated from the diagram. It is interesting to note from the Nernst equation that the reduction potential for the half-reaction is dependent only upon the concentration of the sodium ion, Na^+. Neither the concentration of the hydrogen ion nor the hydroxide ion influences the potential at which the half-reaction occurs since they do not appear in the above equation. Similar calculations may be made for other concentrations of Na^+. It will be found that the horizontal line separating Na^+ and Na moves from -2.71 v at 1.00 M Na^+ to -2.89 v at $10^{-3.0}$ M, to -3.06 v at $10^{-6.0}$ M, to -3.24 v at $10^{-9.0}$ M, and so on.

The pH (horizontal) axis in Figures 1.1 through 1.4 is a reflection of both the hydrogen ion concentration, $[H^+]$, and the hydroxide ion concentration, $[OH^-]$, of the solution. The pH of the solution is related to these values as:

$$pH = -\log [H^+], \text{ or } [H^+] = 10^{-pH}, \tag{2}$$

$$pH = -\log (10^{-14.0}/[OH^-]) = 14.0 + \log [OH^-], \text{ or } [OH^-] = 10^{pH-14.0}. \tag{3}$$

Equation (3) follows from Equation (2), if one recalls the ion product constant of water:

$$[H^+][OH^-] = 10^{-14.0}. \tag{4}$$

3. The Al E–pH Diagram

Figure 1.5 is an E–pH diagram for Al under standard conditions. This means that all soluble species are at 1.00 M, the species being Al^{+3} and $Al(OH)_4^-$. The labels for the four species identify the regions in which they predominate under differing E and pH conditions. These predominance conditions may be seen by examination of the three vertical lines in Figure 1.6 and the three horizontal lines in Figure 1.7. Start at the top of the vertical line at a constant pH of 0.0. As one goes down the line, the predominant species Al^{+3} gives way to the predominant species Al at an E value of about -1.7 v. The reduction half-reaction is written as follows with the $E°$ value as obtained from appropriate tables attached.

$$3e^- + Al^{+3}_{(aq)} \rightarrow Al_{(s)} \quad E° = -1.68 \text{ v}$$

As one scans down the vertical line at a constant pH of 7.0, the predominant species $Al(OH)_3$ is replaced by the predominant Al at an E value of about -1.9 v. The reduction half-reaction along with its $E°$ value is as follows:

$$3e^- + Al(OH)_3 + 3H^+ \rightarrow Al + 3HOH \quad E° = -1.47 \text{ v}$$

Similar observation of the vertical line at a constant pH of 14.0 shows the transformation from $Al(OH)_4^-$ to Al as the predominant species at an E of

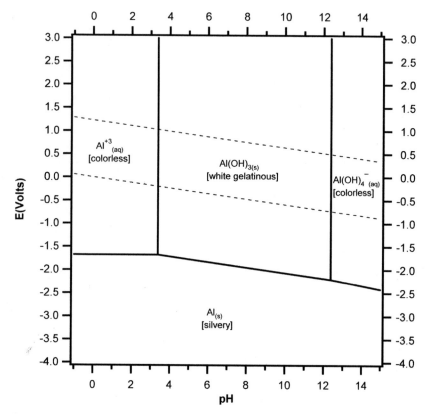

Figure 1.5 E–pH diagram for Al species. Soluble species concentrations (except H^+) = $10^{0.0}$ M. Soluble species and most solids are hydrated. No agents producing complexes or insoluble compounds are present other than HOH and OH^-.

about -2.3 v. The pertinent half-reaction is

$$3e^- + Al(OH)_4^- + 4H^+ \rightarrow Al + 4HOH \quad E° = -1.23 \text{ v.}$$

Values of E° cannot be used for the last two reactions as given because the H^+ concentration (pH) in both cases is not the standard value of 1.00 M. Hence the Nernst equation must be used to ascertain the applicable values of E when $[H^+]$ is $10^{-7.0}$ M in the $Al(OH)_3$ to Al reaction and $10^{-14.0}$ M in the $Al(OH)_4^-$ reaction.

$$E(\text{at pH } 7.0) = E° - (0.0591/3)\log ([Al][HOH]^3/[e^-]^3[Al(OH)_3][H^+]^3)$$

$$= -1.47 - (0.0591/3)\log ([1][1]^3/[1]^3[1][10^{-7.0}]^3) = -1.88 \text{ v}$$

$$E(\text{at pH } 14.0) = E° - (0.0591/3)\log ([Al][HOH]^4/[e^-]^3[Al(OH)_4^-][H^+]^4)$$

$$= -1.23 - (0.0591/3)\log ([1][1]^4/[1]^3[1.00][10^{-14.0}]^4) = -2.33 \text{ v}$$

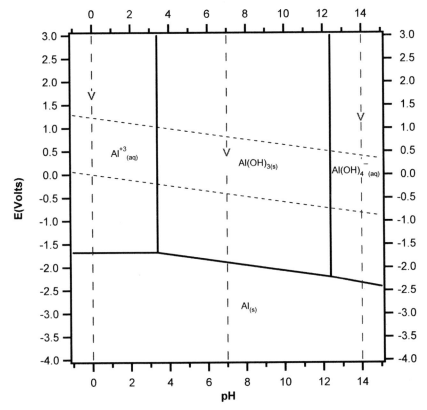

Figure 1.6 E–pH diagram for Al species. Soluble species concentrations (except H^+) = $10^{0.0}$ M. Soluble species and most solids are hydrated. No agents producing complexes or insoluble compounds are present other than HOH and OH^-. Species transformations take place at constant pH.

These values of -1.88 v and -2.33 v match the values of -1.9 v and -2.3 v which were estimated from the diagram. Please note also that the line between $Al(OH)_3$ and Al and the line between $Al(OH)_4^-$ and Al are both sloped. This behavior indicates that the lines are functions of both E and pH. This is obvious by virtue of the presence of $[H^+]$ in both equations.

A scan in Figure 1.7 from right to left of the horizontal line at a constant E of 2.00 v indicates a change from $Al(OH)_4^-$ to $Al(OH)_3$ at a pH of 12.4 and a change from $Al(OH)_3$ to Al^{+3} at a pH of 3.4. These transformations are related to the following reactions, to which are appended equilibrium constants obtained from the literature:

$$Al(OH)_4^- + H^+ \rightarrow Al(OH)_3 + HOH \quad K = 10^{12.4}$$

$$Al(OH)_3 + 3H^+ \rightarrow Al^{+3} + 3HOH \quad K = 10^{10.2}.$$

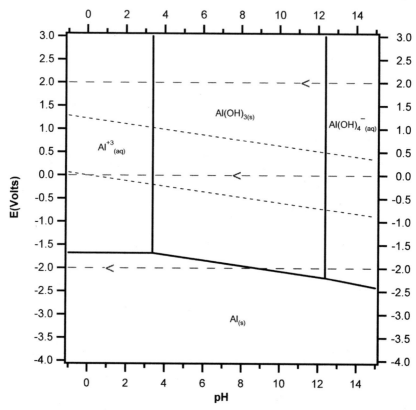

Figure 1.7 E–pH diagram for Al species. Soluble species concentrations (except H^+) = $10^{0.0}$ M. Soluble species and most solids are hydrated. No agents producing complexes or insoluble compounds are present other than HOH and OH^-. Species transformations take place at constant E.

It is to be noticed that both of these equilibrium constants are protonation constants, since they apply to the addition of a proton (H^+) to a given species. The pH values for which these reactions occur can be determined using equilibrium constant expressions and the appropriate equilibrium constant, K. Recall that equilibrium expressions take the general form of:

$$K = \Pi[\text{products}]^x / \Pi[\text{reactants}]^y \tag{5}$$

Applying this relationship to these two reactions and solving for the appropriate pH values gives

$$K = 10^{12.4} = [Al(OH)_3][HOH]/[Al(OH)_4^-][H^+] = [1][1]/[1.00][H^+]$$

$$[H^+] = 10^{-12.4} \quad pH = 12.4$$

$$K = 10^{10.2} = [Al^{+3}][HOH]^3/[Al(OH)_3][H^+]^3 = [1.00][1]^3/[1][H^+]^3$$

$$[H^+] = 10^{-3.4} \quad pH = 3.4$$

The same considerations apply to the horizontal line at a constant E of 0.0 v. However, the horizontal line at a constant E of -2.0 v involves two changes as one proceeds from right to left: $Al(OH)_4^-$ to $Al(OH)_3$ and $Al(OH)_3$ to Al. The first transition occurs at a pH of about 12.4 and the second change occurs at a pH of about 9.0. These are the reactions involved:

$$Al(OH)_4^- + H^+ \rightarrow Al(OH)_3 + 3HOH \quad K = 10^{12.4}$$

$$3e^- + Al(OH)_3 + 3H^+ \rightarrow Al + 3HOH \quad E° = -1.47 \text{ v.}$$

The equilibrium constant is used for the first of these reactions since no electrons are involved in the equation, that is, no change of oxidation state is occurring. But since there are electrons in the second reaction, the Nernst equation which handles reduction reactions must be used.

$$K = 10^{12.4} = [Al(OH)_3][HOH]/[Al(OH)_4^-][H^+] = [1][1]/[1.00][H^+]$$

$$[H^+] = 10^{-12.4} \quad pH = 12.4$$

$$E = E° - (0.0591/n)\log([Al][HOH]^3/[e^-]^3[Al(OH)_3][H^+]^3)$$

$$-2.00 = -1.47 - (0.0591/3)\log([1][1]^3/[1]^3[1][H^+]^3)$$

$$[H^+] = 10^{-9.0} \quad pH = 9.0$$

Notice that the line dividing $Al(OH)_3$ and Al is sloped, this being characteristic of a reaction that is dependent upon both E and pH. Equations of such reactions show both H^+ and electrons.

Figure 1.8 is an E–pH diagram for Al with the soluble species at 0.10 M except for the hydrogen ion concentration. This changed concentration applies to both Al^{+3} and $Al(OH)_4^-$. By observation of Figure 1.9, these transition equations can be seen for the descending vertical lines at constant pH values of 0.0, 7.0, and 14.0:

At pH of 0.0 $3e^- + Al^{+3} \rightarrow Al$ $E° = -1.68$ v

At pH of 7.0 $3e^- + Al(OH)_3 + 3H^+ \rightarrow Al + 3HOH$ $E° = -1.47$ v

At pH of 14.0 $3e^- + Al(OH)_4^- + 4H^+ \rightarrow Al + 4HOH$ $E° = -1.23$ v

Entering the altered concentrations of the soluble species (Al^{+3} and $Al(OH)_4^-$) into the Nernst equation as written for these three reactions will yield values of E which will be found comparable to the ones observed in Figure 1.8. These Nernst relationships are as follows:

At pH of 0.0, $E = -1.68 - (0.0591/3)\log([1]/[1]^3[0.10]) = -1.70$ v

At pH of 7.0, $E = -1.47 - (0.0591/3)\log([1][1]^3/[1]^3[1][10^{-7.0}]^3) = -1.88$ v

At pH of 14.0, $E = -1.23 - (0.0591/3)\log([1][1]^4/[1]^3[0.10][10^{-14.0}]^4)$

$$= -2.35 \text{ v}$$

Further considering Figure 1.10 in which the concentrations of Al^{+3} and $Al(OH)_4^-$ have both been altered to 0.10 M, these reactions reflect the changes

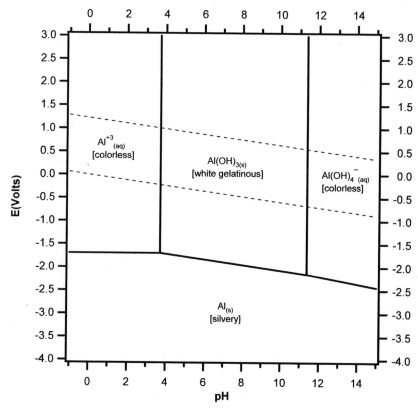

Figure 1.8 E–pH diagram for Al species. Soluble species concentrations (except H^+) = $10^{-1.0}$ M. Soluble species and most solids are hydrated. No agents producing complexes or insoluble compounds are present other than HOH and OH^-.

in species that can be observed as one moves from right to left across the three horizontal lines. Attached to each reaction is its $E°$ value or its K value, whichever is pertinent.

At E of 2.0 v $\quad$ $Al(OH)_4^- + H^+ \rightarrow Al(OH)_3 + HOH \quad K = 10^{12.4}$

At E of 2.0 v $\quad$ $Al(OH)_3 + 3H^+ \rightarrow Al^{+3} + 3HOH \quad K = 10^{10.2}$

At E of 0.0 v $\quad$ The same two equations

At E of − 2.0 v $\quad$ $Al(OH)_4^- + H^+ \rightarrow Al(OH)_3 + HOH \quad K = 10^{12.4}$

At E of − 2.0 v $\quad$ $3e^- + Al(OH)_3 + 3H^+ \rightarrow Al + 3HOH \quad E° = -1.47$ v

The pH values at which the transformations occur may be calculated by use of the equilibrium constant K or the Nernst equation, the latter being applicable to reactions in which the oxidation state changes, that is, reaction equations

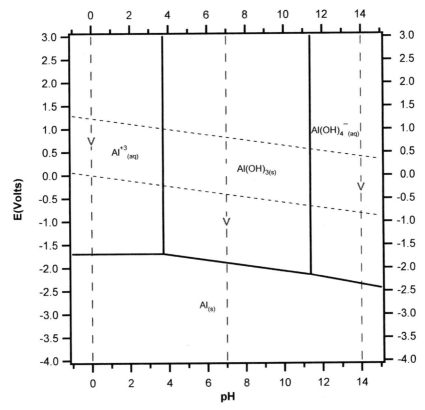

Figure 1.9 E–pH diagram for Al species. Soluble species concentrations (except H^+) $= 10^{-1.0}$ M. Soluble species and most solids are hydrated. No agents producing complexes or insoluble compounds are present other than HOH and OH^-. Species transformations take place at constant pH.

in which electrons appear.

$$K = 10^{12.4} = [1][1]/[0.10][H^+] \qquad [H^+] = 10^{-11.4} \quad pH = 11.4$$

$$K = 10^{10.2} = [0.10][1]^3/[1][H^+]^3 \qquad [H^+] = 10^{-3.7} \quad pH = 3.7$$

$$E = E° - (0.0591/3)\log([1][1]^3/[1]^3[1][H^+]^3)$$

$$-2.00 = -1.47 - (0.0591/3)\log([1][1]^3/[1]^3[1][H^+]^3)$$

$$[H^+] = 10^{-9.0} \quad pH = 9.0$$

The estimated pH values observed on Figure 1.10 for these transformations are 11.5 and 3.8 at a constant E of 2.0 v, the same values at a constant E of 0.0 v, and 11.4 and 9.0 at a constant E of −2.0 v. These fit very well with the calculated values given that the initial values from the figure are estimates. The general procedure presented here can be applied to any number of vertical

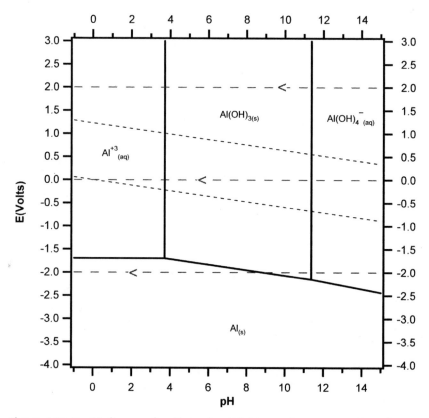

Figure 1.10 E–pH diagram for Al species. Soluble species concentrations (except H^+) = $10^{-1.0}$ M. Soluble species and most solids are hydrated. No agents producing complexes or insoluble compounds are present other than HOH and OH^-. Species transformations take place at constant E.

lines and horizontal lines across the E–pH diagram using any soluble species concentrations.

4. The Fe E–pH Diagram

In Figure 1.11, the E–pH diagram for iron, the predominant species is again determined by the combined effect of the potential E and the pH of the solution. The concentrations of all dissolved species in Figure 1.11 have been adjusted to 0.10 M, except for the hydrogen ion concentration. First, consider the three vertical lines in Figure 1.12 at constant pH values of 0.0, 7.0, and 14.0. The species transformations seen at a constant pH of 0.0 are Fe^{+3} to Fe^{+2} at an E of about 0.8 v, and Fe^{+2} to Fe at an E of about -0.4 v. At a constant pH of 7.0, the transformations are FeO(OH) to $Fe(OH)_2$ at an E of

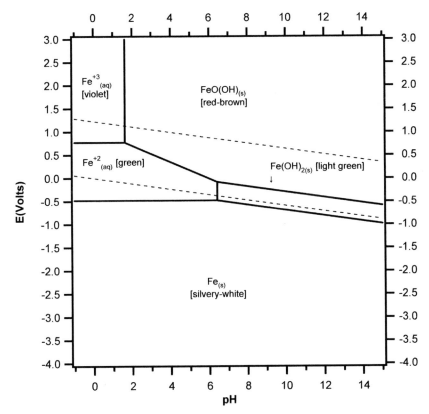

Figure 1.11 E–pH diagram for Fe species. Soluble species concentrations (except H^+) = $10^{-1.0}$ M. Soluble species and most solids are hydrated. No agents producing complexes or insoluble compounds are present other than HOH and OH^-.

about -0.1 v, and $Fe(OH)_2$ to Fe at an E of about -0.5 v. At a constant pH of 14.0, the changes are FeO(OH) to $Fe(OH)_2$ at an E of about -0.5 v, and $Fe(OH)_2$ to Fe at an E of about -0.9 v. The reactions describing these changes are represented by the following equations. Values of $E°$ are attached to the equations.

At a pH of 0.0 $e^- + Fe^{+3} \rightarrow Fe^{+2}$ $E° = 0.77$ v

At a pH of 0.0 $2e^- + Fe^{+2} \rightarrow Fe$ $E° = -0.45$ v

At a pH of 7.0 $e^- + FeO(OH) + H^+ \rightarrow Fe(OH)_2$ $E° = 0.30$ v

At a pH of 7.0 $2e^- + Fe(OH)_2 + 2H^+ \rightarrow Fe + 2HOH$ $E° = -0.10$ v

At a pH of 14.0   Same as the two previous equations

Accurate values of E for comparison with the values estimated from the diagram may be calculated from the Nernst equation as follows. The values

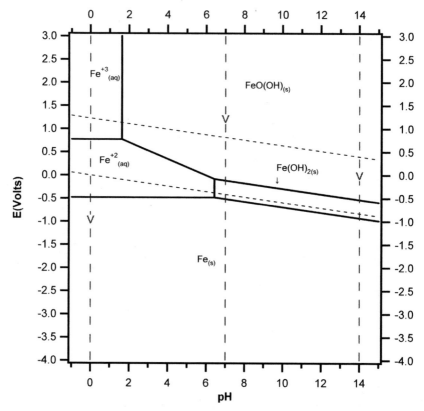

Figure 1.12 E–pH diagram for Fe species. Soluble species concentrations (except H^+) $= 10^{-1.0}$ M. Soluble species and most solids are hydrated. No agents producing complexes or insoluble compounds are present other than HOH and OH^-. Species transformations take place at constant pH.

estimated by observation of the E–pH diagram are presented in brackets.

At a pH of 0.0, $E(Fe^{+3}/Fe^{+2}) = 0.77 - (0.0591/1) \log ([0.10]/[1][0.10])$

$= 0.77$ v   {0.8 v}

At a pH of 0.0, $E(Fe^{+2}/Fe) = -0.45 - (0.0591/2) \log ([1]/[1]^2[0.10])$

$= -0.48$ v   {−0.4 v}

At a pH of 7.0, $E(FeO(OH)/Fe(OH)_2) = 0.30 - (0.0591/1) \log ([1]/[1][1]$

$[10^{-7.0}]) = -0.11$ v   {−0.1 v}

At a pH of 7.0, $E(Fe(OH)_2/Fe) = -0.10 - (0.0591/2) \log ([1][1]^2/[1]^2[1]$

$[10^{-7.0}]^2) = -0.51$ v   {−0.5 v}

At a pH of 14.0, $E(FeO(OH)/Fe(OH)_2) = 0.30 - (0.0591/1)\log([1]/[1][1]$
$[10^{-14.0}]) = -0.53$ v $\{-0.5$ v$\}$

At a pH of 14.0, $E(Fe(OH)_2/Fe) = -0.10 - (0.0591/2)\log([1][1]^2/[1]^2[1]$
$[10^{-14.0}]^2) = -0.93$ v $\{-0.9$ v$\}$

Now, consider the three horizontal lines in Figure 1.13. Moving from right to left, the line at a constant E of 2.0 v shows the change of FeO(OH) to Fe^{+3} at a pH of about 1.5. The line at a constant E of 0.0 v indicates a change from FeO(OH) to Fe^{+2} at a pH of about 6.0. The horizontal line at a constant E of -2.0 v shows no change, the predominant species at all pH values being Fe. The reactions for the transformations are as follows:

At E of 2.0 v $FeO(OH) + 3H^+ \rightarrow Fe^{+3} + 2HOH$ $K = 10^{3.9}$

At E of 0.0 v $e^- + FeO(OH) + 3H^+ \rightarrow Fe^{+2} + 2HOH$ $E° = 1.00$ v

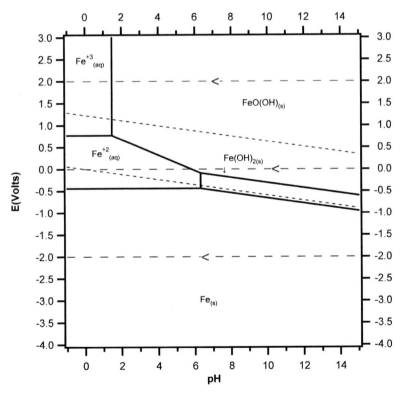

Figure 1.13 E–pH diagram for Fe species. Soluble species concentrations (except H^+) $= 10^{-1.0}$ M. Soluble species and most solids are hydrated. No agents producing complexes or insoluble compounds are present other than HOH and OH^-. Species transformations take place at constant E.

Application of the equilibrium constant equation along with the value for $[Fe^{+3}]$ and then the Nernst equation containing values for E and $E°$ and the concentration changes (0.10 M for all other soluble species) will give accurate pH values for comparison with the estimates from the diagram. The estimated values from the diagram are again presented in brackets.

$$K = 10^{3.9} = [Fe^{+3}][HOH]^2/[FeO(OH)][H^+]^3 = [0.10][1]^2/[1][H^+]^3$$

$$[H^+] = 10^{-1.6} \quad pH = 1.6 \qquad\qquad \{pH = 1.5\}$$

$$E = E° - (0.0591/n)\log([Fe^{+2}][HOH]^2/[e^-][FeO(OH)][H^+]^3)$$

$$0.00 = 1.00 - (0.0591/1)\log([0.10][1]^2/[1][1][H^+]^3)$$

$$[H^+] = 10^{-6.0} \quad pH = 6.0 \qquad\qquad \{pH = 6.0\}$$

5. The V E–pH Diagram

A further, slightly more complicated, E–pH diagram is the one for V which is depicted in Figure 1.14. This diagram is based upon aqueous concentrations of all soluble species (other than H^+) being at $10^{-3.0}$ M. In Figure 1.15, as before, the three vertical lines will be observed. As one goes down from the top of the line at a constant pH of 1.5, VO_2^+ changes to VO^{+2} at about 0.8 v, VO^{+2} changes to V^{+3} at about 0.2 v, V^{+3} goes over to V^{+2} at about -0.3 v, and V^{+2} transforms into V at about -1.3 v. Coming down the line at a pH of 5.0, $H_2VO_4^-$ changes into V_2O_4 at about 0.5 v, V_2O_4 converts to $V(OH)_3$ at about -0.1 v, $V(OH)_3$ is altered to V^{+2} at about -0.5 v, and V^{+2} is replaced by V at about -1.3 v. The vertical line at a constant pH of 11.0 shows these changes as one scans downward: HVO_4^{-2} to $V(OH)_3$ at about -0.4 v, $V(OH)_3$ to $V(OH)_2$ at about -1.2 v, and $V(OH)_2$ to V at about -1.5 v. The half-reactions corresponding to these changes in species along with their $E°$ values are as follows. Please remember that the $E°$ values will need to be altered to E values by proper consideration of the soluble species concentrations and the hydrogen ion concentrations.

At pH of 1.5	$e^- + VO_2^+ + 2H^+ \rightarrow VO^{+2} + HOH$	$E° = 1.00$ v
At pH of 1.5	$e^- + VO^{+2} + 2H^+ \rightarrow V^{+3} + HOH$	$E° = 0.34$ v
At pH of 1.5	$e^- + V^{+3} \rightarrow V^{+2}$	$E° = -0.26$ v
At pH of 1.5	$2e^- + V^{+2} \rightarrow V$	$E° = -1.18$ v
At pH of 5.0	$2e^- + 2H_2VO_4^- + 4H^+ \rightarrow V_2O_4 + 4HOH$	$E° = 1.23$ v
At pH of 5.0	$2e^- + V_2O_4 + 2HOH + 2H^+ \rightarrow 2V(OH)_3;$	$E° = 0.21$ v
At pH of 5.0	$e^- + V(OH)_3 + 3H^+ \rightarrow V^{+2} + 3HOH$	$E° = 0.16$ v
At pH of 5.0	$2e^- + V^{+2} \rightarrow V$	$E° = -1.18$ v

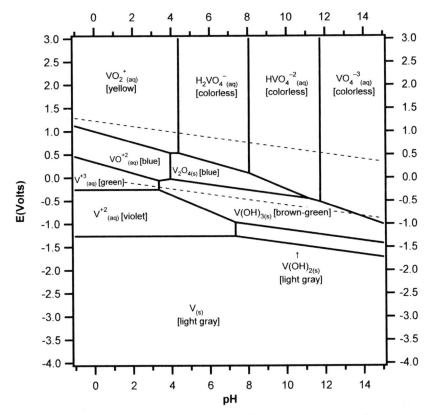

Figure 1.14 E–pH diagram for V species. Soluble species concentrations (except H^+) = $10^{-3.0}$ M. Soluble species and most solids are hydrated. No agents producing complexes or insoluble compounds are present other than HOH and OH^-.

At pH of 11.0 $2e^- + HVO_4^{-2} + 4H^+ \rightarrow V(OH)_3 + HOH$ $E° = 0.96$ v

At pH of 11.0 $e^- + V(OH)_3 + H^+ \rightarrow V(OH)_2 + HOH$ $E° = -0.53$ v

At pH of 11.0 $2e^- + V(OH)_2 + 2H^+ \rightarrow V + 2HOH$ $E° = -0.83$ v

The equations in the previous paragraph may be treated with the Nernst equation to arrive at the calculated values of E at which the changes in predominant species occur. The values estimated from the diagram as given above are shown in brackets.

At pH of 1.5 $E(VO_2^+/VO^{+2}) =$

$1.00 - (0.0591/1)\log([10^{-3.0}][1]/[1][10^{-3.0}][10^{-1.5}]^2) = 0.82$ v {0.8 v}

At pH of 1.5 $E(VO^{+2}/V^{+3}) =$

$0.34 - (0.0591/1)\log([10^{-3.0}][1]/[1][10^{-3.0}][10^{-1.5}]^2) = 0.16$ v {0.2 v}

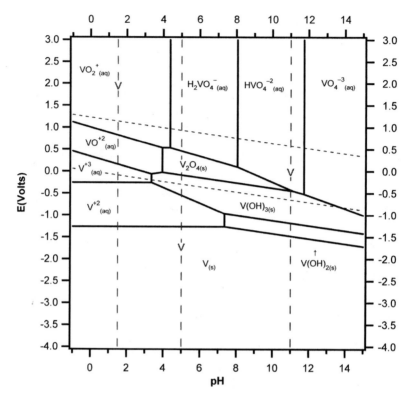

Figure 1.15 E–pH diagram for V species. Soluble species concentrations (except H^+) $= 10^{-3.0}$ M. Soluble species and most solids are hydrated. No agents producing complexes or insoluble compounds are present other than HOH and OH^-. Species transformations take place at constant pH.

At pH of 1.5 $E(V^{+3}/V^{+2}) =$

$- 0.26 - (0.0591/1) \log ([10^{-3.0}]/[1][10^{-3.0}]) = -0.26$ v $\{-0.3 \text{ v}\}$

At pH of 1.5 $E(V^{+2}/V) =$

$- 1.18 - (0.0591/2) \log ([1]/[1]^2[10^{-3.0}]) = -1.27$ v $\{-1.3 \text{ v}\}$

At pH of 5.0 $E(H_2VO_4^-/V_2O_4) =$

$1.23 - (0.0591/2) \log ([1][1]^4/[1]^2[10^{-3.0}]^2[10^{-5.0}]^4) = 0.46$ v $\{0.5 \text{ v}\}$

At pH of 5.0 $E(V_2O_4/V(OH)_3) =$

$0.21 - (0.0591/2) \log ([1]^2/[1]^2[1][1]^2[10^{-5.0}]^2) = -0.09$ v $\{-0.1 \text{ v}\}$

At pH of 5.0 $E(V(OH)_3/V^{+2}) =$

$0.16 - (0.0591/1) \log ([10^{-3.0}][1]^3/[1][1][10^{-5.0}]^3) = -0.55$ v $\{-0.5 \text{ v}\}$

At pH of 1.5 $E(V^{+2}/V) =$

$-1.18 - (0.0591/2)\log([1]/[1]^2[10^{-3.0}]) = -1.27$ v   {−1.3 v}

At pH of 11.0 $E(HVO_4^{-2}/V(OH)_3) =$

$0.96 - (0.0591/2)\log([1][1]/[1]^2[10^{-3}][10^{-11.0}]^4) = -0.43$ v {−0.4 v}

At pH of 11.0 $E(V(OH)_3/V(OH)_2) =$

$-0.53 - (0.0591/1)\log([1][1]/[1][1][10^{-11.0}]) = -1.18$ v   {−1.2 v}

At pH of 11.0 $E(V(OH)_2/V) =$

$-0.83 - (0.0591/2)\log([1][1]^2/[1]^2[1][10^{-11.0}]^2) = -1.48$ v {-1.5 v}

There are two horizontal lines in Figure 1.16 which represent changes in pH at constant E values. Observation of the topmost one which occurs at a constant E of 1.50 v shows the following changes as one proceeds from right to left: from VO_4^{-3} to HVO_4^{-2} at a pH of about 11.8, from HVO_4^{-2} to $H_2VO_4^{-}$ at a pH of about 8.1, and from $H_2VO_4^{-}$ to VO_2^{+} at a pH of about 4.3. Consideration of the other horizontal line at a constant E of -0.60 v shows a transformation of VO_4^{-3} to $V(OH)_3$ at a pH of about 12.5 and of $V(OH)_3$ to V^{+2} at a pH of about 5.2.

At an E of 1.50 v $VO_4^{-3} + H^+ \rightarrow HVO_4^{-2}$ $K = 10^{11.8}$

At an E of 1.50 v $HVO_4^{-2} + H^+ \rightarrow H_2VO_4^{-}$ $K = 10^{8.1}$

At an E of 1.50 v $H_2VO_4^{-} + 2H^+ \rightarrow VO_2^+ + 2HOH$ $K = 10^{8.8}$

At an E of -0.60 v $2e^- + VO_4^{-3} + 5H^+ \rightarrow V(OH)_3 + HOH$ $E° = 1.31$ v

At an E of -0.60 v $e^- + V(OH)_3 + 3H^+ \rightarrow V^{+2} + 3HOH$ $E° = 0.16$ v

The first three equations above lend themselves to treatment by equilibria expressions and the last two by the Nernst expression. In all five cases it is the pH that is being solved for. The estimated values taken from the E–pH diagram are presented in brackets for comparison.

At an E of 1.50 v $K = 10^{11.8} = [10^{-3.0}]/[10^{-3.0}][H^+]$

$[H^+] = 10^{-11.8}$ pH $= 11.8$ {pH $= 11.8$}

At an E of 1.50 v $K = 10^{8.1} = [10^{-3.0}]/[10^{-3.0}][H^+]$

$[H^+] = 10^{-8.1}$ pH $= 8.1$ {pH $= 8.1$}

At an E of 1.50 v $K = 10^{8.8} = [10^{-3.0}][1]^2/[10^{-3.0}][H^+]^2$

$[H^+] = 10^{-4.4}$ pH $= 4.4$ {pH $= 4.3$}

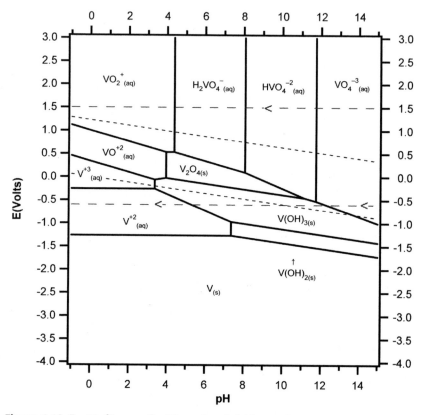

Figure 1.16 E–pH diagram for V species. Soluble species concentrations (except H^+) = $10^{-3.0}$ M. Soluble species and most solids are hydrated. No agents producing complexes or insoluble compounds are present other than HOH and OH^-. Species transformations take place at constant E.

At an E of -0.60 v $\quad -0.60 = 1.31 - (0.0591/2)\log\,([1][1]/[1]^2[10^{-3.0}][H^+]^5)$

$$[H^+] = 10^{-12.3} \quad pH = 12.3 \qquad \{pH = 12.5\}$$

At an E of -0.60 v $\quad -0.60 = 0.16 - (0.0591/1)\log\,([10^{-3.0}][1]^3/[1][1][H^+]^3)$

$$[H^+] = 10^{-5.3} \quad pH = 5.3 \qquad \{pH = 5.2\}$$

At this juncture, four different E–pH diagrams have been empirically examined in detail. Several important points are to be noted. First, an E–pH diagram shows the regions of species predominance under conditions of E and pH. Second, the regions of predominance are separated by lines which show the transformations of predominant species. Third, these lines are reflections of transformation reactions which can be represented by equations. Fourth, when the transformation equation does not show the hydrogen ion, the line is horizontal. Fifth, when the transformation equation does not show electrons, the line is vertical. Sixth, when the transformation equation shows both the hydrogen ion and electrons, the line is sloped. Seventh, E values for transformation equations which contain electrons may be ascertained through use of the Nernst relationship along with pertinent $E°$ values. Eighth, pH values for transformation equations which do not contain electrons may be ascertained through use of equilibrium expressions and appropriate equilibrium constant values.

With regard to the lines which show transformations of predominant species, it is of interest to know what the concentration gradients are on each side of the line. For example, consider the line which separates Fe^{+3} from Fe^{+2} in Figure 1.11. On the line the molar amounts of iron are equal: 50% Fe^{+3} and 50% Fe^{+2}. Just 0.07 v below the line, there is 10% Fe^{+3} and 90% Fe^{+2}. For a second illustration, consider the line which separates VO_4^{-3} and HVO_4^{-2} in Figure 1.14. On the line, the molar amounts of vanadium are equal: 50% HVO_4^{-2} and 50% VO_4^{-3}. One pH unit to the right of the line, there is 10% HVO_4^{-2} and 90% VO_4^{-3}. These illustrations indicate that species change is quite sensitive to voltage but less sensitive to pH.

6. The HOH E–pH Diagram

The reader will have seen in all the E–pH diagrams described so far that there are two dashed lines running respectively from 1.29 v to −0.34 v and from 0.06 v to −0.89 v on every diagram. These are the species change lines for water. And since the E–pH diagrams are for aqueous solutions, water is involved in all cases and its behavior under all E and pH conditions must be taken into account. Figure 1.17 shows only the water E–pH diagram. In between the dotted lines the species H^+ and HOH appear, represented hereafter in the text as $HOH{\equiv}H^+$. The HOH represents water and the H^+ represents the hydrogen ion. Since water ionizes slightly into H^+ and OH^-, and the two ions are interdependent, then OH^- is tacitly included in this area. If one observes any vertical line (constant pH) and moves down from the top O_2 is seen to transform into $HOH{\equiv}H^+$ and then $HOH{\equiv}H^+$, is seen to transform into H_2. The equation for the first transformation may be written in two ways

$$4e^- + 4H^+ + O_2 \rightarrow 2HOH, \text{ or } \quad E° = 1.23 \text{ v}$$

$$4e^- + 2HOH + O_2 \rightarrow 4OH^-. \quad E(\text{with } OH^- \text{ at } 1.00 \text{ M and } H^+$$

$$\text{at } 10^{-14.0}M) = 0.40 \text{ v}$$

Since E–pH diagrams involve the pH which is a representation of the hydrogen ion concentration, the first of these equations is preferred.

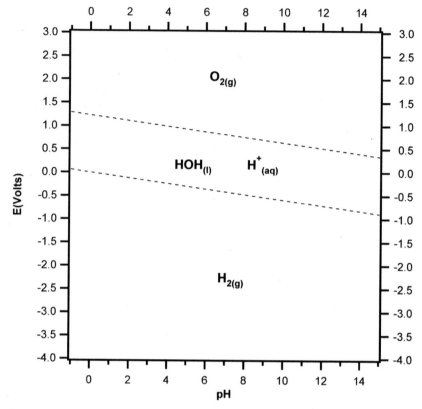

Figure 1.17 E–pH diagram for HOH.

The second transition from HOH≡H$^+$ may also be written in two ways:

$$2e^- + 2H^+ \rightarrow H_2 \quad E° = 0.00 \text{ v}$$

$$2e^- + 2HOH \rightarrow H_2 + 2OH^- \quad E(\text{with OH}^- \text{ at } 1.00 \text{ M and H}^+$$
$$\text{at } 10^{-14.0} \text{ M}) = -0.83 \text{ v}$$

Once again, the equation showing H$^+$ is preferred because one of the variables for the E–pH diagram is the pH.

As is the pattern with other E–pH diagrams the area above the top dotted line shows the predominance of O_2, the area below the lower dashed line shows the predominance of H_2, and the area in between the dashed lines shows the predominance of HOH. In other words, at high potentials water decomposes into O_2 and at low potentials water decomposes into H_2. Values of E at H$^+$ concentrations other than 1.00 M may be readily calculated by introducing the above equations and the appropriate E° values into the Nernst relation.

2

The Construction of E–pH Diagrams

1. Introduction

In order to construct an E–pH diagram one needs to follow eight basic steps:

(1) Select the species of the element involved which contain one or more of the following entities: the element, oxygen, and hydrogen. This is best done by reading the descriptive chemistry of the element in a good inorganic text and identifying the species, both soluble and insoluble, which persist, at least for several minutes, in aqueous solution.

(2) Starting at the lower left-hand corner of an E–pH framework, arrange the selected species in vertical order of increasing oxidation number of the element. Then, if there are different species with the same oxidation number, arrange them in horizontal order of decreasing protonation (increasing hydroxylation). If there is only one species of a given oxidation number, this species extends across the entire pH range for the purposes of diagram construction.

(3) Draw in border lines between the species, that is, the lines representing the transformation of a species to another species. You will not know exactly where these lines occur but the approximate regions are sufficient for the purposes of diagram construction.

(4) Write equations for the transformations that have been indicated. Some of them will involve electrons and therefore will be half-reactions. Such equations must always be written as reductions, that is, with the electrons on the left. In addition, no reaction should contain the OH^- ion; only the H^+ and/or HOH instead.

(5) From appropriate tabulations, obtain the standard free energy values ($\Delta G°$ in kJ/mole) of every species in the equations. These $\Delta G°$ values are to be employed in the following relationship which applies to each of the above equations.

$$\Delta G° \text{ (reaction)} = \Sigma \Delta G°(\text{products}) - \Sigma \Delta G° \text{ (reactants)} \quad (6)$$

(6) The $\Delta G°$ (reaction) values for each equation are to be converted into $E°$ values for those equations containing electrons and into K values for those equations which do not. This is done by use of the following expressions:

$$E° = \Delta G°/-96.49n \qquad \log K = \Delta G°/-5.7 \quad (7/8)$$

where n represents the number of electrons in an equation.

(7) For each reduction half-reaction, the Nernst equation is written with the proper $E°$, and the relationship is solved for E as a function of pH and the concentrations of the soluble species. For each reaction which does not involve electrons, the equilibrium expression is written with the pertinent K and is solved for pH as a function of the concentrations of the soluble species.

(8) The above equations of E or K are straight line relationships and either are or can be rewritten in the familiar $y = ax + b$ or $y = b + ax$ form. The lines can then be drawn on the E–pH framework and portions of them can be erased when overlaps occur.

2. Constructing the Ga E–pH Diagram

The eight steps described in the previous section will now be applied to the construction of a Ga E–pH diagram.

Step (1). *Species identification.* Perusal of descriptive inorganic chemistry texts will lead to the discovery of the Ga-, O-, and H-containing species which persist in water. These species consist of the solids Ga and $Ga(OH)_3$ and the soluble ions Ga^{+3} and $Ga(OH)_4^-$. It should be noted that $Ga(OH)_3$ should occur in the basic region and that Ga will sit low on the E–pH diagram because it is the most highly reduced species.

Step (2). *Species placement.* In accordance with the procedure, the most reduced species Ga (oxidation number 0) is placed in the lower left-hand corner. There are no other species with a 0 oxidation number, and therefore Ga is to be assumed to extend all across the bottom of the diagram. The other

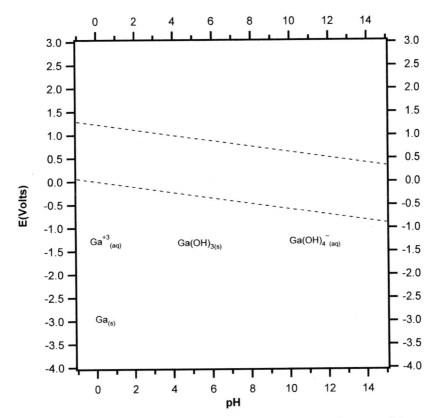

Figure 2.1 E–pH diagram for Ga species. Soluble species and most solids are hydrated. No agents producing complexes or insoluble compounds are present other than HOH and OH⁻. Species placement approximation.

three species are all seen to have a Ga oxidation number of III. Ga^{+3} is likely to exist in the acidic region (low pH) and the other two toward the basic region. Of these other two, $Ga(OH)_4^-$ will exist farthest toward the basic region because it involves four OH^- whereas $Ga(OH)_3$ involves only three. These species have been placed on the E–pH framework in Figure 2.1.

Step (3). *Drawing in transformation lines.* A line is now drawn in between every contiguous pair of species. These lines are preliminary placements of transformations from one predominant species to another. See Figure 2.2.

Step (4). *Writing equations.* Each line on the diagram symbolizes a species transformation. The following equations represent these reactions.

$$3e^- + Ga^{+3} \rightarrow Ga$$

$$3e^- + Ga(OH)_3 + 3H^+ \rightarrow Ga + 3HOH$$

$$3e^- + Ga(OH)_4^- + 4H^+ \rightarrow Ga + 4HOH$$

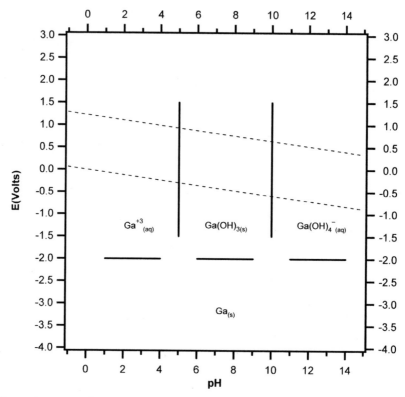

Figure 2.2 E–pH diagram for Ga species. Soluble species and most solids are hydrated. No agents producing complexes or insoluble compounds are present other than HOH and OH⁻. Species with initial line separations.

$$Ga(OH)_4^- + H^+ \rightarrow Ga(OH)_3 + HOH$$

$$Ga(OH)_3 + 3H^+ \rightarrow Ga^{+3} + 3HOH$$

Step (5). *Free energy calculations.* Consultation of standard free energy value tables gives the following $\Delta G°$ values in kJ/mole for the four species being considered: Ga (0.0), Ga^{+3} (−159.4), $Ga(OH)_3$ (−831.4), and $Ga(OH)_4^-$ (−982.4). These values are attached to the above equations and $\Delta G°$ (reaction) values have been calculated using Equation (6). The $\Delta G°$ values for the electron and for H^+ are both 0.00 kJ/mole and that for HOH is −237.2 kJ/mole.

$$3e^- + Ga^{+3} \rightarrow Ga$$

0   − 159.4   0 $\Delta G°(Ga^{+3}/Ga) = 159.4\,kJ/mole$

$$3e^- + Ga(OH)_3 + 3H^+ \rightarrow Ga + 3HOH$$

0   − 831.4   0   0   3(−237.2) $\Delta G°(Ga(OH)_3/Ga)$

$$= 119.8\,kJ/mole$$

$$3e^- + Ga(OH)_4^- + 4H^+ \rightarrow Ga + 4HOH$$

$$0 \quad -982.4 \quad 0 \quad 0 \quad 4(-237.2) \quad \Delta G°(Ga(OH)_4^-/Ga)$$
$$= 33.6 \, kJ/mole$$

$$Ga(OH)_4^- + H^+ \rightarrow Ga(OH)_3 + HOH$$

$$-982.4 \quad 0 \quad -831.4 \quad -237.2 \quad \Delta G°(Ga(OH)_4^-/Ga(OH)_3)$$
$$= -86.2 \, kJ/mole$$

$$Ga(OH)_3 + 3H^+ \rightarrow Ga^{+3} + 3HOH$$

$$-831.4 \quad 0 \quad -159.4 \quad 3(-237.2) \quad \Delta G°(Ga(OH)_3/Ga^{+3})$$
$$= -39.6 \, kJ/mole$$

Step (6). *Conversions into E° or K values.* Values of $\Delta G°$ for the first three equations are converted into E° values using Equation (7). Values of $\Delta G°$ for the last two equations are converted into K values using Equation (8). These conversions are appropriate because the first three equations contain electrons whereas the last two do not.

$$3e^- + Ga^{+3} \rightarrow Ga \quad E° = \Delta G°/-96.49n = 159.4/-96.49(3) = -0.55 \, v$$

$$3e^- + Ga(OH)_3 + 3H^+ \rightarrow Ga + 3HOH \quad E° = \Delta G°/-96.49n$$
$$= 119.8/-96.49(3) = -0.41 \, v$$

$$3e^- + Ga(OH)_4^- + 4H^+ \rightarrow Ga + 4HOH \quad E° = \Delta G°/-96.49n$$
$$= 33.6/-96.49(3) = -0.12 \, v$$

$$Ga(OH)_4^- + H^+ \rightarrow Ga(OH)_3 + HOH \quad \log K = \Delta G°/-5.7$$
$$= -86.2/-5.7 = 15.1 \quad K = 10^{15.1}$$

$$Ga(OH)_3 + 3H^+ \rightarrow Ga^{+3} + 3HOH \quad \log K = \Delta G°/-5.7$$
$$= -39.6/-5.7 = 6.9 \quad K = 10^{6.9}$$

Step (7). *Writing E and pH equations.* For each of the above three equations, the Nernst expression is written out, then the equation is solved for E as a function of pH. For each of the last two equations, the equilibrium constant is written out, the equation is put into logarithmic form, and is then solved for the pH.

$$E = E° - (0.0591/n) \log(\Pi[products]^x/\Pi[reactants]^y) \tag{1}$$

$$E(Ga^{+3}/Ga) = -0.55 - (0.0591/3) \log ([1]/[1]^3[Ga^{+3}])$$
$$= -0.55 + 0.020 \log [Ga^{+3}]$$

$$E(Ga(OH)_3/Ga) = -0.41 - (0.0591/3) \log ([1][1]^3/[1]^3[1][H^+]^3)$$

$$= -0.41 - 0.059 \, pH$$

$$E(Ga(OH)_4^-/Ga) = -0.12 - (0.0591/3) \log ([1][1]^4/[1]^3[Ga(OH)_4^-][H^+]^4)$$

$$= -0.12 - 0.079pH + 0.020 \log [Ga(OH)_4^-]$$

$$K(Ga(OH)_4^-/Ga(OH)_3) = 10^{15.1} = [1][1]/[Ga(OH)_4^-][H^+]$$

$$pH = 15.1 + \log [Ga(OH)_4^-]$$

$$K(Ga(OH)_3/Ga^{+3}) = 10^{6.9} = [Ga^{+3}][1]^3/[1][H^+]^3$$

$$3pH = 6.9 - \log [Ga^{+3}]$$

Step (8). *Drawing lines.* The five expressions above represent straight lines. They have been drawn in Figure 2.3. The lines have not been extended

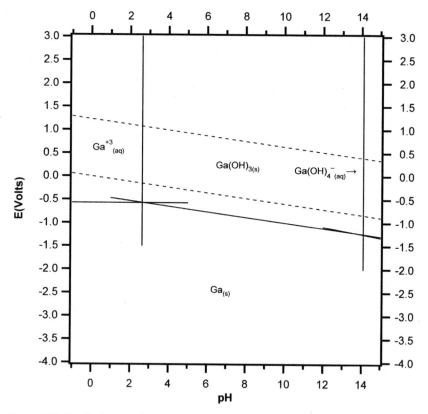

Figure 2.3 E–pH diagram for Ga species. Soluble species concentrations (except H^+) $= 10^{-1.0}$ M. Soluble species and most solids are hydrated. No agents producing complexes or insoluble compounds are present other than HOH and OH^-. Preliminary line placements.

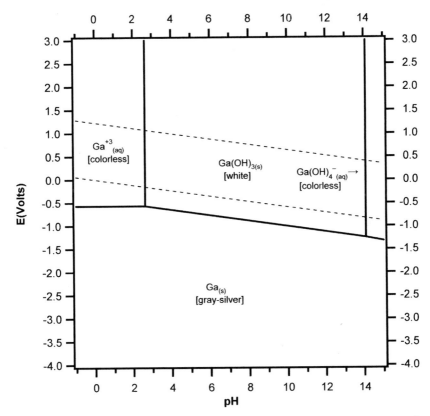

Figure 2.4 E–pH diagram for Ga species. Soluble species concentrations (except H^+) = $10^{-1.0}$ M. Soluble species and most solids are hydrated. No agents producing complexes or insoluble compounds are present other than HOH and OH^-. Final diagram.

into regions where the species involved are not pertinent. For example, the Ga^{+3}/Ga line has not been extended into the basic region. The portions of the lines which invade regions of predominant species then have been erased to produce the completed E–pH diagram as displayed in Figure 2.4. Note that the five expressions given above can be used to make detailed calculations of the exact values of E and pH for species transformations.

3. Constructing the Ga E–pH Diagram with Concentration Variations

By using the above procedures, Ga E–pH diagrams at soluble species concentrations of $10^{-1.0}$ M, $10^{-4.0}$ M, and $10^{-7.0}$ M can be made. Individual diagrams for the three concentrations are presented in Figures 2.4–2.6. These three figures have been combined to give Figure 2.7. Notice that E transition

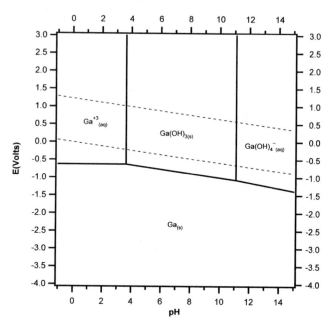

Figure 2.5 E–pH diagram for Ga species. Soluble species concentrations (except H$^+$) = 10$^{-4.0}$ M. Soluble species and most solids are hydrated. No agents producing complexes or insoluble compounds are present other than HOH and OH$^-$.

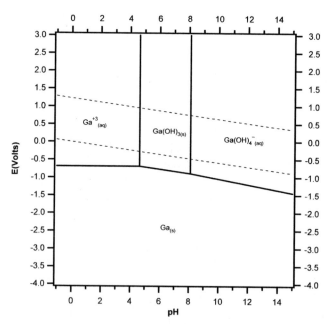

Figure 2.6 E–pH diagram for Ga species. Soluble species concentrations (except H$^+$) = 10$^{-7.0}$ M. Soluble species and most solids are hydrated. No agents producing complexes or insoluble compounds are present other than HOH and OH$^-$.

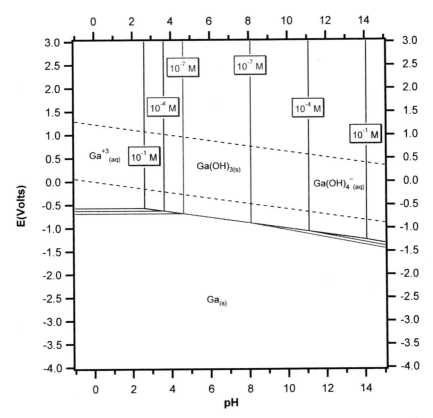

Figure 2.7 E–pH diagram for Ga species. Soluble species concentrations (except H^+) = $10^{-1.0}$, $10^{-4.0}$, and $10^{-7.0}$ M. Soluble species and most solids are hydrated. No agents producing complexes or insoluble compounds are present other than HOH and OH^-. Combined diagram.

values are not highly sensitive to concentration variations, for example the Ga^{+3}/Ga line moves from -0.57 v to -0.63 v to -0.69 v with the above concentration changes. However, the pH transition values are quite sensitive to concentration variations, as the values go from pH = 14.1 to pH = 11.1 to pH = 8.1 for the $Ga(OH)_4^-/Ga(OH)_3$ line when the above concentrations decrease.

4. Constructing the Mn E–pH Diagram

The eight steps described in the first section of this chapter will now be applied to the construction of an Mn E–pH diagram.

Step (1). *Species identification.* Consultation of a good descriptive inorganic chemistry textbook will indicate that the major species of Mn which

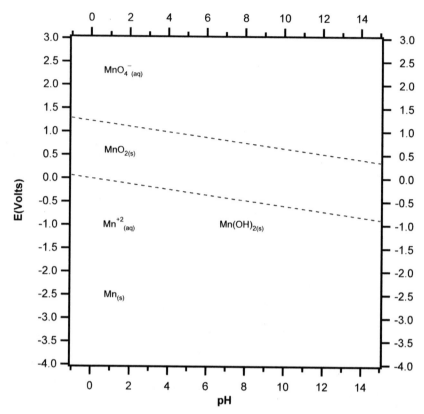

Figure 2.8 E–pH diagram for Mn species. Soluble species and most solids are hydrated. No agents producing complexes or insoluble compounds are present other than HOH and OH⁻. Species placement approximation.

need to be considered for an E–pH diagram are Mn, Mn^{+2}, $Mn(OH)_2$, MnO_2, and MnO_4^-. Several points should be noted: Mn and MnO_2 are solids and Mn^{+2} and MnO_4^- are solution species; Mn is the only species with an Mn oxidation number of 0 and MnO_4^- is the only species with an Mn oxidation number of VII; both Mn^{+2} and $Mn(OH)_2$ have Mn oxidation numbers of II; and Mn^{+2} is likely to be an acidic-region species whereas $Mn(OH)_2$ is a basic-region species.

Step (2). *Species placement.* In accordance with the procedure, the most reduced species Mn (oxidation number 0) is placed on Figure 2.8 in the lower left-hand corner. There are no other species with a 0 oxidation number, and therefore Mn is to be assumed to extend all across the bottom of the diagram. The two species with an Mn oxidation number of II are next added to the figure, with Mn^{+2} above Mn and $Mn(OH)_2$ placed to the right (the basic direction). Then MnO_2 is placed above the Mn^{+2} and extended all the

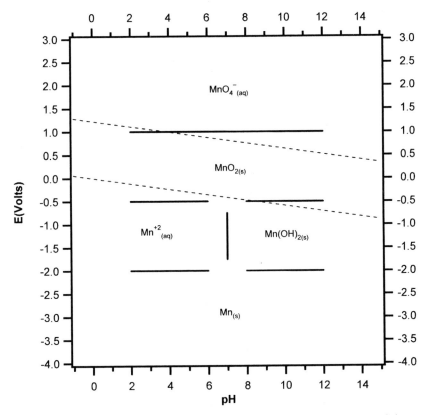

Figure 2.9 E–pH diagram for Mn species. Soluble species and most solids are hydrated. No agents producing complexes or insoluble compounds are present other than HOH and OH⁻. Species with initial line separations.

way across the diagram, and MnO_4^- above the MnO_2 and all across the diagram.

Step (3). *Drawing in transformation lines.* A line is now drawn in between every contiguous pair of species. These lines are preliminary placements of transformations from one predominant species to another. See Figure 2.9.

Step (4). *Writing equations.* Each line on the diagram symbolizes a species transformation. The following equations represent these reactions.

$$2e^- + Mn^{+2} \rightarrow Mn$$

$$2e^- + MnO_2 + 4H^+ \rightarrow Mn^{+2} + 2HOH$$

$$2e^- + MnO_2 + 2H^+ \rightarrow Mn(OH)_2$$

$$3e^- + MnO_4^- + 4H^+ \rightarrow MnO_2 + 2HOH$$

$$2e^- + Mn(OH)_2 + 2H^+ \rightarrow Mn + 2HOH$$

$$2H^+ + Mn(OH)_2 \rightarrow Mn^{+2} + 2HOH$$

Step (5). *Free energy calculations.* Consultation of standard free energy value tables gives the following $\Delta G°$ values in kJ/mole for the five species being considered: Mn (0.0), Mn^{+2} (−228.4), $Mn(OH)_2$ (−615.0), MnO_2 (−465.3), and MnO_4^- (−447.3). These values are attached to the above equations and $\Delta G°$(reaction) values have been calculated using Equation (6). The $\Delta G°$ values for the electron and for H^+ are both 0.00 kJ/mole and that for HOH is −237.2 kJ/mole.

$$2e^- + Mn^{+2} \rightarrow Mn$$

$$0 \quad -228.4 \quad\quad 0 \quad\quad \Delta G°(Mn^{+2}/Mn) = 228.4\,kJ/mole$$

$$2e^- + MnO_2 + 4H^+ \rightarrow Mn^{+2} + 2HOH$$

$$0 \quad -465.3 \quad 0 \quad\quad -228.4 \quad 2(-237.2) \quad \Delta G°(MnO_2/Mn^{+2})$$

$$= -237.5\,kJ/mole$$

$$2e^- + MnO_2 + 2H^+ \rightarrow Mn(OH)_2$$

$$0 \quad -465.3 \quad 0 \quad\quad -615.0 \quad\quad \Delta G°(MnO_2/Mn(OH)_2)$$

$$= -149.7\,kJ/mole$$

$$3e^- + MnO_4^- + 4H^+ \rightarrow MnO_2 + 2HOH$$

$$0 \quad -447.3 \quad\quad 0 \quad\quad -465.3 \quad 2(-237.2) \quad \Delta G°(MnO_4^-/MnO_2)$$

$$= -492.4\,kJ/mole$$

$$2e^- + Mn(OH)_2 + 2H^+ \rightarrow Mn + 2HOH$$

$$0 \quad -615.0 \quad\quad 0 \quad\quad 0 \quad 2(-237.2) \quad \Delta G°(Mn(OH)_2/Mn)$$

$$= 140.6\,kJ/mole$$

$$2H^+ + Mn(OH)_2 \rightarrow Mn^{+2} + 2HOH$$

$$0 \quad -615.0 \quad\quad -228.4 \quad 2(-237.2) \quad\quad \Delta G°(Mn(OH)_2Mn^{+2})$$

$$= -87.8\,kJ/mole$$

Step (6). *Conversions into E° or K values.* Values of $\Delta G°$ for the first five equations are converted into E° values using Equation (7). The values of $\Delta G°$ for the last equation is converted into a K value using Equation (8). These conversions are appropriate because the first five equations contain electrons whereas the last one does not.

$$2e^- + Mn^{+2} \rightarrow Mn \qquad\qquad E° = 228.4/-96.49(2) = -1.18 \text{ v}$$

$$2e^- + MnO_2 + 4H^+ \rightarrow Mn^{+2} + 2HOH \qquad E° = -237.5/-96.49(2) = 1.23 \text{ v}$$

$$2e^- + MnO_2 + 2H^+ \rightarrow Mn(OH)_2 \qquad E° = -149.7/-96.49(2) = 0.78 \text{ v}$$

$$3e^- + MnO_4^- + 4H^+ \rightarrow MnO_2 + 2HOH \quad E° = -492.4/-96.49(3) = 1.70 \text{ v}$$

$$2e^- + Mn(OH)_2 + 2H^+ \rightarrow Mn + 2HOH \qquad E° = 140.6/-96.49(2) = -0.73 \text{ v}$$

$$2H^+ + Mn(OH)_2 \rightarrow Mn^{+2} + 2HOH \qquad \log K = -87.8/-5.7 = 15.4$$

$$K = 10^{15.4}$$

Step (7). *Writing E and pH equations.* For each of the first five equations above, the Nernst expression is written out, then the equation is solved for E as a function of pH. For the last equation, the equilibrium constant is written out, the equation is put into logarithmic form, and is then solved for the pH.

$$E(Mn^{+2}/Mn) = -1.18 - (0.0591/2) \log [1]/[1][Mn^{+2}]$$

$$= -1.18 + 0.030 \log [Mn^{+2}]$$

$$E(MnO_2/Mn^{+2}) = 1.23 - (0.0591/2) \log ([Mn^{+2}][1]^2/[1]^2[1][H^+]^4)$$

$$= 1.23 - 0.030 \log [Mn^{+2}] - 0.118 \text{ pH}$$

$$E(MnO_2/Mn(OH)_2) = 0.78 - (0.0591/2) \log ([1]/[1]^2[1][H^+]^2)$$

$$= 0.78 - 0.059 \text{ pH}$$

$$E(MnO_4^-/MnO_2) = 1.70 - (0.0591/3) \log ([1][1]^2/[1]^3[MnO_4^-][H^+]^4)$$

$$= 1.70 + 0.020 \log [MnO_4^-] - 0.079 \text{ pH}$$

$$E(Mn(OH)_2/Mn) = -0.73 - (0.0591/2) \log ([1][1]^2/[1]^2[1][H^+]^2)$$

$$= -0.73 - 0.059 \text{ pH}$$

$$K(Mn(OH)_2/Mn^{+2}) = 10^{15.4} = [Mn^{+2}][1]^2/[1][H^+]^2$$

$$2pH = 15.4 - \log [Mn^{+2}]$$

Step (8). *Drawing lines.* The six expressions above represent straight lines. They have been drawn into Figure 2.10. The lines have not been extended into regions where the species involved are not pertinent. For example, the Mn^{+2}/Mn line has not been extended into the basic region. The portions of the lines which invade regions of predominant species then have been erased to produce the completed E–pH diagram as displayed in Figure 2.11. Note that the six expressions given above can be used to make detailed calculations of the exact values of E and pH for species transformations.

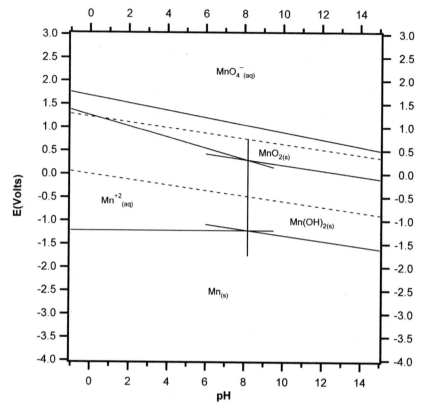

Figure 2.10 E–pH diagram for Mn species. Soluble species concentrations (except H^+) = $10^{-1.0}$ M. Soluble species and most solids are hydrated. No agents producing complexes or insoluble compounds are present other than HOH and OH^-. Preliminary line placements.

5. E–pH Computer Programs

The above calculations, being algebraic, can be readily incorporated into a computer program which quickly makes them, generates the lines, and then presents the resulting E–pH diagram. However, the species must be specified. A further advantage of a computer program is that it can be set up to minimize the free energy of the complete system involving any number of species. Such free energy minimizations by hand are tedious and time consuming, but the program can carry them out rapidly. Hence, species which do not exist under competition with other species will be omitted. Several such programs are available commercially, some from universities, and some in the public domain.

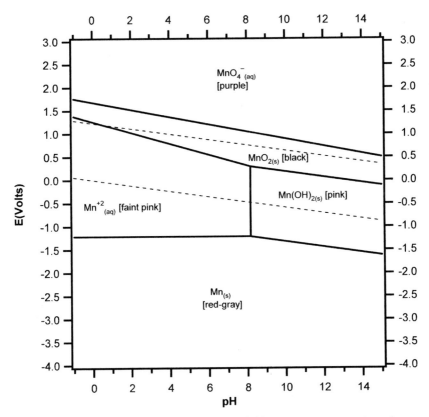

Figure 2.11 E–pH diagram for Mn species. Soluble species concentrations (except H^+) = $10^{-1.0}$ M. Soluble species and most solids are hydrated. No agents producing complexes or insoluble compounds are present other than HOH and OH^-. Final diagram.

6. Species Selection and Treatment

All of the solution and solid Mn/O/H species which persist in water (for at least several minutes) and for which standard free energy values are available are as follows (with the $\Delta G°$ values in kJ/mole in parentheses): Mn (0.0), MnO (−362.8), MnO_2 (−465.3), Mn_2O_3 (−878.9), Mn_3O_4 (−1283.0), $Mn(OH)_2$ (−615.0), MnO(OH) (−567.1), $HMnO_2^-$ (−506.1), Mn^{+3} (−85.0), Mn^{+2} (−228.4), MnO_2^{-2} (−429.0), MnO_4^- (−447.3), MnO_4^{-2} (−503.7), and $Mn(OH)^+$ (−407.0). The very fact that $\Delta G°$ values are available for these species is an indicator that they are probably important to the aqueous chemistry of Mn. When all these species are employed to construct an E–pH diagram, Figure 2.12 results. Notice that free energy minimization has

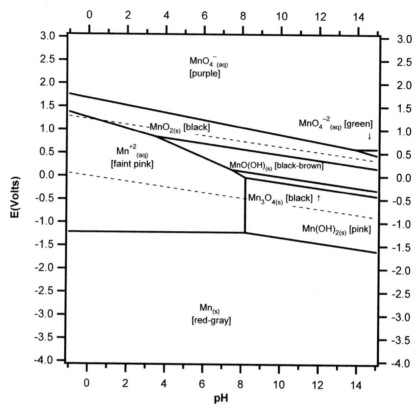

Figure 2.12 E–pH diagram for Mn species. Soluble species concentrations (except H^+) = $10^{-1.0}$ M. Soluble species and most solids are hydrated. No agents producing complexes or insoluble compounds are present other than HOH and OH$^-$. Species include Mn, Mn^{+2}, $Mn(OH)_2$, Mn_3O_4, $MnO(OH)$, MnO_2, MnO_4^-, and MnO_4^{-2}.

excluded several of the species, indicating that they are not as stable as the competing species which remain.

The species remaining in Figure 2.12 are Mn, Mn^{+2}, $Mn(OH)_2$, Mn_3O_4, $MnO(OH)$, MnO_2, MnO_4^{-2}, and MnO_4^-. Notice that Mn_3O_4 has a very narrow range of existence on the E scale, so it might be difficult to maintain under experimental conditions. If it is omitted, the diagram presented as Figure 2.13 is obtained. However, it will be important to include it if geochemical systems are being treated, since it occurs in nature as the mineral hausmannite. Further, the species MnO_4^{-2} also has a very narrow area of existence, and again might be difficult to maintain. Should it be omitted, Figure 2.14 is obtained.

Attention is now directed to the species MnO(OH). This compound, in which Mn is in an oxidation state of III, in a water system could be written as Mn_2O_3, MnO(OH), or $Mn(OH)_3$, the only differences between them being

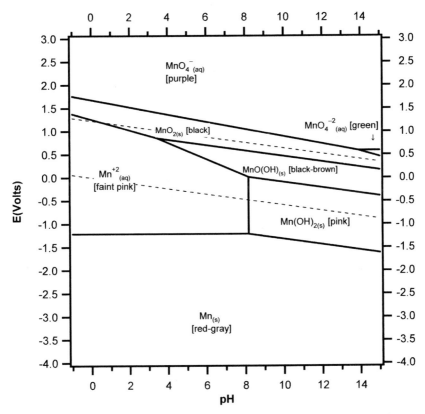

Figure 2.13 E–pH diagram for Mn species. Soluble species concentrations (except H^+) = $10^{-1.0}$ M. Soluble species and most solids are hydrated. No agents producing complexes or insoluble compounds are present other than HOH and OH^-. Species include Mn, Mn^{+2}, $Mn(OH)_2$, $MnO(OH)$, MnO_2, MnO_4^-, and MnO_4^{-2}.

the state of hydration [$Mn(OH)_3 - HOH = MnO(OH)$ and $2Mn(OH)_3 - 3HOH = Mn_2O_3$]. Among inorganic hydroxides of metals carrying a III or greater charge, there is a tendency for a freshly precipitated metal ion to be the hydroxide, for example $Mn(OH)_3$. As this compound ages or is heated, it tends to dehydrate, going first to $MnO(OH)$, then perhaps to Mn_2O_3. In general, the rate of this transformation increases as the charge on the metal ion gets larger (IV or greater). Many investigators believe that for many of the more-highly charged cations, the species with hydroxide ions never occur, the oxide being formed even upon fresh precipitation. For the Mn compounds, it is instructive to consider the $\Delta G°$ values of the following reactions:

$$2Mn(OH)_3 \rightarrow 2MnO(OH) + 2HOH$$

$2(-757.3)\qquad 2(-567.1)\qquad 2(-237.2)\quad \Delta G°(\text{reaction}) = -94.0\,\text{kJ}$

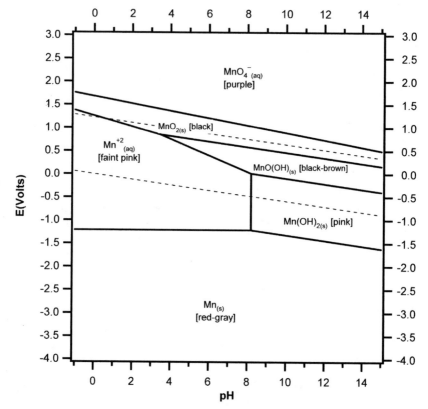

Figure 2.14 E–pH diagram for Mn species. Soluble species concentrations (except H^+) = $10^{-1.0}$ M. Soluble species and most solids are hydrated. No agents producing complexes or insoluble compounds are present other than HOH and OH^-. Species include Mn, Mn^{+2}, $Mn(OH)_2$, $MnO(OH)$, MnO_2, and MnO_4^-.

$$2Mn(OH)_3 \rightarrow Mn_2O_3 + 3HOH$$

2(−757.3)   − 878.9   3(−237.2) $\Delta G°$(reaction) = −75.9 kJ

$$2MnO(OH) \rightarrow Mn_2O_3 + HOH$$

2(−567.1)   − 878.9   − 237.2 $\Delta G°$(reaction) = +18.1 kJ (Note : +)

Since the $Mn(OH)_3$ system seeks the lowest (most stable) free energy, the preferred form of the Mn(III) oxide/hydroxide is $MnO(OH)$ as the E–pH diagram shows. In other words, the most stable is the partially dehydrated species. In other systems, for example, the Al system, possible species would be $Al(OH)_3$, $AlO(OH)$, and Al_2O_3. Analysis of the free energy relations of these species indicates that Al_2O_3 is the preferred species at equilibrium, but chemical experimentation shows that $Al(OH)_3$ precipitates. The $Al(OH)_3$ species then undergoes transformation to Al_2O_3 over time or with heating.

In many other cases, free energy values will lead to conclusions regarding which species is the most stable. A further illustration of the importance of species which differ by hydration is AlO_2^- and $Al(OH)_4^-$, the latter being preferred by free energy calculations ($AlO_2^- + 2HOH \rightarrow Al(OH)_4^-$).

Another important aspect of E–pH diagrams is the oversimplification of species. For example, the Mn^{+2} species in water solution is more properly written as $Mn(HOH)_6^{+2}$ which reflects the coordination number of the ion. The coordination numbers of $+1$, $+2$, and $+3$ cations for water in solution are generally 6, except for those which show 4 (Li^+, Be^{+2}, Ag^+, Pd^{+2}, Pt^{+2}, Sn^{+2}), those which show 8 (K^+ and Rb^+ sometimes, Cs^+, Sr^{+2}, Ba^{+2}, Y^{+3}, Gd^{+3} to Lu^{+3}), and those which show 9 (La^{+3} to Eu^{+3}, Ac^{+3}). In a similar simplification, $Al(OH)_4^-$ is more accurately written as $Al(HOH)_2(OH)_4^-$, again the coordination number of 6 for the cation being recognized. Another example is afforded by the Be^{+2} ion which undergoes hydrolysis (addition of hydroxide ion) at low pH values to give the predominant species $Be_3(HOH)_6(OH)_3^{+3}$. Details on actual species in aqueous solution can be found in D. T. Richens, The Chemistry of Aqua Ions, Wiley, New York, NY, 1997 and in the chemical literature as indexed on SciFinder, a facility provided to libraries by the American Chemical Society. Another oversimplification that will be made is that solid species will be written in the anhydrous form even though in many cases they commonly exist in the hydrated form. Details on the hydration of commonly existing compounds may be obtained from catalogs of suppliers. For example, $CuSO_4$ commonly comes as the pentahydrate $CuSO_4 \cdot 5H_2O$, but it will be treated by writing only $CuSO_4$. It is very important that these oversimplifications be constantly borne in mind when dealing with E–pH diagrams and when reading listings of compounds.

A further characteristic of E–pH diagrams is one which involves species which are thermodynamically unstable but kinetically are slow to decompose. Consider a piece of Mn metal placed in acid. The reaction is slow because of the limited surface area. However, if the Mn is powdered the reaction is faster due to a marked increase in surface area. Or, consider the MnO_4^- ion in water solution. This ion is known to decompose under these conditions, however, the reaction is ordinarily quite slow. Further, a piece of Al metal will not readily dissolve in dilute acid even though a thermodynamic prediction indicates it will. The reason is that there usually is a thin coating of refractory (non-reactive) Al_2O_3 on the surface. A similar situation is seen, at least in a portion of the pH range, for numerous other metals, particularly those that form highly stable oxides, such as Ti, Zr, Sn, Hf, Be, In, and Cr. Another consideration with regard to rates relates to the following ions: Ir^{+3}, Rh^{+3}, Cr^{+3}, Ru^{+3}, and Pt^{+2}. These ions are referred to as inert, because they and their complexes tend to react slowly, in contradistinction to labile ions which consist of most other ions and which usually react rapidly.

Finally, some considerations with regard to standard free energy values must be taken into account. Many $\Delta G°$ values are well known, some have sizable error limits, and some are estimated from chemical trends and behavior. When an E–pH diagram is derived and then seen to be in contradiction to chemical behavior, the cause may be various including kinetics (slow decomposition), erroneous species identification, or an erroneous $\Delta G°$ value. Certain aspects of an E–pH diagram are especially sensitive to $\Delta G°$ values, particularly pH values at which a hydroxide precipitates. Consider, for example, the precipitation of $La(OH)_3$ from a 1.00 M solution of La^{+3}. The accepted value of the $\Delta G°$ for $La(OH)_3$ is -1272.8 kJ/mole, that for La^{+3} is -686.1 kJ/mole, and that for HOH is -237.2 kJ/mole. These values lead to the following equation for the line on the E–pH diagram which separates $La(OH)_3$ from La^{+3}: $3pH = 21.9 - \log [La^{+3}]$. This gives a pH value of 7.3 for the precipitation of the hydroxide. Now let us suppose there is a 13 kJ error (about 1%) in the $\Delta G°$ for the $La(OH)_3$. Using a value of -1259.8 kJ/mole gives this relationship: $3pH = 24.2 - \log [La^{+3}]$, which predicts a precipitation pH value of 8.1. Similar calculations show that a difference of 2 kJ (about 0.16%) in the $\Delta G°$ of $La(OH)_3$ results in a 0.1 change in the pH value for precipitation. It is to be noted that many $\Delta G°$ values are derived from measured $E°$ values, and many $E°$ values are derived from measured $\Delta G°$ values. Equation (7) is used to convert one to the other.

3

Reactions and Applications

1. Introduction

E–pH diagrams involve two types of reactions: (1) Non-redox full reactions and (2) Redox half-reactions. Non-redox full reactions are exemplified by ones such as $Mn(OH)_2 + 2H^+ \rightarrow Mn^{+2} + 2HOH$. This is not a redox (reduction–oxidation) reaction, since there are no changes in oxidation numbers of the elements. Such reactions are reflected as vertical lines on E–pH diagrams. An example of a redox half-reaction is $2e^- + Mn^{+2} \rightarrow Mn$. As can be seen, this is a redox reaction, since electrons appear in the equation, and there is an oxidation number change (II to 0 for Mn). Such reactions are represented by horizontal or sloped lines in an E–pH diagram. In order to write complete reactions in which oxidation numbers change, two half-reactions must be combined. One half reaction will represent a reduction ($2e^- + Mn^{+2} \rightarrow Mn$) and the other will represent an oxidation ($Mg \rightarrow Mg^{+2} + 2e^-$). These half-reactions are combined such that the electrons cancel out and a complete redox equation is obtained ($Mn^{+2} + Mg \rightarrow Mn + Mg^{+2}$). Each of the two half-reactions has an E value, and the E value of the resulting complete redox equation is obtained by the difference in the E values of the contributing half-reactions. E–pH diagrams may be employed to predict non-redox full reactions and complete redox reactions and to ascertain E values of the latter. This will be the subject matter of the next few sections.

2. Reactions with HOH

Dashed lines in every E–pH diagram represent the E values for changes in HOH-related species (HOH, H^+, H_2, O_2, and implicitly OH^-). The upper dashed sloped line represents the reaction $4e^- + 4H^+ + O_2 \rightarrow 2HOH$ and is described by the equation $E = 1.23 - 0.059$ pH. The lower dashed sloped line represents the reaction $2e^- + 2H^+ \rightarrow H_2$ and is described by the equation $E = 0.00 - 0.059$ pH. Figure 3.1 shows the E–pH diagram for HOH, and Figure 3.2 shows the E–pH diagram for Mg with all soluble species at 1.00 M except H^+. The solid horizontal line represents the reaction $2e^- + Mg^{+2} \rightarrow Mg$, and the equation for the line is $E = -2.36 + 0.030 \log [Mg^{+2}]$.

In order to predict the interaction of Mg with HOH at a pH of 2.0, a vertical cut is made at this pH on the Mg diagram and another at this pH on the HOH diagram. See the dashed vertical lines on Figures 3.1 and 3.2.

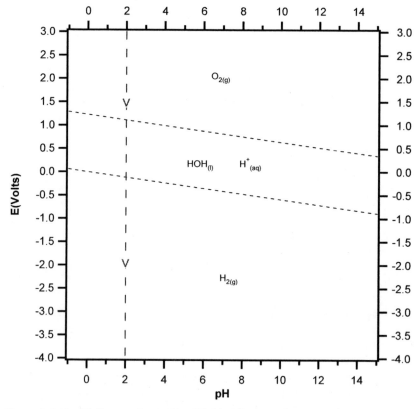

Figure 3.1 E–pH diagram for HOH. Soluble species and most solids are hydrated. No agents producing complexes or insoluble compounds are present other than HOH and OH^-. Species transformations take place at constant pH.

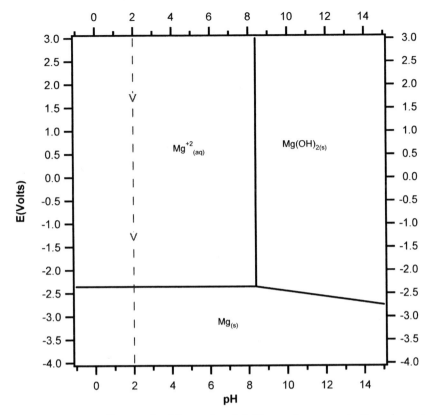

Figure 3.2 E–pH diagram for Mg species. Soluble species concentrations (except H^+) = $10^{0.0}$ M. Soluble species and most solids are hydrated. No agents producing complexes or insoluble compounds are present other than HOH and OH^-. Species transformations take place at constant pH.

These vertical cuts are called electron ladders. Rungs are placed on them at the proper E values, and proper species are indicated below and above the rungs. These are shown in Figure 3.3. The exact values of E for the HOH ladder have been calculated from the equations given above, and that for the Mg ladder from $E = -2.36 + 0.030 \log [Mg^{+2}]$, which is the equation that describes the line between Mg^{+2} and Mg on the Mg E–pH diagram.

Once these electron ladders have been written out (using the E–pH diagrams and the equations that describe the lines), the two ladders are combined as in Figure 3.4. The interaction between Mg and HOH $\equiv H^+$ at a pH of 2.0 can now be described using these rules: (1) Identify the two reacting species, Mg and HOH $\equiv H^+$, (2) Identify the lowest rung which contains one of the reacting species, -2.36 v with Mg, (3) Identify the rung above it which contains the other reacting species, -0.12 v with HOH $\equiv H^+$, (4) Write an equation showing the reaction of the lower species of the lower

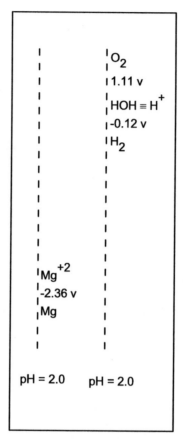

Figure 3.3 Electron ladders for Mg and H_2O at pH $= 2.0$. Soluble species concentrations (except H^+) $= 10^{0.0}$ M.

rung (Mg) with the upper species of the upper rung (HOH $\equiv H^+$) to give the two intermediate species, Mg^{+2} and H_2, (5) Balance the equation, (6) Subtract the E value of the lower rung from the E value of the upper rung to give the E value of the complete equation. These steps lead to this complete redox reaction:

$$Mg + 2H^+ \rightarrow Mg^{+2} + H_2 \quad E = -0.12 - (-2.36) = 2.24 \text{ v}$$

When the E for a reaction is positive, as this one is, the reaction will proceed to the right. A positive E value corresponds to a negative ΔG value as indicated by $\Delta G = -96.49nE$.

A similar procedure can be carried out at any desired pH. When such is done for a pH of 10.0, the following equations can be used to define the values

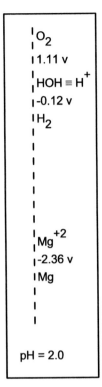

Figure 3.4 Combined electron ladder for Mg and H_2O at pH = 2.0. Soluble species concentrations (except H^+) = $10^{0.0}$ M.

and species at the three rungs:

$$E(O_2/HOH \equiv H^+) = 1.23 - 0.059 \text{ pH} = 0.64 \text{ v},$$

$$E(HOH \equiv H^+/H_2) = 0.00 - 0.059 \text{ pH} = -0.59 \text{ v, and}$$

$$E(Mg(OH)_2/Mg) = -1.86 - 0.059 \text{ pH} = -2.45 \text{ v}.$$

The consequent electron ladder is shown in Figure 3.5. The reaction which results by the combination of the lower species at the lower rung with the upper species at the upper rung to give the intermediate species is:

$$Mg + 2HOH \rightarrow Mg(OH)_2 + H_2 \quad E = -0.59 - (-2.45) = 1.86 \text{ v}$$

Another variation on the theme would be to alter the concentration of the Mg in the Mg E–pH diagram. This would give a different value for the rung separating the two Mg species. Let us suppose that the Mg soluble species concentrations be designated as $10^{-4.0}$ M and that the pH be specified as 4.0.

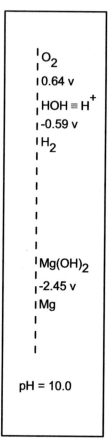

Figure 3.5 Combined electron ladder for Mg and H_2O at pH $= 10.0$. Soluble species concentrations (except H^+) $= 10^{0.0}$ M.

The pertinent equations for the rungs on the electron ladder are as follows:

$$E(O_2/HOH \equiv H^+) = 1.23 - 0.059 \, pH = 1.23 - 0.059(4.0) = 0.99 \text{ v}$$

$$E(HOH \equiv H^+/H_2) = 0.00 - 0.059 \, pH = 0.00 - 0.059(4.0) = -0.24 \text{ v}$$

$$E(Mg^{+2}/Mg) = -2.36 + 0.03 \log [Mg^{+2}] = -2.36 + 0.03 \log [10^{-4.0}]$$
$$= -2.48 \text{ v}$$

The electron ladder which shows these interrelationships is pictured in Figure 3.6. And the reaction which takes place along with its potential is

$$Mg + 2H^+ \rightarrow Mg^{+2} + H_2 \quad E = -0.24 - (-2.48) = 2.24 \text{ v}$$

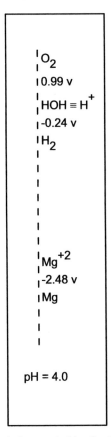

Figure 3.6 Combined electron ladder for Mg and H_2O at pH = 4.0. Soluble species concentrations (except H^+) = $10^{-4.0}$ M.

3. The Mn–Fe–HOH System (Excess Mn Species)

Consideration will now be paid to the Mn–Fe–HOH system. The E–pH diagrams for Mn and Fe at soluble species concentrations of 1.00 M (except H^+) are given in Figures 3.7 and 3.8. Remember that the HOH E–pH diagram is involved in each of these. Electron ladders have been abstracted at a pH of 1.0, the Mn soluble species concentrations have been left at 1.00 M, and the Fe species have been set at $10^{-3.0}$ M. These diagrams appear as Figures 3.9 and 3.10, with a combined diagram as Figure 3.11. Recognizing the soluble species concentrations as 1.00 M for Mn, and $10^{-3.0}$ M for Fe, and the pH as 1.0, the exact values of E for each of the rungs on the ladders have been calculated from these equations:

$$E(MnO_4^-/MnO_2) = 1.70 - 0.079 \text{ pH} + 0.020 \log [MnO_4^-] = 1.62 \text{ v}$$

$$E(MnO_2/Mn^{+2}) = 1.23 - 0.118 \text{ pH} - 0.030 \log [Mn^{+2}] = 1.11 \text{ v}$$

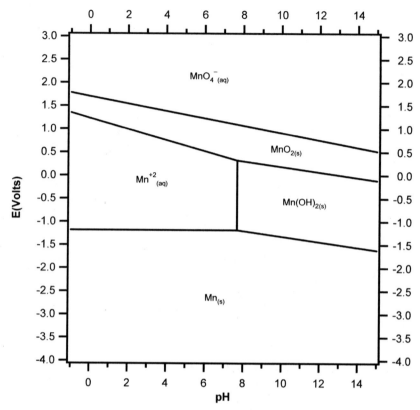

Figure 3.7 E–pH diagram for Mn species. Soluble species concentrations (except H^+) = $10^{0.0}$ M. Soluble species and most solids are hydrated. No agents producing complexes or insoluble compounds are present other than HOH and OH^-.

$$E(Mn^{+2}/Mn) = -1.18 + 0.030 \log [Mn^{+2}] = -1.18 \text{ v}$$

$$E(O_2/HOH \equiv H^+) = 1.23 - 0.059 \text{ pH} = 1.17 \text{ v}$$

$$E(HOH \equiv H^+/H_2) = 0.00 - 0.059 \text{ pH} = -0.06 \text{ v}$$

$$E(Fe^{+3}/Fe^{+2}) = 0.77 \text{ v}$$

$$E(Fe^{+2}/Fe) = -0.45 + 0.030 \log [Fe^{+2}] = -0.54 \text{ v}$$

Now numerous possible reactions can be predicted. Take special note that the Mn species are in considerable excess of the Fe species. Consider the situation in which a large amount of Mn metal is placed in the solution with Fe metal. Obviously nothing happens because the Fe is as far reduced as it can be. In other words, Fe is not an upper species on a rung above

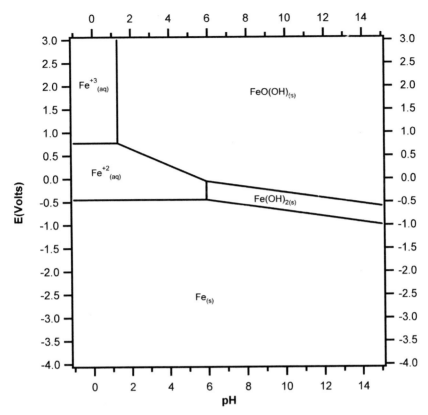

Figure 3.8 E–pH diagram for Fe species. Soluble species concentrations (except H^+) $= 10^{0.0}$ M. Soluble species and most solids are hydrated. No agents producing complexes or insoluble compounds are present other than HOH and OH⁻.

the Mn. Next, consider Mn metal placed in a solution containing Fe^{+2}. The Mn, being a lower species on a lower rung, will react with the Fe^{+2} to give the intermediate species Mn^{+2} and Fe, that is $Mn + Fe^{+2} \rightarrow Fe + Mn^{+2}$. Now notice that the next rung is the $HOH \equiv H^+/H_2$ rung. This shows that Mn will react with H^+ to give Mn^{+2} and H_2 (lower + upper → intermediate species), or $Mn + 2H^+ \rightarrow Mn^{+2} + H_2$. In fact, Mn will react with any species on an upper rung above it: Fe^{+2}, H^+, Fe^{+3}, O_2. Mn^{+2} will be produced and the Fe^{+2} would end up as Fe, except it in turn will react with H^+ to produce Fe^{+2} and H_2. Hence in this system the final result of the reaction of excess Mn with a small amount of Fe^{+2} will be Mn^{+2}, H_2, and Fe^{+2}. By similar reasoning, the Fe^{+3} will end up as the same products.

Now consider starting with Mn^{+2} put into solution with various Fe species. In order to react Mn^{+2} must function as the lower species on a lower rung (a reductant) or as an upper species on an upper rung (an oxidant).

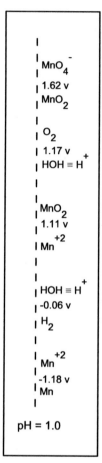

Figure 3.9 Combined electron ladder for Mn and H_2O at pH = 1.0. Soluble species concentrations (except H^+) = $10^{0.0}$ M.

If Mn^{+2} is placed with Fe^{+2}, there will be no reaction, because it is not the lower species with regard to Fe^{+2}. If Mn^{+2} is placed with Fe^{+3}, again there will be no reaction since the lower/lower–upper/upper relationship is not fulfilled. Further, take notice that Mn^{+2} does not react with $HOH \equiv H^+$ for the same reason. However, Mn^{+2} sits as the lower species on the MnO_2/Mn^{+2} rung, which indicates that it can react with the upper species O_2 on the $O_2/HOH \equiv H^+$ rung as follows: $2Mn^{+2} + O_2 + 2HOH \rightarrow 2MnO_2 + 4H^+$. Thus, if the system is in air, O_2 would be available, and the process would occur.

Next, take a look at the situation in which MnO_2 is placed in solution with different Fe species. MnO_2 appears as the upper species on the MnO_2/Mn^{+2} rung, therefore it is well to look at lower Fe and HOH species on rungs below.

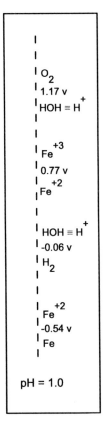

Figure 3.10 Combined electron ladder for Fe and H_2O at pH = 1.0. Soluble species concentrations (except H^+) = $10^{0.0}$ M.

It can be seen that MnO_2 will oxidize Fe^{+2} to Fe^{+3}, H_2 (if any is around) to H^+, and Fe to Fe^{+2}, then on to Fe^{+3}. Appropriate equations may be written for these reactions, and voltages calculated by differences. MnO_2 appears as the lower species on the MnO_4^-/MnO_2 rung, which means that it will react with the upper species on any rung above it, and there is none.

As a final consideration in this system, look at MnO_4^- placed in solution with several Fe species. The anion MnO_4^- is the upper species on the upper rung of Figure 3.11. This implies that it has the potential to react with all lower species on rungs below it. These include $HOH \equiv H^+$, Fe^{+2}, H_2 (probably not present), and Fe. These reactions are described by the following equations, with the calculated E values appended:

$$4MnO_4^- + 4H^+ \rightarrow 3O_2 + 4MnO_2 + 2HOH$$

$$E = 1.62 - 1.17 = 0.45 \text{ v}$$

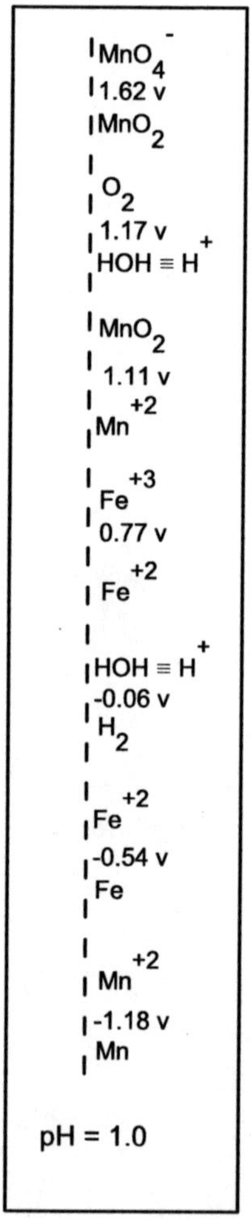

$$MnO_4^-$$
$$1.62 \text{ v}$$
$$MnO_2$$

$$O_2$$
$$1.17 \text{ v}$$
$$HOH \equiv H^+$$

$$MnO_2$$
$$1.11 \text{ v}$$
$$Mn^{+2}$$

$$Fe^{+3}$$
$$0.77 \text{ v}$$
$$Fe^{+2}$$

$$HOH \equiv H^+$$
$$-0.06 \text{ v}$$
$$H_2$$

$$Fe^{+2}$$
$$-0.54 \text{ v}$$
$$Fe$$

$$Mn^{+2}$$
$$-1.18 \text{ v}$$
$$Mn$$

pH = 1.0

Figure 3.11 Combined electron ladder for Mn, Fe, and H_2O at pH = 1.0. Mn soluble species concentrations = $10^{0.0}$ M. Fe soluble species concentrations = $10^{-3.0}$ M.

$$MnO_4^- + 3Fe^{+2} + 4H^+ \rightarrow MnO_2 + 2HOH + 3Fe^{+3}$$

$$E = 1.62 - 0.77 = 0.85 \text{ v}$$

$$2MnO_4^- + 3Fe + 8H^+ \rightarrow 2MnO_2 + 4HOH + 3Fe^{+2}$$

$$E = 1.62 - (-0.54) = 2.16 \text{ v}$$

The Fe^{+2} produced in the last reaction will proceed to be oxidized according to the second reaction.

The electron ladders also give predictions with regard to the coexistence of species of the same element. For example, consider an excess of MnO_4^- introduced to a solution containing Mn^{+2}. By observation of the MnO_4^-/MnO_2 and the MnO_2/Mn^{+2} rungs, the following reaction can be foreseen:

$$2MnO_4^- + 3Mn^{+2} + 2HOH \rightarrow 5MnO_2 + 4H^+ \quad E = 1.62 - 1.11 = 0.51 \text{ v}$$

It is assumed that the reader recognizes that this equation is a subtractive combination of the half-reactions that apply to the two rungs being considered:

$$3e^- + MnO_4^- + 4H^+ \rightarrow MnO_2 + 2HOH \qquad E = 1.62 \text{ v}$$

minus   minus

$$2e^- + MnO_2 + 4H^+ \rightarrow Mn^{+2} + 2HOH \qquad E = 1.11 \text{ v}$$

4. The Mn–Fe–HOH System (Excess Fe Species)

The previous exercise will now be repeated except that the soluble Fe species will be put in excess (1.00 M) with the soluble Mn species being at $10^{-3.0}$ M. Various Fe species will be introduced in the presence of smaller concentrations of Mn species. When the previous equations are applied to obtain precise values of E, the electron ladder at a pH $= 1.0$ is as shown in Figure 3.12. Consider the introduction of excess Fe into a solution containing Mn species. The key consideration is that Fe acting as a lower species on the second rung from the bottom will react with all upper species on rungs above it: $HOH \equiv H^+, O_2, MnO_2,$ and MnO_4^-. Fe does not appear as an upper species, and therefore it cannot react with any species below it. The possible reactions are as follows:

$$Fe + 2H^+ \rightarrow Fe^{+2} + H_2$$

$$E = -0.06 \text{ v} - (-0.45) = 0.39 \text{ v}$$

$$2Fe + O_2 + 4H^+ \rightarrow 2Fe^{+2} + 2HOH$$

$$E = 1.17 - (-0.45) = 1.62 \text{ v}$$

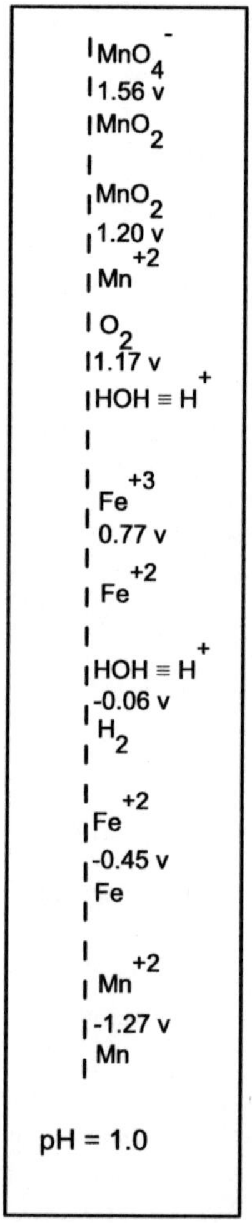

$$
\begin{array}{l}
| \; MnO_4^- \\
| \; 1.56 \; v \\
| \; MnO_2 \\
| \\
| \; MnO_2 \\
| \; 1.20 \; v \\
 \quad \;^{+2} \\
| \; Mn \\
\\
| \; O_2 \\
| \; 1.17 \; v \\
| \; HOH \equiv H^+ \\
| \\
| \quad \;^{+3} \\
| \; Fe \\
| \; 0.77 \; v \\
| \quad \;^{+2} \\
| \; Fe \\
| \\
| \; HOH \equiv H^+ \\
| \; -0.06 \; v \\
| \; H_2 \\
| \\
| \quad \;^{+2} \\
| \; Fe \\
| \; -0.45 \; v \\
| \; Fe \\
| \\
| \quad \;^{+2} \\
| \; Mn \\
| \; -1.27 \; v \\
| \; Mn \\
| \\
\end{array}
$$

pH = 1.0

Figure 3.12 Combined electron ladder for Mn, Fe, and H_2O at pH $=$ 1.0. Mn soluble species concentrations $=$ $10^{-3.0}$ M. Fe soluble species concentrations $= 10^{0.0}$ M.

$$Fe + MnO_2 + 4H^+ \rightarrow Fe^{+2} + Mn^{+2} + 2HOH$$

$$E = 1.20 - (-0.45) = 1.65 \text{ v}$$

$$3Fe + 2MnO_4^- + 8H^+ \rightarrow 3Fe^{+2} + 2MnO_2 + 4HOH$$

$$E = 1.56 - (-0.45) = 2.01 \text{ v}$$

The MnO_2 produced in the last reaction is subject to further reduction as will be pointed out below.

Next, excess Fe^{+2} is to be introduced to a solution which might contain various Mn species. Fe^{+2} functions as an upper species on the Fe^{+2}/Fe rung and as the lower species on the Fe^{+3}/Fe^{+2} rung. Hence it can react with any lower species on a rung below it, or with any upper species on a rung above it. This indicates that Fe^{+2} will react with Mn, and that Fe^{+2} will react with O_2, MnO_2, and MnO_4^-. Appropriate equations for these reactions are

$$Fe^{+2} + Mn \rightarrow Fe + Mn^{+2}$$

$$E = -0.45 - (-1.27) = 0.82 \text{ v}$$

$$4Fe^{+2} + O_2 + 4H^+ \rightarrow 4Fe^{+3} + 2HOH$$

$$E = 1.17 - 0.77 = 0.40 \text{ v}$$

$$2Fe^{+2} + MnO_2 + 4H^+ \rightarrow 2Fe^{+3} + Mn^{+2} + 2HOH$$

$$E = 1.20 - 0.77 = 0.43 \text{ v}$$

$$3Fe^{+2} + MnO_4^- + 4H^+ \rightarrow 3Fe^{+3} + MnO_2 + 2HOH$$

$$E = 1.56 - 0.77 = 0.79 \text{ v}$$

Now, consider the addition of an excess of Fe^{+3} to a solution which contains small concentrations of various Mn species. Fe^{+3} appears as only an upper species which indicates that it cannot be oxidized further, that is, it reacts with no species above it. However, being an upper species on the Fe^{+3}/Fe^{+2} rung, it can attack lower species on lower rungs, such as H_2 and Mn. H_2 is usually not present and thus no reaction occurs, but the attack on Mn is described by this equation:

$$2Fe^{+3} + Mn \rightarrow 2Fe^{+2} + Mn^{+2} \quad E = 0.77 - (-1.27) = 2.04 \text{ v}$$

It is well to realize that Fe^{+3} can also interact with Fe to give Fe^{+2}. This is recognized by noting that the Fe^{+3}/Fe^{+2} rung rests above the Fe^{+2}/Fe rung.

5. Combining Redox Reactions

Consider the electron ladder in Figure 3.13. This ladder was abstracted from the E–pH diagrams of HOH and Cr with all soluble species at $10^{0.0}$ M and a

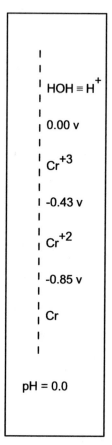

Figure 3.13 Combined electron ladder for Cr and H_2O at pH $= 0.0$. Soluble species concentrations (except H^+) $= 10^{0.0}$ M.

pH of 0.0. The interaction of Cr metal with H^+ is described by this equation

$$Cr + 2H^+ \rightarrow Cr^{+2} + H_2$$

and the E value is obtained by subtracting the value on the lower rung from the value on the upper rung $0.00 - (-0.85) = 0.85$ v. Likewise, the interaction of Cr^{+2} with H^+ is described by

$$2Cr^{+2} + 2H^+ \rightarrow 2Cr^{+3} + H_2$$

and the E is calculated to be $0.00 - (-0.43) = 0.43$ v. It can be seen that combining the two equations will lead to an equation describing the oxidation of Cr metal to Cr^{+3} : $Cr \rightarrow Cr^{+2} \rightarrow Cr^{+3}$. This overall oxidation is given by the following equation:

$$2Cr + 6H^+ \rightarrow 2Cr^{+3} + 3H_2$$

Caution must be exercised with regard to calculating the E value of this overall oxidation reaction. Calculations which convert the E values for the $Cr \rightarrow Cr^{+2}$ and $Cr^{+2} \rightarrow Cr^{+3}$ equations into ΔG values are required. These conversions follow:

$$2Cr + 4H^+ \rightarrow 2Cr^{+2} + 2H_2 \quad E = 0.85 \text{ v} \quad \Delta G = -96.49(4)0.85 = -328.1 \text{ kJ}$$

$$2Cr^{+2} + 2H^+ \rightarrow 2Cr^{+3} + H_2 \quad E = 0.43 \text{ v} \quad \Delta G = -96.49(2)0.43 = -83.0 \text{ kJ}$$

The equations are now added and the ΔG values are added to give the following:

$$2Cr + 6H^+ \rightarrow 2Cr^{+3} + 3H_2 \quad \Delta G = -411.1 \text{ kJ}$$

This value of ΔG is now converted to the value of E, as follows: $E = -411.1/-96.49(6) = 0.71$ v. ΔG values must be employed since the number of electrons in the final equation differs from those in the equations which are combined. This is because E values are potentials, whereas equations involve ΔG values which are energies.

6. The HOH E–pH Diagram Revisited

Thermodynamically, any species residing above the upper water line in an E–pH diagram should react to produce O_2, and any species below the lower water line should react to produce H_2 (See Figure 3.14). In a number of cases, these species do not react as predicted when they are investigated experimentally. There can be several reasons for this: (1) slow kinetics, (2) passivity, (3) reaction surface.

Experimentally, in some cases, the $E°$ for water decomposition to O_2 seems to be 1.48 ± 0.25 v or thereabouts rather than the 1.23 v that is predicted theoretically from free-energy considerations. Some reactions in the range of 1.23 to 1.73 v, and even above, go slowly enough that there is some pseudo-stability. This gives an added region of short-term, non-equilibrium stability in water solutions. For example, MnO_4^- can persist in water solution for several days. The $E°$ for the MnO_4^-/MnO_2 transformation is 1.69 v. The persistence of the MnO_4^- is because the reaction $4MnO_4^- + 4H^+ \rightarrow 4MnO_2 + 2HOH + O_2$ is slow, not reaching thermodynamic equilibrium rapidly. It is observed that as the pH is increased, the reaction goes even more slowly. Other examples of this pseudo-stability are afforded by H_2O_2, BaO_2, RhO_2, and ClO_4^-.

In practice, in some cases, the $E°$ for water decomposition into H_2 appears to be -0.25 ± 0.25 v or thereabouts rather than 0.00 v as is predicted thermodynamically. Examples are Yb^{+2}, Ti^{+3}, V^{+2}, and Cr^{+2}. Often the reason is a slow reaction as treated above. The pseudo-stability of a solid species may often be attributed to one or both of two causes: (1) passivity or (2) reaction surface. Many metals, especially those which show high oxidation numbers, develop very strong, thin, refractory (non-reactive) oxide layers

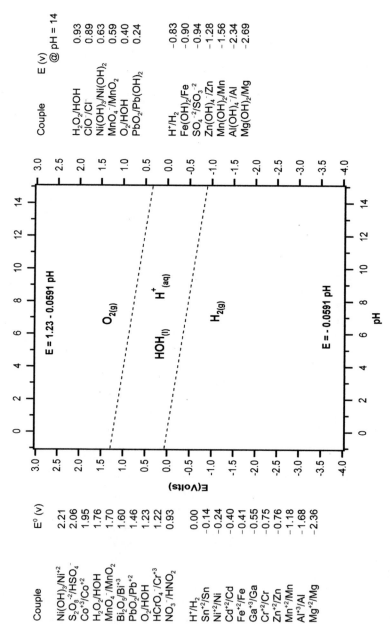

Figure 3.14 E–pH diagram for water system with commonly used oxidants and reductants. Soluble species and most solids are hydrated. No agents producing complexes or insoluble compounds are present other than HOH and OH⁻.

when exposed to air or O_2 or HOH or oxidizing agents, particularly acids, such as HNO_3. Examples are Al, Ti, Zr, Hf, V, Nb, Ta, Cr, Mo, W, and somewhat by Fe, Co, and Ni. These coatings often make the metals resistant to further attack by HOH and H^+, and to other reactions. The degree of resistance can vary considerably depending upon the previous treatment of the metal. With regard to reaction surface, consider a cube of Zn 1.00 cm on a side. This chunk will dissolve slowly in acid, giving a portion of the Zn a relatively long-term existence. Should the same cube of Zn be powdered, the greatly increased surface area will shorten the existence of the metal.

Another important consideration with regard to reaction surface is illustrated by the behavior of oxides of higher-oxidation-number metals. Take for example, the oxide La_2O_3. This oxide can be produced in several ways, but one of the most common methods is by the precipitation of $La(OH)_3$ and then heating the precipitate to drive off the water. The rate and final temperature of this dehydration can strongly affect the reactivity of the oxide with acid. High temperatures and slow rates lead to an oxide which is resistant to dissolution by acids. This phenomenon is due to the character of the metal to oxide bridges in the compound, as well as the crystal structure. It is also to be noticed that different forms of compounds often show different standard free-energy values in addition to evidencing kinetic differences. Again, this is a crystal structure phenomenon. An example is afforded by the several forms of the oxide of Al. These are given here with the standard free energies in kJ/mole shown in parentheses: α-corundum (-1582.3), boehmite (-1588.2), diaspore (-1603.8), gibbsite (-1598.2), amorphous (-1532.0).

7. General Conclusions

In general then, reactions may be thermodynamically predicted by using the E–pH diagrams of the two elements involved. A pH is selected, and a vertical cut (an electron ladder) in both diagrams is made at that pH. Then a soluble-species concentration for each of the two elements is selected, and the rungs on the ladders (vertical cuts) are adjusted in E value by using the equations which describe the transformation lines on the E–pH diagrams. Following this, all rungs from both electron ladders, plus the HOH rungs, are combined into one electron ladder in decreasing order of E value. Then reactions can be predicted by combining lower species on lower rungs with upper species on upper rungs to give the intermediate species. The equations describing these reactions are readily balanced by consideration of the two half-reactions involved. Finally, attention is to be paid to the factors which might make the reactions go slowly. All of these considerations will lead to reactions which usually correlate with experimental observations.

Figure 3.14 displays commonly used oxidants and reductants along with their potentials. Accompanying the redox couples is the HOH E-pH diagram since it represents the medium to which they are referred.

4

Precipitation and Complexation

1. Introduction

In Chapter 2, a method for the construction of single-element E–pH diagrams has been presented. No agent which could produce an insoluble compound nor any agent which could complex with any of the simple ions in the single-element diagrams except OH^- has been included. However, it is of interest in many cases to derive E–pH diagrams for systems which involve the precipitation or complexation of one or more of the simple ions present in the single-element diagram. One method of deriving such diagrams is to recognize that precipitation or complexation of a simple cation or anion reduces the concentration of the simple ion in solution considerably. If this reduced simple-ion concentration is calculated, it can then be used to construct a new E–pH diagram for the simple ion. Therefore, since the predominant species in the region labeled as the simple ion is no longer the simple cation or anion due to the precipitation or complexation, the region is re-labeled with the precipitated compound or the complex. E–pH diagrams which involve the precipitation or complexation of one or more of the simple ions present in the single-element diagram may also be obtained using one of the available computer programs. For complicated systems, the hand calculations become time consuming and it is often better to employ a computer program.

2. Precipitation and E–pH Diagrams

In order to determine the changes in a single element E–pH diagram that occur due to the addition of a precipitating species and the resulting formation of an insoluble compound, the following five steps may be followed.

(1) Select an element of interest showing a simple cation or anion and construct the E–pH diagram for the element at a soluble species equilibrium concentration of $10^{-4.0}$ M.

(2) Select a precipitating agent which will form an insoluble compound with a simple cation or anion of the element of interest. In order to determine which species will form insoluble compounds with the simple ions of an element, it is often useful to consult a list of solubility rules. These rules are a summary of empirical observations that certain combinations of cations and anions in HOH react to produce insoluble or very slightly soluble compounds. The solubility rules given below are to be used in the order given, with careful recognition of exceptions where indicated.

a. Most NH_4^+, Li^+, Na^+, K^+, Rb^+, and Cs^+ salts are soluble.
b. Most NO_3^-, NO_2^-, $C_2H_3O_2^-$, ClO_4^-, ClO_3^-, BrO_3^-, MnO_4^-, ReO_4^-, and many F^- salts are soluble. (F^- salts of Ca^{+2}, Sr^{+2}, Ba^{+2}, Sc^{+3}, Th^{+4}, and lanthanoid(III) ions are insoluble.)
c. Most Ag^+, Hg_2^{+2}, Pb^{+2}, and Tl^+ salts are insoluble.
d. Most Cl^-, Br^-, and I^- salts are soluble.
e. Most CO_3^{-2}, S^{-2}, OH^-, SO_3^{-2}, PO_4^{-3}, AsO_4^{-3}, CrO_4^{-2}, $C_2O_4^{-2}$, VO_4^{-3}, MoO_4^{-2}, and WO_4^{-2} salts are insoluble.
f. Most SO_4^{-2} salts are soluble. (SO_4^{-2} salts of Ca^{+2}, Sr^{+2}, Ba^{+2}, Pb^{+2}, and Hg_2^{+2} are insoluble.)

The solubility of a specific compound may also be found in literature references such as Lange's Handbook of Chemistry[5] or the CRC Handbook of Chemistry and Physics[6]. These references have extensive listings of inorganic compounds and their aqueous solubilities.

(3) Write the equation for the slight dissolution of the insoluble compound and calculate the $\Delta G°$(reaction) using Equation (6) from Chapter 2. Once the $\Delta G°$(reaction) value is obtained, the K value is calculated using Equation (8) also from Chapter 2. Keeping in mind that the concentration (activity) of the insoluble compound is defined as 1, it is recognized that the K value for the dissolution of the insoluble compound is the solubility product constant, K_{sp}. Alternatively, the K_{sp} value may be available from compilations of such values as presented in Lange's Handbook of Chemistry[5] or the CRC Handbook of Chemistry and Physics.[6]

(4) With the original soluble-species equilibrium concentration of $10^{-4.0}$ M, consider the addition of $10^{0.0}$ M (1.00 M) concentration of the precipitating agent to the system. Then, using the K_{sp}, calculate the resulting equilibrium concentration of the original soluble ion. The E–pH diagram is now plotted using this equilibrium concentration. Then the label for the original simple ion is changed to the formula of the precipitate.

(5) Repeat the procedure in step (4) using $10^{-2.0}$ M as the concentration of the precipitating agent. Then repeat using $10^{-4.0}$ M, $10^{-6.0}$ M, $10^{-8.0}$ M, etc. as the concentrations of the precipitant until the resulting equilibrium concentration of the original ion is greater than $10^{-4.0}$ M. Such a situation indicates that no precipitate will form, and the resulting E–pH diagram is no longer different than the original.

3. Addition of Cl⁻ to the Ag E–pH System

Using the five steps outlined in the previous section, the effects of the addition of Cl^- on the Ag E–pH system will now be determined.

(1) Element selection.   The Ag E–pH diagram is selected for study because silver exhibits a single soluble cation species: the Ag^+ ion. The Ag E–pH diagram is then constructed for a dissolved species concentration of $10^{-4.0}$ M using the method discussed in Chapter 2. This diagram appears as Figure 4.1.

(2) Precipitating agent selection.   From consideration of the solubility rules, it is noted that most Ag^+ compounds (other than those in rule [b]) are insoluble. Therefore, except for the anions listed in rule [b], almost any anion could be used as a precipitating agent for the Ag^+ ion. For this example, Cl^- is selected as the precipitating agent and the resulting insoluble compound is $AgCl_{(s)}$. It is of importance to recognize that the Cl^- is not stable over the entire E range given on E–pH diagrams. As Figure 4.2 illustrates, Cl^- is not a pertinent species above the bold solid line. Above this line the Cl^- has been oxidized to Cl_2 and/or ClO_4^-. Therefore, the E–pH diagrams for the Ag^+/Cl^- system will show that the shaded region above the bold solid line is irrelevant to chloride precipitation. This bold solid line is described by the following equations, the first applying to the oxidation to Cl_2 in the low pH region, and the second applying to the oxidation to ClO_4^- over the remaining expanse.

$$E = 1.40 - 0.0591 \log [Cl^-] + 0.030 \log [Cl_2] \quad \text{for the } Cl_2/Cl^- \text{ line, and}$$

$$E = 1.39 - 0.0591 \text{ pH} \qquad\qquad\qquad \text{for the } ClO_4^-/Cl^- \text{ line}$$

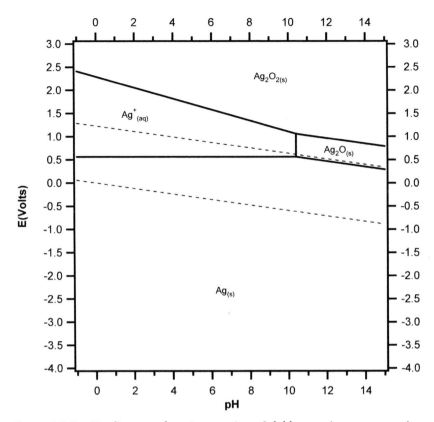

Figure 4.1 E–pH diagram for Ag species. Soluble species concentrations (except H^+) $= 10^{-4.0}$ M. Soluble species and most solids are hydrated. No agents producing complexes or insoluble compounds are present other than HOH and OH^-.

(3) Calculation of K_{sp} value. The slight dissolution of the insoluble compound is given by the equation:

$$AgCl_{(s)} \rightarrow Ag^+_{(aq)} + Cl^-_{(aq)}$$

The pertinent ΔG° values (in kJ/mole) are $AgCl_{(s)}$ (-109.9), $Ag^+_{(aq)}$ (77.1), and $Cl^-_{(aq)}$ (-131.4). Calculation of the ΔG°(reaction) is carried out using Equation (6) from Chapter 2.

$$\Delta G^\circ(\text{reaction}) = \Sigma \Delta G^\circ(\text{products}) - \Sigma \Delta G^\circ(\text{reactants}) \tag{6}$$

$$\Delta G^\circ(\text{reaction}) = [77.1 + (-131.4)] - [-109.9] = 55.6 \text{ kJ/mole}$$

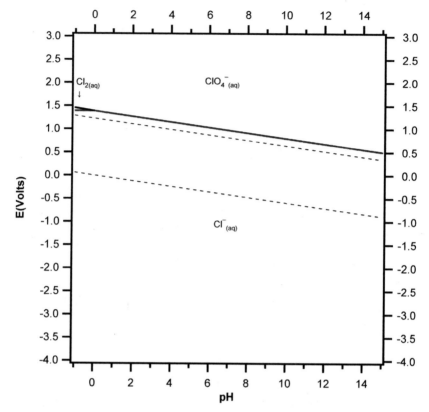

Figure 4.2 E–pH diagram for Cl species. Soluble species concentrations (except H^+) = $10^{0.0}$ M. Soluble species and most solids are hydrated. No agents producing complexes or insoluble compounds are present other than HOH and OH^-.

Once the $\Delta G°$(reaction) value is obtained, the K value can be calculated using Equation (8) from Chapter 2.

$$\log K = \Delta G° / - 5.7 = 55.6 / - 5.7 = -9.75 \tag{8}$$

$$K = 10^{-9.8} = [Ag^+][Cl^-]/[AgCl] = K_{sp} = [Ag^+][Cl^-]$$

Noting that the concentration of the insoluble compound is defined as 1, the K value is re-labeled as K_{sp}.

(4) Flooding the system with the precipitant.   With the Ag^+ in the system at $10^{-4.0}$ M, enough Cl^- is added to give an equilibrium concentration of $10^{0.0}$ or 1.00 M. A calculation using the K_{sp} is now made to ascertain the equilibrium concentration of Ag^+.

$$[Ag^+] = K_{sp}/10^{0.0} = 10^{-9.8}/10^{0.0} = 10^{-9.8} \text{ M}$$

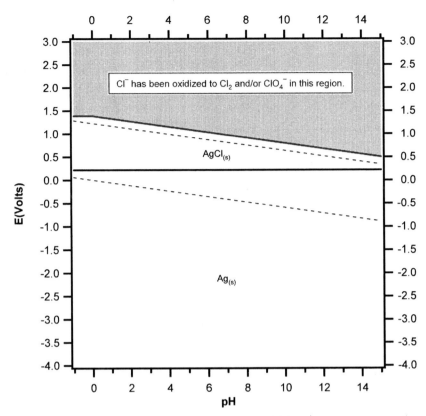

Figure 4.3 E–pH diagram for Ag species. Soluble species concentrations (except H^+) = $10^{-9.8}$ M. Soluble species and most solids are hydrated. No agents producing complexes or insoluble compounds are present other than HOH, OH^-, and Cl^-. Ag^+ label changed to AgCl (see text).

The E–pH diagram for Ag is now plotted using $[Ag^+] = 10^{-9.8}$ M. Since the majority of the Ag^+ has now been precipitated, the predominant species in the Ag^+ region of the diagram is no longer Ag^+, but AgCl. In accordance with this, the Ag^+ label is changed to AgCl. The resulting diagram is depicted in Figure 4.3.

(5) Varying the precipitant concentration. Next, it is instructive to add only enough Cl^- to give an equilibrium value of $10^{-2.0}$ M. The calculation of the $[Ag^+]$ follows the previous pattern.

$$[Ag^+] = K_{sp}/10^{-2.0} = 10^{-9.8}/10^{-2.0} = 10^{-7.8} \text{ M}$$

Again, the E–pH diagram for Ag is plotted using $[Ag^+]$ and the label Ag^+ is changed to AgCl. This diagram is shown in Figure 4.4. By using an equilibrium value of $10^{-4.0}$ M for the Cl^-, the $[Ag^+]$ is found to be $10^{-5.8}$ M. The resultant

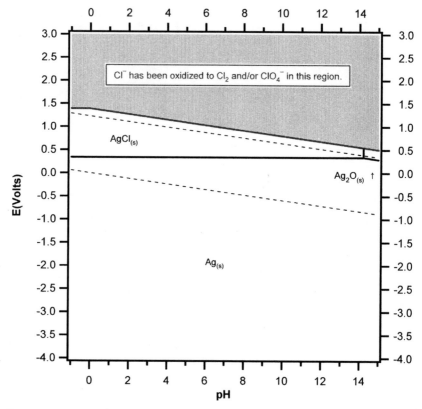

Figure 4.4 E–pH diagram for Ag species. Soluble species concentrations (except H^+) = $10^{-7.8}$ M. Soluble species and most solids are hydrated. No agents producing complexes or insoluble compounds are present other than HOH, OH^-, and Cl^-. Ag^+ label changed to AgCl (see text).

diagram is depicted in Figure 4.5. The result for the addition of a much smaller amount of Cl^- to give an equilibrium value of $10^{-6.0}$ M yields an equilibrium value for $[Ag^+]$ of $10^{-3.8}$ M. This value is greater than the initial value of $[Ag^+]$ = $10^{-4.0}$ M which means that no precipitate will form. Hence, the diagram is unaltered by the addition of this small amount of Cl^-. Carefully consider that with $10^{-4.0}$ M Ag^+, an equilibrium chloride concentration greater than $10^{-5.8}$ M will yield a precipitate, whereas a value below $10^{-5.8}$ M will not produce a precipitate.

Important Note: In the above considerations, for the purpose of explanatory simplification, the region above the Cl^- ion oxidation has been represented as a shaded area. The more complete diagram is presented in Figure 4.6. It is to be noted that above the AgCl area, Cl^- has been oxidized to Cl_2 then to ClO_4^- or directly to ClO_4^-. Hence, Ag^+ appears, since $AgClO_4$ is soluble in accordance with rule [b] of the solubility rules.

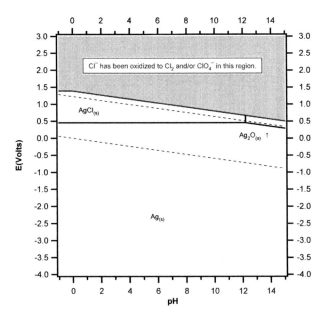

Figure 4.5 E–pH diagram for Ag species. Soluble species concentrations (except H$^+$) = $10^{-5.8}$ M. Soluble species and most solids are hydrated. No agents producing complexes or insoluble compounds are present other than HOH, OH$^-$, and Cl$^-$. Ag$^+$ label changed to AgCl (see text).

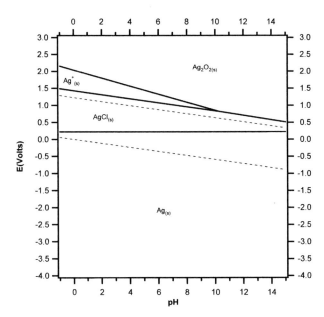

Figure 4.6 E–pH diagram for Ag/Cl system. Soluble Ag species = $10^{-4.0}$ M. Soluble Cl species = $10^{0.0}$ M. Soluble species and most solids are hydrated. No agents producing complexes or insoluble compounds are present other than HOH, OH$^-$, and Cl$^-$. Generated by computer including Cl$^-$ oxidation.

4. Addition of S^{-2} to the Zn E–pH System

The effects of the addition of S^{-2} on the Zn E–pH system will now be considered.

(1) Element selection. The Zn E–pH diagram is selected for study because zinc exhibits a single soluble cation species in the acidic region Zn^{+2} and a single soluble anionic species in the strongly basic region $Zn(OH)_4^{-2}$. The Zn E–pH diagram is then constructed for a dissolved species concentration of $10^{-4.0}$ M using the method discussed in Chapter 2. This system is represented by Figure 4.7.

(2) Precipitating agent selection. From consideration of the solubility rules, it is noted that most S^{-2} compounds are insoluble except for those noted

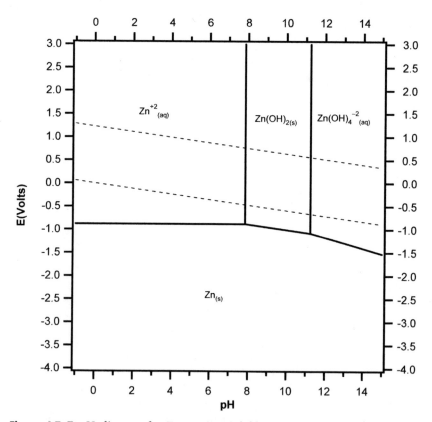

Figure 4.7 E–pH diagram for Zn species. Soluble species concentrations (except H^+) $= 10^{-4.0}$ M. Soluble species and most solids are hydrated. No agents producing complexes or insoluble compounds are present other than HOH and OH^-.

in rule [a]. For this example, S^{-2} is selected as the precipitating agent for the Zn^{+2} ion and the resulting insoluble compound is $ZnS_{(s)}$. It is of importance to recognize that the S^{-2}, HS^-, and H_2S are not stable over the entire E range given in E–pH diagrams. For $[S]total = 10^{-1.0}$ M, as Figure 4.8 illustrates, S^{-2}, HS^-, and H_2S are not pertinent species above the gray line that runs from 0.25 v on the left to -0.75 v on the right. Above this line the S^{-2}, HS^-, and H_2S have been oxidized to S_8 and/or SO_4^{-2} and/or HSO_4^-, and then to $S_2O_8^{-2}$. Therefore, the E–pH diagrams for this Zn/S system will show that the shaded region above this line is irrelevant to sulfide precipitation. It should be noted that the S^{-2} ion is a weak base. Therefore, it is necessary to take into account that as the pH of the solution is lowered, the concentration of the S^{-2} ion will be lowered due to protonation. The pertinent protonation reactions and corresponding $\Delta G°$ values and $\Delta G°$ (reaction) values (in kJ/mole) are

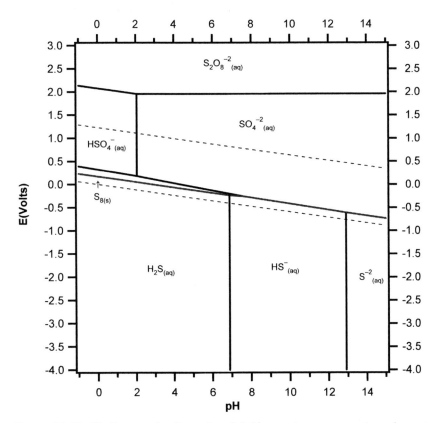

Figure 4.8 E–pH diagram for S species. Soluble species concentrations (except H^+) = $10^{-1.0}$ M. Soluble species and most solids are hydrated. No agents producing complexes or insoluble compounds are present other than HOH and OH^-.

given below:

$$S^{-2} + H^+ \rightarrow HS^- \quad \Delta G°(\text{reaction}) = -73.7 \quad \text{Protonation 1}$$
$$85.8 \quad 0.0 \quad 12.1$$
$$HS^- + H^+ \rightarrow H_2S \quad \Delta G°(\text{reaction}) = -39.3 \quad \text{Protonation 2}$$
$$12.1 \quad 0.0 \quad -27.2$$

From the $\Delta G°(\text{reaction})$ values, the K values for the reactions can be calculated as follows:

$$\text{Protonation 1} \quad \log K = \Delta G°/-5.7 = -73.7/-5.7 = 12.93$$
$$K = 10^{12.9} = [HS^-]/[S^{-2}][H^+] = K_{p1}$$
$$\text{Protonation 2} \quad \log K = \Delta G°/-5.7 = -39.3/-5.7 = 6.89$$
$$K = 10^{6.9} = [H_2S]/[HS^-][H^+] = K_{p2}$$

These K values are often designated as K_p values (protonation constants) and are related to the commonly encountered acid dissociation constants, K_a, by the following relationship: $K_p = 1/K_a$. From the K_p values, it can be seen that below pH values of 12.9 the S^{-2} ion is protonated to HS^- which lowers the concentration of S^{-2} ion in solution. The concentration of the S^{-2} ion is further influenced as the HS^- ion is protonated to H_2S below pH values of 6.9. It is therefore important to take into account the protonation reactions when determining the concentration of sulfide in solution. This is most readily done by recognizing that the total sulfur concentration in solution [S]total will be divided among the three species: S^{-2}, HS^-, and H_2S. This relationship can be expressed as:

$$[S]\text{total} = [S^{-2}] + [HS^-] + [H_2S]$$

By applying(combining) this equation and the two protonation expressions from above, the $[S^{-2}]$ can be determined for any [S]total as a function of the $[H^+]$ using the following equation:

$$[S^{-2}] = [S]\text{total}/\{1 + K_{p1}[H^+] + K_{p1}K_{p2}[H^+]^2\}$$

As an example, for a $10^{-1.0}$ M [S]total solution, the $[S^{-2}]$ changes as the pH is lowered. The variation in the $[S^{-2}]$ over the pH range from 15.0 to -1.0 for a [S]total of $10^{-1.0}$ M is listed below and is shown in graphical form in Figure 4.9. The value of $10^{-1.0}$ is being used because H_2S is soluble only to that extent.

pH	$[S^{-2}]$	pH	$[S^{-2}]$	pH	$[S^{-2}]$
15	$10^{-1.0}$	9	$10^{-4.9}$	3	$10^{-14.8}$
14	$10^{-1.0}$	8	$10^{-5.9}$	2	$10^{-16.8}$
13	$10^{-1.3}$	7	$10^{-7.2}$	1	$10^{-18.8}$
12	$10^{-2.0}$	6	$10^{-8.9}$	0	$10^{-20.8}$
11	$10^{-2.9}$	5	$10^{-10.8}$	-1	$10^{-21.8}$
10	$10^{-3.9}$	4	$10^{-12.8}$		

(3) Calculation of K_{sp} value. The slight dissolution of the insoluble compound is given by the equation:

$$ZnS_{(s)} \rightarrow Zn^{+2}{}_{(aq)} + S^{-2}{}_{(aq)}$$

The pertinent $\Delta G°$ values (in kJ/mole) are $ZnS_{(s)}$ (-181.0), $Zn^{+2}{}_{(aq)}$ (-146.9), and $S^{-2}{}_{(aq)}$ (85.8). Calculation of the $\Delta G°$(reaction) is carried out using Equation (6) from Chapter 2.

$$\Delta G°(\text{reaction}) = \Sigma \Delta G°(\text{products}) - \Sigma \Delta G°(\text{reactants}) \qquad (6)$$

$$\Delta G°(\text{reaction}) = [85.8 + (-146.9)] - [-181.0] = 119.9 \text{ kJ/mole}$$

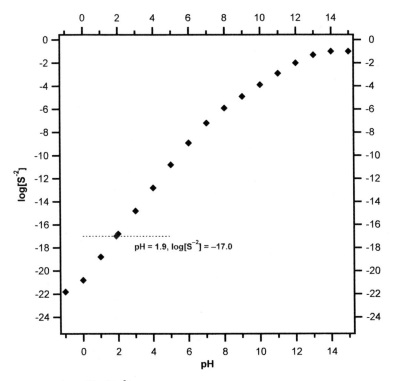

Figure 4.9 Plot of $\log[S^{-2}]$ versus pH.

Once the $\Delta G°$(reaction) value is obtained, the K value can be calculated using Equation (8) from Chapter 2.

$$\log K = \Delta G°/ - 5.7 = 119.9/ - 5.7 = -21.04 \tag{8}$$

$$K = 10^{-21.0} = [Zn^{+2}][S^{-2}]/[ZnS] = K_{sp} = [Zn^{+2}][S^{-2}]$$

Noting that the concentration of the insoluble compound is defined as 1, the K value is re-labeled as K_{sp}.

(4) Flooding the system with precipitant.   Using the calculated K_{sp} value and the initial concentration of the Zn^{+2} ion ($10^{-4.0}$ M), an initial concentration of the S^{-2} ion can be calculated from the solubility product expression.

$$K_{sp} = [Zn^{+2}][S^{-2}] = 10^{-21.0} \quad [10^{-4.0}][S^{-2}] = 10^{-21.0} \quad [S^{-2}] = 10^{-17.0} \text{ M}$$

This S^{-2} ion concentration is the minimum concentration necessary for a precipitate to form. From initial considerations, it would appear that it should be relatively easy to precipitate the Zn^{+2} ion from solution using the S^{-2} ion. However, one must take into account the weak base character of the S^{-2} ion. For [Zn^{+2}] of $10^{-4.0}$ M and [S]total of $10^{-1.0}$ M, the concentration of the S^{-2} ion drops below the minimum concentration ($10^{-17.0}$ M) value as the pH is lowered from 2 to 1 (see above values and Figure 4.9). Therefore at pH values below 2, the concentration of S^{-2} ion in solution is insufficient to cause the precipitation of ZnS. This observation must be kept in mind when selecting a concentration range for the S^{-2} ion, since even relatively concentrated sulfide solutions do not contain sufficient S^{-2} ion to cause precipitation at low pH values. By careful and detailed consideration of the changing $[S^{-2}]$ value combined with the procedures elucidated in Chapter 2, E–pH diagrams for the Zn–S system can be constructed. However, due to the variation in the $[S^{-2}]$ as a function of pH, the calculation of the E–pH diagram can be carried out most easily by using one of the computer programs. Figure 4.10 depicts this system with Zn^{+2} at an initial concentration of $10^{-4.0}$ M and the total sulfur species at an equilibrium concentration of $10^{-1.0}$ M (0.10 M).

(5) Varying the precipitant concentration.   The procedure of step (4) can be carried out with different total concentrations of the sulfur species. Figures 4.11–4.14 show the results for total sulfur species concentrations of $10^{-4.0}$ M, $10^{-7.0}$ M, $10^{-10.0}$ M, and $10^{-12.0}$ M. Note the interesting progression of predominant species in Figure 4.13.

Important Note: In the above considerations, for the purpose of explanatory simplification, in the region above the S^{-2}, HS^-, and H_2S species, oxidation has been represented as a shaded area. The more complete diagram for Figure 4.12 is presented in Figure 4.15. Note that above the ZnS area the

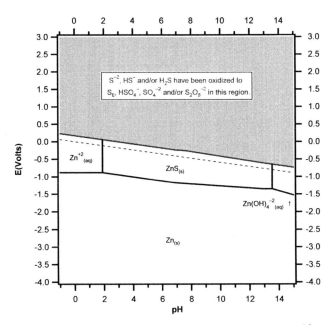

Figure 4.10 E–pH diagram for Zn/S system. Soluble Zn species $= 10^{-4.0}$ M. Soluble S species $= 10^{-1.0}$ M. Soluble species and most solids are hydrated. No agents producing complexes or insoluble compounds are present other than HOH, OH⁻, and S⁻². Generated by computer.

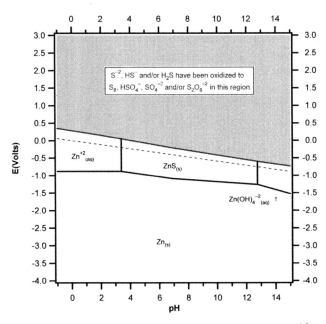

Figure 4.11 E–pH diagram for Zn/S system. Soluble Zn species $= 10^{-4.0}$ M. Soluble S species $= 10^{-4.0}$ M. Soluble species and most solids are hydrated. No agents producing complexes or insoluble compounds are present other than HOH, OH⁻, and S⁻². Generated by computer.

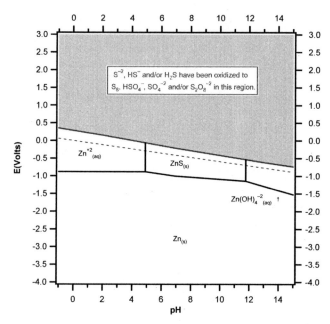

Figure 4.12 E–pH diagram for Zn/S system. Soluble Zn species $= 10^{-4.0}$ M. Soluble S species $= 10^{-7.0}$ M. Soluble species and most solids are hydrated. No agents producing complexes or insoluble compounds are present other than HOH, OH$^-$, and S^{-2}. Generated by computer.

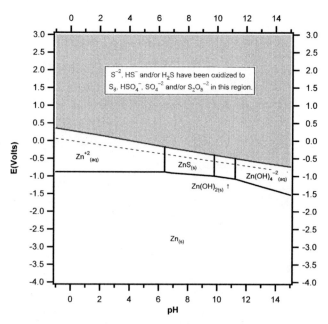

Figure 4.13 E–pH diagram for Zn/S system. Soluble Zn species $= 10^{-4.0}$ M. Soluble S species $= 10^{-10.0}$ M. Soluble species and most solids are hydrated. No agents producing complexes or insoluble compounds are present other than HOH, OH$^-$, and S^{-2}. Generated by computer.

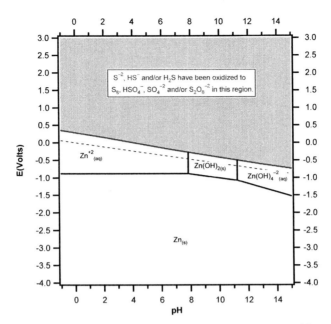

Figure 4.14 E–pH diagram for Zn/S system. Soluble Zn species = $10^{-4.0}$ M. Soluble S species = $10^{-12.0}$ M. Soluble species and most solids are hydrated. No agents producing complexes or insoluble compounds are present other than HOH, OH⁻, and S⁻². Generated by computer.

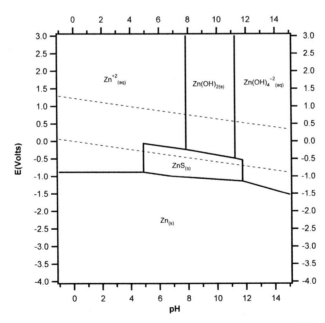

Figure 4.15 E–pH diagram for Zn/S system. Soluble Zn species = $10^{-4.0}$ M. Soluble S species = $10^{-7.0}$ M. Soluble species and most solids are hydrated. No agents producing complexes or insoluble compounds are present other than HOH, OH⁻ and S⁻². Generated by computer including S⁻², HS⁻, and H₂S oxidation.

sulfide species have been oxidized to S_8, SO_4^{-2}, HSO_4^-, and/or $S_2O_8^{-2}$. In other words, there are no predominant sulfide species at the higher E values.

5. Review of Complexation and Complex Species

A complex or coordination compound generally consists of one or more groups (molecules or anions) surrounding and bonded to a central metal (cation or neutral atom). Depending upon the charges on the central metal and bonded groups, the complex can be cationic, anionic, or neutral. Examples are:

$$Cu(NH_3)_4^{+2} \quad PdCl_4^{-2} \quad NiBr_2(NH_3)_2$$

The groups bonded to the central metal are called coordinating species or ligands. Ligands contain one or more atoms with a lone pair of electrons with which it forms a bond to the central metal. These atoms are often referred to as donor or ligating atoms. Typical donor atoms include: N, P, O, S, F, Cl, Br, and I. A ligand which forms only one bond to the central metal is called a monodentate ligand, for example, H_2O, NH_3, F^-, and OH^-. Ligands which form two, three, four, five, or six bonds are designated as bidentate, tridentate, tetradentate, pentadentate, or hexadentate ligands, or simply polydentate ligands. Examples of polydentate ligands include: ethylenediamine (bidentate), oxalate (bidentate), and EDTA (hexadentate). A more extensive list of common ligands is given in Table 4.1. Ligands which contain a minimum of two donor atoms that can simultaneously bond to the central metal are called chelating ligands, for example ethylenediamine and EDTA, and the complexes are called chelates. These complexes are usually made up of 5- or 6-membered rings which contain the metal atom or ion. The number of donor atoms bonded to the central metal is referred to as the coordination number of the central metal. Coordination numbers typically range from 1 to 9 with 4 and 6 being the most common.

The stability (or tendency of the donor atoms to bond to the central metal) of a complex is expressed in terms of a formation constant, β_n (where n is the number of ligands attached to the central metal). Formation constants are equilibrium constants, K, written for the formation of a complex from the free ligands and central metal. Consider the following example of the formation of several Ni^{+2} complexes with ammonia:

$$Ni^{+2} + NH_3 \rightarrow Ni(NH_3)^{+2} \quad \beta_1 = [Ni(NH_3)^{+2}]/[Ni^{+2}][NH_3] = 10^{2.7}$$

$$Ni^{+2} + 2NH_3 \rightarrow Ni(NH_3)_2^{+2} \quad \beta_2 = [Ni(NH_3)_2^{+2}]/[Ni^{+2}][NH_3]^2 = 10^{4.9}$$

$$Ni^{+2} + 3NH_3 \rightarrow Ni(NH_3)_3^{+2} \quad \beta_3 = [Ni(NH_3)_3^{+2}]/[Ni^{+2}][NH_3]^3 = 10^{6.6}$$

$$Ni^{+2} + 4NH_3 \rightarrow Ni(NH_3)_4^{+2} \quad \beta_4 = [Ni(NH_3)_4^{+2}]/[Ni^{+2}][NH_3]^4 = 10^{7.7}$$

Table 4.1
Common Polydentate Ligands

Ligand	Structure	Abbreviation	Denticity
Carbonato			bidentate
Oxalato		ox	bidentate
Ethylenediamine	H_2N ... NH_2	en	bidentate
1,2-Propanediamine	H_2N ... NH_2	pn	bidentate
Acetylacetonato		acac	bidentate
8-Hydroxyquinolinato		oxine	bidentate
2,2′-Dipyridyl		dipy	bidentate
1,10-Phenanthroline		phen	bidentate
Glycinato	H_2N ... O^-	gly	bidentate
Diethylenetriamine	H_2N ... NH_2	dien	tridentate
Tartrate anion		tar	tetradentate

Table 4.1
(Continued)

Ligand	Structure	Abbreviation	Denticity
Citrate anion		cit	tetradentate
Nitrilotriacetate anion		nta	tetradentate
2,2,'2″ -Triaminotriethylamine		trien	tetradentate
Triethylenetetramine		tren	tetradentate
Ethylenediaminetetraacetate anion		edta	hexadentate

It should be noted that the larger the formation constant the more stable the complex. Consider the following example of Ni^{+2} complexes with cyanide and ammonia:

$$Ni^{+2} + 4CN^- \rightarrow Ni(CN)_4^{-2} \quad \beta_4 = [Ni(CN)_4^{-2}]/[Ni^{+2}][CN^-]^4 = 10^{30.2}$$

$$Ni^{+2} + 4NH_3 \rightarrow Ni(NH_3)_4^{+2} \quad \beta_4 = [Ni(NH_3)_4^{+2}]/[Ni^{+2}][NH_3]^4 = 10^{7.7}$$

Cyanide forms a much more stable complex with Ni^{+2} than ammonia.

It is also observed that chelating agents (polydentate ligands) form more stable complexes than a comparable number of donor atoms in monodentate ligands. Consider the following example of Ni^{+2} complexes with four ammonia molecules (each monodentate) and two ethylenediamine molecules (each bidentate):

$$Ni^{+2} + 4NH_3 \rightarrow Ni(NH_3)_4^{+2} \quad \beta_4 = [Ni(NH_3)_4^{+2}]/[Ni^{+2}][NH_3]^4 = 10^{7.7}$$

$$Ni^{+2} + 2\,en \rightarrow Ni(en)_2^{+2} \quad \beta_2 = [Ni(en)_2^{+2}]/[Ni^{+2}][en]^2 = 10^{13.5}$$

6. Complexation and E–pH Diagrams

In order to determine the changes in a single element E–pH diagram that occur due to the addition of a complexing agent and the resulting formation of a complex, the following five steps may be carried out.

(1) Select an element of interest showing a simple cation and construct the E–pH diagram for the element at a soluble species concentration of $10^{-4.0}$ M.

(2) Select a complexing agent which will form a complex with the simple cation of the element of interest. Unlike the solubility rules for selecting precipitating agents, there is no list of general complexation rules for the selection of complexing agents. This is due to the large variation in the types of ligands and metal centers available to form complexes and the potential for many exceptions to any given rule to exist. With this in mind, some general statements regarding complex formation with specific ligands and/or metals can be made.

    a. The cations, Pd^{+2}, Pt^{+2}, Pt^{+4}, Cu^+, Cu^{+2}, Ag^+, Au^+, Zn^{+2}, Cd^{+2}, Hg_2^{+2}, Hg^{+2}, Tl^+, Tl^{+3}, Sn^{+2}, Pb^{+2}, Sb^{+3}, and Bi^{+3} usually form their most stable complexes with ligands containing S, P, Cl, Br, and I as donor atoms.

    b. Most other cations usually form their most stable complexes with ligands containing F, O, and N as donor atoms.

    c. Alkali metals generally form the weakest complexes, while transition metals form a wide variety of relatively stable complexes.

    d. The ligands, ClO_4^- and NO_3^-, ordinarily show little tendency to form complexes.

    e. Chelating agents which form 5- and 6-membered rings with the central metal form the most stable complexes.

    f. Generally, as the number of bonds a chelating agent forms with the central metal increases, the stability of the resulting complex increases.

The reader will want to consult a list of stability constants before selecting a specific complexing agent. The stability of specific complexes may be found in literature references such as Lange's Handbook of Chemistry[5] or the CRC Handbook of Chemistry and Physics[6] or the Handbook of Chemical Equilibria in Analytical Chemistry[7] or NIST Critically Selected Stability Constants of Metal Complexes.[8]

These references have extensive listings of inorganic complexes and their aqueous stabilities. Special note: If the complexing agent is one which protonates, the concentrations of the non-protonated agent (as a function of pH) must be calculated. These values are to be used in the construction of the E-pH diagram.

(3) Write the equation for the formation of the complex corresponding to the β_n with the largest value, and then calculate the $\Delta G°$(reaction) using Equation (6) from Chapter 2. Once the $\Delta G°$(reaction) value is obtained, the K value is calculated using Equation (8) also from Chapter 2. It should be recognized that the K value for the complex formation is generally referred to as the formation constant, β_n, n being the number of ligands attaching.

(4) Flood the system with the complexing agent, putting it in to give a concentration of $10^{1.0}$ M (or $10^{0.0}$ M, if more appropriate). Then employing the β_n value arrived at in step (3), calculate the equilibrium concentration of the simple (uncomplexed) ion. Plot an E–pH diagram using this uncomplexed ion concentration. Then replace the label for the simple (uncomplexed) ion with the symbol for the complexed ion.

(5) By the use of the same β_n and other values of the concentrations of the complexing agent, data for the equilibrium concentration of the simple ion can be found. As before, these simple ion values can be used to construct E–pH plots and then the simple-ion labels can be replaced with the complex labels.

7. Addition of Cl⁻ to the Pd E–pH System to Form PdCl₄⁻²

Using the five steps outlined in the previous section, the effects of the addition of excess Cl^- on the Pd E–pH system to form $PdCl_4^{-2}$ will now be determined.

(1) Element selection.   The Pd E–pH diagram is selected for study because palladium exhibits a single soluble cation species: the Pd^{+2} ion. The Pd E–pH diagram is then constructed for a dissolved species concentration of $10^{-4.0}$ M using the method discussed in Chapter 2. See Figure 4.16.

(2) Complexing agent selection.   From consideration of the complexation statements, it is noted that Pd^{+2} has a tendency to form complexes with ligands containing the donor atoms: P, S, Cl, and Br. For this example, Cl^- is selected as the complexing agent. (Note that Cl^- is a non-protonating complexing agent.) The Pd^{+2} ion has a coordination number of 4, so a more accurate description of its aqueous form is $Pd(HOH)_4^{+2}$. When it complexes with the Cl^- ion, different numbers of the waters may be replaced by Cl^- to give the following chloride complexes: $Pd(HOH)_3Cl^+$, $Pd(HOH)_2Cl_2$,

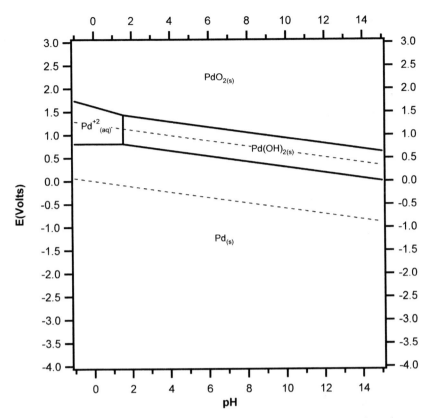

Figure 4.16 E–pH diagram for Pd species. Soluble species concentrations (except H^+) = $10^{-4.0}$ M. Soluble species and most solids are hydrated. No agents producing complexes or insoluble compounds are present other than HOH and OH^-.

$Pd(HOH)Cl_3^-$, $PdCl_4^{-2}$, or to write the species more simply, $PdCl^+$, $PdCl_2$, $PdCl_3^-$, $PdCl_4^{-2}$. When no Cl^- is added to the Pd^{+2} system, the predominant ion is Pd^{+2}. When a relatively small amount of Cl^- is added, the predominant species becomes $PdCl^+$. Then as successively larger amounts of Cl^- are added, the predominant form becomes: $PdCl_2$, then $PdCl_3^-$, then finally $PdCl_4^{-2}$. In this example, a large amount or a considerable excess of Cl^- will be added to ensure that $PdCl_4^{-2}$ is the predominant species. This complex is characterized by a formation constant β_4. In the next section of this chapter, the entire array of complexes ($PdCl^+$, $PdCl_2$, $PdCl_3^-$, $PdCl_4^{-2}$) will be dealt with. But for here the focus will be solely upon $PdCl_4^{-2}$.

As in the previous insoluble AgCl system, it is important to recall that Cl^- is not stable over the entire E range given on E–pH diagrams. As Figure 4.2 illustrated, Cl^- is not the predominant species above the bold solid line. Above this line, the Cl^- has been oxidized to Cl_2 and/or ClO_4^-, therefore the E–pH diagrams for the Pd^{+2}/Cl^- system will involve Cl^- only below the shaded area. Pd at high E values shows a IV oxidation state, but it does not have to be considered, since its occurrence is in the shaded area above the bold line which describes the oxidation of Cl^-.

(3) Calculation of β_4 value. The expression for the formation of the $PdCl_4^{-2}$ complex follows. Values of ΔG° in kJ/mole for the species involved have been provided, and the ΔG° of the reaction calculated from Equation (6).

$$Pd^{+2} \quad + \quad 4Cl^- \quad \rightarrow PdCl_4^{-2} \quad \Delta G^\circ(\text{reaction}) = -69.3 \text{ kJ}$$

$$+177.8 \quad 4(-131.4) \quad -417.1$$

From the ΔG°(reaction) value, the corresponding K (β_4) value for the reaction can be calculated from Equation (8) as follows:

$$\log K = \Delta G^\circ/-5.7 = -69.3/-5.7 = 12.16$$

$$K = \beta_4 = 10^{12.2} = [PdCl_4^{-2}]/[Pd^{+2}][Cl^-]^4$$

Quite often the values of β_n will be given in the chemical literature so that they will not need to be calculated from ΔG° values.

(4) Flooding the system with complexing agent. Put the concentration of Cl^- as $10^{1.0}$ M (10.0 M) into the expression for β_4 and calculate the equilibrium concentration of the Pd^{+2} ion. Assume that practically all of the Pd^{+2} forms $PdCl_4^{-2}$ such that approximately $10^{-4.0}$ M can be put in for the equilibrium concentration of $PdCl_4^{-2}$.

$$\beta_4 = 10^{12.2} = [PdCl_4^{-2}]/[Pd^{+2}][Cl^-]^4 = [10^{-4.0}]/[Pd^{+2}][10.0]^4$$

$$[Pd^{+2}] = 10^{-20.2}$$

An E–pH diagram is now constructed using a Pd concentration of $10^{-20.2}$ M. Then the Pd^{+2} label on the diagram is replaced with the label $PdCl_4^{-2}$ since this species is in equilibrium with the $10^{-20.2}$ M Pd^{+2}, but is now the predominant species. The resulting E–pH diagram is shown in Figure 4.17.

(5) Concentration variations. Using the β_4 and other concentrations of Cl^- ($10^{0.0}$ and $10^{-1.0}$ M) to make calculations comparable to the ones in step (4), the following results are obtained:

$$\beta_4 = 10^{12.2} = [PdCl_4^{-2}]/[Pd^{+2}][Cl^-]^4 = [10^{-4.0}]/[Pd^{+2}][10^{0.0}]^4$$

$$[Pd^{+2}] = 10^{-16.2} \text{ M}$$

$$\beta_4 = 10^{12.2} = [PdCl_4^{-2}]/[Pd^{+2}][Cl^-]^4 = [10^{-4.0}]/[Pd^{+2}][10^{-1.0}]^4$$

$$[Pd^{+2}] = 10^{-12.2} \text{ M}$$

These results are used to derive E–pH diagrams which are presented in Figures 4.18 and 4.19. Take careful note of the movement of the line dividing the precipitate $Pd(OH)_2$ and the soluble complex $PdCl_4^{-2}$ as the concentration of the Cl^- changes.

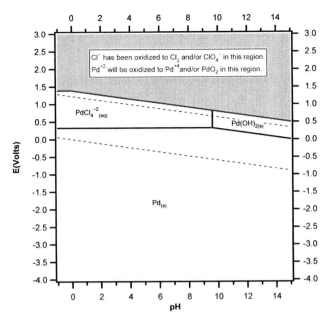

Figure 4.17 E–pH diagram for Pd species. Soluble species concentrations (except H$^+$) = 10$^{-20.2}$ M. Soluble species and most solids are hydrated. No agents producing complexes or insoluble compounds are present other than HOH, OH$^-$, and Cl$^-$. Pd^{+2} label changed to PdCl$_4{}^{-2}$ (see text).

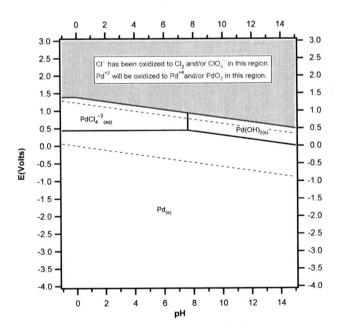

Figure 4.18 E–pH diagram for Pd species. Soluble species concentrations (except H+) = 10$^{-16.2}$ M. Soluble species and most solids are hydrated. No agents producing complexes or insoluble compounds are present other than HOH, OH$^-$, and Cl$^-$. Pd^{+2} label changed to PdCl$_4{}^{-2}$ (see text).

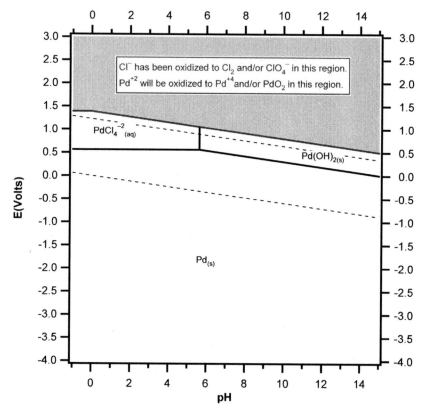

Figure 4.19 E–pH diagram for Pd species. Soluble species concentrations (except H^+) $= 10^{-12.2}$ M. Soluble species and most solids are hydrated. No agents producing complexes or insoluble compounds are present other than HOH, OH^-, and Cl^-. Pd^{+2} label changed to $PdCl_4^{-2}$ (see text).

8. Multiple Complexes and E–pH Diagrams

A detailed treatment of all the complexes formed in the Pd–Cl system will now be made. Recall that Pd^{+2} forms four separate complexes with Cl^-: $PdCl^+$, $PdCl_2$, $PdCl_3^-$, and $PdCl_4^{-2}$. The pertinent complex formation reactions and corresponding $\Delta G°$ values (in kJ/mole) along with the $\Delta G°$(reaction) values (in kJ) are given below:

Formation reaction 1 $Pd^{+2} + Cl^- \rightarrow PdCl^+$ $\Delta G°$(reaction) $= -23.8$ kJ

177.8 -131.4 22.6

Formation reaction 2 $Pd^{+2} + 2Cl^- \rightarrow PdCl_2$ $\Delta G°$(reaction) $= -43.9$ kJ

177.8 $2(-131.4)$ -128.9

Formation reaction 3 $Pd^{+2} + 3Cl^- \rightarrow PdCl_3^-$ $\Delta G°(\text{reaction}) = -59.7 \text{ kJ}$

$\quad\quad\quad$ 177.8   3(-131.4) -276.1

Formation reaction 4 $Pd^{+2} + 4Cl^- \rightarrow PdCl_4^{-2}$ $\Delta G°(\text{reaction}) = -69.3 \text{ kJ}$

$\quad\quad\quad$ 177.8   4(-131.4) -417.1

From the $\Delta G°(\text{reaction})$ values, the corresponding K (β_n) values for the reactions can be calculated as follows:

Formation reaction 1 $\log K = \Delta G°/-5.7 = -23.8/-5.7 = 4.18$

$$K = \beta_1 = 10^{4.2} = [PdCl^+]/[Pd^{+2}][Cl^-]$$

Formation reaction 2 $\log K = \Delta G°/-5.7 = -43.9/-5.7 = 7.70$

$$K = \beta_2 = 10^{7.7} = [PdCl_2]/[Pd^{+2}][Cl^-]^2$$

Formation reaction 3 $\log K = \Delta G°/-5.7 = -59.7/-5.7 = 10.47$

$$K = \beta_3 = 10^{10.5} = [PdCl_3^-]/[Pd^{+2}][Cl^-]^3$$

Formation reaction 4 $\log K = \Delta G°/-5.7 = -69.3/-5.7 = 12.16$

$$K = \beta_4 = 10^{12.2} = [PdCl_4^{-2}]/[Pd^{+2}][Cl^-]^4$$

Note these K's are the formation constants for the specific complex species indicated and are re-labeled as the appropriate β_n values. Values of β_n may be calculated from $\Delta G°$ values, as has been done, or they may often be obtained from the chemical literature. Depending upon adjustment of the relative concentrations of the Pd^{+2} and Cl^- ions, each of the four complex species above can be formed as the predominant species in solution.

Because multiple Pd–Cl complex species can co-exist in solution, the Pd^{+2} ion concentration in solution after the addition of a given amount of the Cl^- ion can be calculated using the same type of treatment employed to determine the concentration of the S^{-2} ion as a function of pH. The total concentration of Pd, [Pd]total, found in solution is expressed by the following equation:

$$[Pd]total = [Pd^{+2}] + [PdCl^+] + [PdCl_2] + [PdCl_3^-] + [PdCl_4^{-2}]$$

By applying(combining) this equation with the four β_n values from above, the $[Pd^{+2}]$ can be determined as a function of the $[Cl^-]$ for any [Pd]total using the following equation:

$$[Pd^{+2}] = [Pd]total/(1 + \beta_1[Cl^-] + \beta_2[Cl^-]^2 + \beta_3[Cl^-]^3 + \beta_4[Cl^-]^4)$$

or substituting $[Pd]total = 10^{-4.0}$ and the β_n values

$$[Pd^{+2}] = 10^{-4.0}/(1 + 10^{4.2}[Cl^-] + 10^{7.7}[Cl^-]^2 + 10^{10.5}[Cl^-]^3 + 10^{12.2}[Cl^-]^4)$$

This equation permits the construction of a table which will identify the predominant species at a given Cl^- concentration, and will facilitate the construction of an E–pH diagram through use of the simple Pd^{+2} concentration in equilibrium with the predominant species. The table will contain columns for $[Cl^-]$, each of the terms in the denominator of the above equation, and the resulting $[Pd^{+2}]$. The initial $[Pd^{+2}]$ is $10^{-4.0}$ M, so a range of increasing $[Cl^-]$ values will be used starting with $[Cl^-]$ less than the $[Pd^{+2}]$ at $10^{-5.0}$ M. See Table 4.2.

The largest number among the second through sixth columns representing the terms in the denominator is the major influence upon the $[Pd^{+2}]$. Note carefully that the column labeled 1 represents Pd^{+2}, the column labeled $10^{4.2}[Cl^-]$ represents $PdCl^+$, the column labeled $10^{7.7}[Cl^-]^2$ represents $PdCl_2$, the column labeled $10^{10.5}[Cl^-]^3$ represents $PdCl_3^-$, and the column labeled $10^{12.2}[Cl^-]^4$ represents $PdCl_4^{-2}$. Therefore, the species represented by the largest number is the predominant species and the concentration of the Pd^{+2} ion in equilibrium with it is given in the last column. The largest number in each row is indicated by a # sign. Hence at $[Cl^-]$ $= 10^{-5.0}$ M, the predominant species is Pd^{+2} and its concentration is $10^{-4.1}$ M. At $[Cl^-] = 10^{-4.0}$ M, the predominant species is $PdCl^+$ and $[Pd^{+2}] = 10^{-4.5}$ M. At $[Cl^-] = 10^{-3.0}$ M, the predominant species is $PdCl_2$ and $[Pd^{+2}] = 10^{-6.0}$ M. At $[Cl^-] = 10^{-2.3}$ M, the predominant species is $PdCl_3^-$ and $[Pd^{+2}] = 10^{-7.8}$ M. And at $[Cl^-] = 10^{-1.3}$ M, the predominant species is $PdCl_4^{-2}$ and $[Pd^{+2}] = 10^{-11.1}$ M.

In order to plot an E–pH diagram at any given Cl^- concentration, the following procedure is to be carried out. (1) Identify the equilibrium

Table 4.2
Pd complexation with chloride $[Pd^{+2}] = 10^{-4}$ M

$[Cl^-]$	1	$10^{4.2}[Cl^-]$	$10^{7.7}[Cl^-]^2$	$10^{10.5}[Cl^-]^3$	$10^{12.2}[Cl^-]^4$	$[Pd^{+2}]$
$10^{-5.0}$	1#	$10^{-0.8}$	$10^{-2.3}$	$10^{-4.5}$	$10^{-7.8}$	$10^{-4.1}$
$10^{-4.6}$	1#	$10^{-0.4}$	$10^{-1.5}$	$10^{-3.3}$	$10^{-6.2}$	$10^{-4.2}$
$10^{-4.3}$	1#	$10^{-0.1}$	$10^{-0.9}$	$10^{-2.4}$	$10^{-5.0}$	$10^{-4.3}$
$10^{-4.0}$	1	$10^{0.2}$#	$10^{-0.3}$	$10^{-1.5}$	$10^{-3.8}$	$10^{-4.5}$
$10^{-3.6}$	1	$10^{0.6}$#	$10^{0.5}$	$10^{-0.3}$	$10^{-2.2}$	$10^{-4.9}$
$10^{-3.3}$	1	$10^{0.9}$	$10^{1.1}$#	$10^{0.6}$	$10^{-1.0}$	$10^{-5.4}$
$10^{-3.0}$	1	$10^{1.2}$	$10^{1.7}$#	$10^{1.5}$	$10^{-0.2}$	$10^{-6.0}$
$10^{-2.6}$	1	$10^{1.6}$	$10^{2.5}$	$10^{2.7}$#	$10^{1.8}$	$10^{-7.0}$
$10^{-2.3}$	1	$10^{1.9}$	$10^{3.1}$	$10^{3.6}$#	$10^{3.0}$	$10^{-7.8}$
$10^{-2.0}$	1	$10^{2.2}$	$10^{3.7}$	$10^{4.5}$#	$10^{4.2}$	$10^{-8.7}$
$10^{-1.6}$	1	$10^{2.6}$	$10^{4.5}$	$10^{5.7}$	$10^{5.8}$#	$10^{-10.1}$
$10^{-1.3}$	1	$10^{2.9}$	$10^{5.1}$	$10^{6.6}$	$10^{7.0}$#	$10^{-11.1}$
$10^{-1.0}$	1	$10^{3.2}$	$10^{5.7}$	$10^{7.5}$	$10^{8.2}$#	$10^{-12.2}$

sign indicates the largest number in each row.

[Cl⁻], say for example $10^{-2.0}$ M. (2) Read the corresponding equilibrium [Pd⁺²] from the last column, that is, $10^{-8.7}$ M. (3) Plot a Pd E–pH diagram using $[Pd^{+2}] = 10^{-8.7}$ M. (4) Identify the predominant species from the chart, namely, $PdCl_3^-$. (5) Substitute the label $PdCl_3^-$ for the label Pd⁺² on the E–pH diagram. The resultant diagram is presented in Figure 4.20. Figures 4.21–4.24 present diagrams for these [Cl⁻]: $10^{-5.0}$, $10^{-3.6}$, $10^{-3.0}$, and $10^{-1.0}$ M at which the predominant species are respectively Pd⁺², PdCl⁺, PdCl₂, and $PdCl_4^{-2}$. Precisely the same diagrams may be obtained from appropriate computer programs.

Once again, it is important to recall that a complexing agent which undergoes protonation (unlike Cl⁻) requires more consideration. This is because the concentration of the non-protonated complexing species varies as the pH changes. An example would be the acetate ion $C_2H_3O_2^-$ which protonates

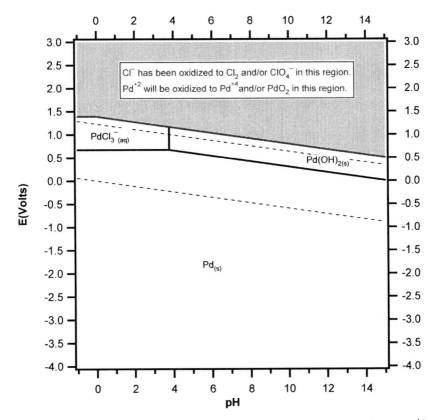

Figure 4.20 E–pH diagram for Pd species. Soluble species concentrations (except H⁺) $= 10^{-8.7}$ M. Soluble species and most solids are hydrated. No agents producing complexes or insoluble compounds are present other than HOH, OH⁻, and Cl⁻. Pd⁺² label changed to $PdCl_3^-$ (see text).

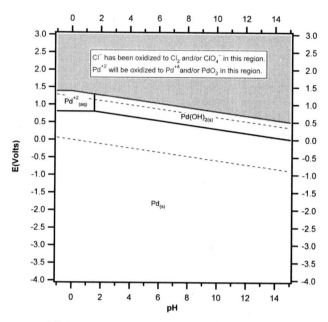

Figure 4.21 E–pH diagram for Pd species. Soluble species concentrations (except H^+) $= 10^{-4.1}$ M. Soluble species and most solids are hydrated. No agents producing complexes or insoluble compounds are present other than HOH, OH^-, and Cl^-.

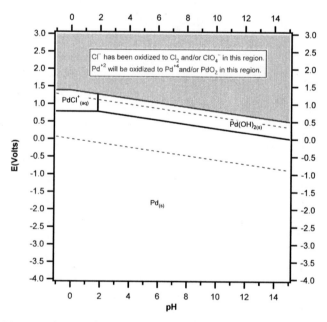

Figure 4.22 E–pH diagram for Pd species. Soluble species concentrations (except H^+) $= 10^{-4.9}$ M. Soluble species and most solids are hydrated. No agents producing complexes or insoluble compounds are present other than HOH, OH^-, and Cl^-. Pd^{+2} label changed to $PdCl^+$ (see text).

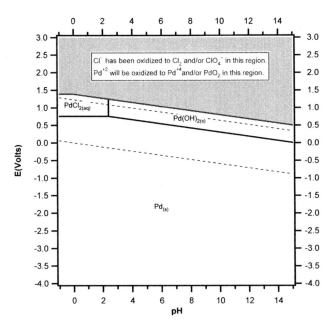

Figure 4.23 E–pH diagram for Pd species. Soluble species concentrations (except H^+) $= 10^{-6.0}$ M. Soluble species and most solids are hydrated. No agents producing complexes or insoluble compounds are present other than HOH, OH^-, and Cl^-. Pd^{+2} label changed to $PdCl_2$ (see text).

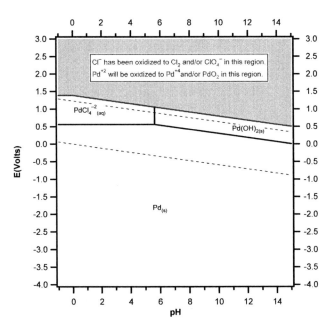

Figure 4.24 E–pH diagram for Pd species. Soluble species concentrations (except H^+) $= 10^{-12.2}$ M. Soluble species and most solids are hydrated. No agents producing complexes or insoluble compounds are present other than HOH, OH^-, and Cl^-. Pd^{+2} label changed to $PdCl_4^{-2}$ (see text).

to $HC_2H_3O_2$. The situation becomes complicated when the non-protonated complexing agent can undergo multiple protonations, as is the case with $edta^{-4}$ which protonates to $Hedta^{-3}$, H_2edta^{-2}, H_3edta^-, and H_4edta. It is possible to make these calculations by hand, but they are better treated by computer.

5

The Lithium Group

1. Introduction

The elements which constitute Group 1 of the Periodic Table are known as the alkali metals. They are lithium Li, sodium Na, potassium K, rubidium Rb, cesium Cs, and francium Fr. (Sometimes the NH_4^+ ion is included among these since it resembles K^+ or Rb^+ in many of its reactions.) All six of the elements have atoms characterized by an outer electron structure of ns^1 with n representing the principal quantum number. The elements exhibit marked resemblances to each other with Li deviating the most. This deviation is assignable to the small size of Li which causes the positive charge of Li^+ to be concentrated, that is, the charge density is high. All of the elements exhibit oxidation numbers of 0 and I, with exceptions being rare, such that their chemistries are dominated by the oxidation state I. The six metals are exceptionally reactive, being strong reductants, reacting with HOH at all pH values to give H_2 and M^+, and having hydroxides MOH which are strong and soluble. Ionic sizes in pm for the members of the group are as follows: Li (76), Na (102), K (139), Rb (152), Cs (167), and Fr (180). The $E°$ values for the M^+/M couples are as follows: Li (-3.04 v), Na (-2.71 v), K (-2.93 v), Rb (-2.92 v), Cs (-2.92 v), and Fr (about -3.03 v).

2. Lithium (Li) 2s^1

a. E–pH diagram. The E–pH diagram for $10^{-1.0}$ M Li is presented in Figure 5.1. The figure legend provides an equation for the line that separates Li$^+$ and Li. The horizontal line appears at an E value of -3.10 v. Considerably above the Li$^+$/Li line, the HOH $\equiv$ H$^+$/H$_2$ line appears, which indicates that Li metal is unstable in HOH, reacting with it to produce H$_2$ and Li$^+$. Note further that Li$^+$ dominates the diagram reflecting that the aqueous chemistry of Li is largely that of the ion Li$^+$. In the region above the Li$^+$/Li line, no

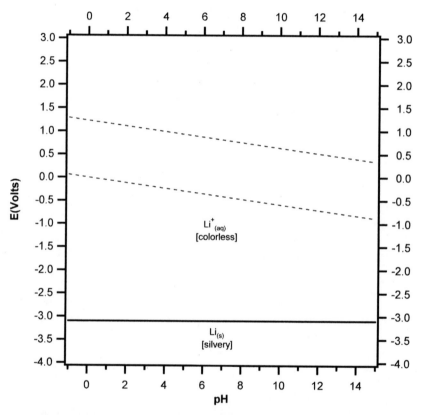

Figure 5.1 E–pH diagram for Li species. Soluble species concentrations (except H$^+$) $= 10^{-1.0}$ M. Soluble species and most solids are hydrated. No agents producing complexes or insoluble compounds are present other than HOH and OH$^-$.

Species ($\Delta G°$ in kJ/mol): Li (0.0), Li$^+$ (-293.3), HOH (-237.2), H$^+$ (0.0), and OH$^-$ (-157.3).

Equations for the lines:

$$\text{Li}^+/\text{Li} \qquad E = -3.04 + 0.059 \log [\text{Li}^+]$$

oxide or hydroxide precipitates as the pH is increased, denoting that neither is insoluble in water. The region of Li^+ extends all the way to the top of the diagram, pointing out that further oxidation is not possible. And the reaction of Li with HOH indicates that the metal cannot be prepared in water solution.

b. *Discovery.*   The element Li was discovered in the mineral petalite $LiAl(Si_2O_5)_2$ in 1817 by Johan August Arfvedson of Sweden. His professor Jöns Jacob Berzelius named it after the Greek word for stone (lithos). Shortly thereafter, in 1818, both Humphry Davy and William Thomas Brandé isolated the metal by electrolysis of molten Li_2O.

c. *Extraction.*   Lithium occurs in the form of several minerals and in brines and brine deposits. A brine is water with sizable concentrations of halide salts. The most common source of lithium is the mineral spodumene which has the formula $LiAl(SiO_3)_2$. Two treatments are used to extract the Li from it. In one, the mineral is ignited to enhance its solubility, is digested with H_2SO_4 to put Li^+ and a small amount of Al^{+3} into solution, the pH is raised to precipitate $Al(OH)_3$, then the solution of Li_2SO_4 is evaporated to give the solid sulfate, or it is treated with Na_2CO_3 to precipitate Li_2CO_3. In the other treatment, the powdered spodumene is mixed with powdered $CaCO_3$, calcined (roasted), cooled, and treated with HOH to put LiOH into solution. Concentration of the solution by evaporation gives $LiOH\cdot HOH$. The Li_2CO_3 obtained in the first treatment and the $LiOH\cdot HOH$ in the second are the materials from which other Li compounds can be prepared.

d. *The element.* Li_2CO_3 or $LiOH\cdot HOH$ is treated with HCl solution followed by evaporation to produce LiCl. This compound is mixed with KCl, melted, and electrolyzed at a controlled voltage. Li metal comes out at the cathode and Cl_2 gas evolved at the anode. This soft, silvery Li is exceptionally active as indicated by the E–pH diagram. It reacts violently with HOH to give Li^+, H_2, and OH^-, and it oxidizes upon heating in O_2 to produce the oxide Li_2O. The reaction with water may be written in two ways:

$$2Li + 2HOH \rightarrow 2Li^+ + 2OH^- + H_2, \text{ or}$$

$$2Li + 2H^+ \rightarrow 2Li^+ + H_2.$$

The first equation indicates an increase in $[OH^-]$ and thus a decrease in $[H^+]$, whereas the second equation indicates a decrease in $[H^+]$ and thus an increase in $[OH^-]$. Hence they are signifying essentially the same thing. The first equation can be obtained by addition of the second equation to this equation

$$2HOH \rightarrow 2H^+ + 2OH^-,$$

this being a reaction that always occurs in HOH solution.

e. Oxides and hydroxides. White solid Li_2O is formed upon heating Li metal in O_2. The white solid $LiOH \cdot HOH$ can be made by adding Li_2O to water and then evaporating the solution. The reaction in solution is $Li_2O + HOH \rightarrow 2Li^+ + 2OH^-$, and when the solution is evaporated $LiOH \cdot HOH$ solid remains. Mild heating of the $LiOH \cdot HOH$ drives the HOH off to yield anhydrous LiOH. $LiOH \cdot HOH$ and LiOH are very soluble, are completely dissociated in HOH, and act as a strong base. The peroxide compound Li_2O_2 can be made by addition of H_2O_2 to a solution of LiOH.

f. Compounds and solubilities. Table 5.1 presents the names, formulas, states, solubilities, and standard free energies of a number of the more important inorganic species of Li. A few organic salts are also shown. The species Li, Li_2O, LiOH, and Li_2CO_3 can be treated with appropriate acids to produce the corresponding salts. Note that almost all Li^+ inorganic salts are soluble, with the notable exceptions of Li_2CO_3, LiF, and Li_3PO_4. These compounds have solubilities of approximately 13.3, 1.3, and 0.4 g/L respectively. Further, most Li compounds are white, except for those which have a colored anion such as Li_2CrO_4 and $LiMnO_4$. When Li salts of weak-acid anions are dissolved they give basic solutions by freeing OH^- when the anions take H^+ away from HOH. A typical reaction of this type is

$$LiCN \text{ (solid)—dissolve in water} \rightarrow Li^+ + CN^-, \text{ then}$$

$$CN^- + HOH \rightarrow HCN + OH^-.$$

The most common strong-acid anions are Cl^-, Br^-, I^-, ClO_3^-, BrO_3^-, IO_3^-, ClO_4^-, BrO_4^-, IO_4^-, HSO_4^-, $HSeO_4^-$, NO_3^-, $H_2PO_4^-$, and $H_2AsO_4^-$. Almost all other anions are weak-acid anions, and their salts will give basic solutions when they are dissolved. The extent to which reactions of the type $CN^- + HOH \rightarrow HCN + OH^-$ occur can be determined by a free-energy calculation of the reaction energy of this equation, followed by conversion of the reaction free energy to a K value, followed by solving for the pH.

g. Redox reactions. As can be seen from the E–pH diagram, Li ordinarily shows only one valence $+1$ other than 0, an apparent exception being Li_2O_2, a highly unstable compound in which the O oxidation number is $-I$. The metal Li is a powerful reducing agent, reducing water readily by virtue of its $E°$ value of -3.04 v. The E–pH diagram also indicates that Li^+ is highly stable, resisting both oxidation and reduction.

h. Complexes. The Li^+ ion probably exists in HOH solution in the form of an exceptionally stable complex with four HOH molecules attached, $Li(HOH)_4^+$, the oxygens on the water molecules being the points of attachment. The ion is said to have a coordination number of 4, this value referring to the number of points of attachment that Li^+ provides to molecules

Table 5.1
Lithium Species

Name	Formula	State	Color	Solubility	$\Delta G°$ (kJ/mole)
Lithium	Li	s	Silvery	Decomp	0.0
Lithium ion	Li$^+$	aq	Colorless		−293.8
Lithium(I)					
acetate	LiC$_2$H$_3$O$_2$	s	White	Sol	
arsenate	Li$_3$AsO$_4$	s	White	Sl sol	
bromate	LiBrO$_3$	s	White	Sol	
bromide	LiBr	s	White	Sol	−341.6
carbonate	Li$_2$CO$_3$	s	White	Insol	−1132.4
chlorate	LiClO$_3$	s	White	Sol	
chloride	LiCl	s	White	Sol	−384.0
chromate	Li$_2$CrO$_4$	s	Yellow	Sol	
citrate	Li$_3$C$_6$H$_5$O$_7$	s	White	Sol	
cyanate	LiOCN	s	White	Sol	
cyanide	LiCN	s	White	Sol	
fluoride	LiF	s	White	Insol	−584.7
formate	LiOOCH	s	White	Sol	
germanate	Li$_2$GeO$_3$	s	White	Insol	
hydrogen carbonate	LiHCO$_3$	s	White	Sol	
hydroxide	LiOH	s	White	Sol	−441.9
iodate	LiIO$_3$	s	White	Sol	
iodide	LiI	s	White	Sol	−266.9
laurate	LiC$_{12}$H$_{23}$O$_2$	s	White	Insol	
metaborate	LiBO$_2$	s	White	Sol	−961.8
molybdate	Li$_2$MoO$_4$	s	White	Sol	
nitrate	LiNO$_3$	s	White	Sol	−389.5
nitrite	LiNO$_2$	s	White	Sol	−332.6
oxalate	Li$_2$C$_2$O$_4$	s	White	Sol	
oxide	Li$_2$O	s	White	Decomp	−562.1
perchlorate	LiClO$_4$	s	White	Sol	−254.0
permanganate	LiMnO$_4$	s	Purplish	Sol	
peroxide	Li$_2$O$_2$	s	White	Sol	−571.0
perrhenate	LiReO$_4$	s	White	Sol	
phosphate	Li$_3$PO$_4$	s	White	Insol	−1947.7
sulfate	Li$_2$SO$_4$	s	White	Sol	−1324.7
sulfide	Li$_2$S	s	White-yellow	Sol	−438.9
sulfite	Li$_2$SO$_3$	s	White	Sol	
tartrate	LiC$_4$H$_4$O$_6$	s	White	Sol	
thiocyanate	LiSCN	s	White	Sol	
vanadate	Li$_3$VO$_4$	s	White	Sol	
tungstate	Li$_2$WO$_4$	s	White	Sol	

or ions (ligands) which can attach. Some investigations indicate that this coordination number may be 6 instead of 4. This complex with attachments to oxygen is so very stable that displacement of the water molecules by other ligands (attaching molecules or ions) is difficult. As a result, Li^+ forms very few other stable complexes, and when it does the complexes are weak. An interesting complex is $Li(Ph_3PO)_4NO_3$ which is insoluble. Some complexation constants ($\log \beta_1$) for ligands with Li^+ are as follows: HPO_4^{-2} (0.2), $P_2O_7^{-4}$ (3.4), OH^- (0.3), SO_4^{-2} (0.6), $C_2H_3O_2^-$ (0.3), citrate^{-3} (0.6), iminodiacetate^{-2} (1.0), nitrilotriacetate^{-3} (2.5), edta^{-4} (2.8), diethylenetrinitrilopentaacetate^{-5} (3.1).

i. Analysis. The main methods used for the analyses of most elements at present are colorimetric methods using appropriate color-developing agents, atomic absorption spectroscopy (AAS), electrothermal atomic absorption spectroscopy (ETAES), inductively-coupled plasma atomic emission spectroscopy (ICPAES), inductively-coupled mass spectroscopy (ICPMS), and ion chromatogaphy (IC). These methods are prominent because their sample preparation and treatment is minimized, even though some of them require expensive instrumentation. The methods are for the analysis of the elements only, except for IC and colorimetry, which can usually discriminate among various ionic species. Colorimetry is often limited by the presence of other species which react similarly to the species being measured.

    Li or a Li compound in the flame gives a bright crimson color due to its emission of 670.8 nm photons produced by the short-lived species LiOH. This is the property that allows for the spectrophotometric determination of Li by atomic absorption spectroscopy (AAS) down to 20 ppb. Inductively-coupled plasma emission spectroscopy (ICPAES), inductively-coupled plasma mass spectroscopy (ICPMS), and ion chromatography (IC) improve this limit to about 0.1 ppb. A spot test for detection of Li down to 2 ppm is provided by basic KIO_4 plus $FeCl_3$.

j. Health aspects. Li_2CO_3 is medically administered as an antidepressant at dosages of 1.00 g/day or less. Doses in excess of this often begin to affect the central nervous system, with resulting tremors, nausea, and diarrhea. Ingestion of much larger amounts (over 5.0 g) can lead to death. The LD50 of ingested Li_2CO_3 for a rat is about 0.5 g/kg. Other simple compounds of Li have values as high as 1.8 g/kg. LD50 (lethal dose 50) stands for the dose which will lead to death in 50% of a sizable population. Further details will be found in Section 3j below.

3. Sodium (Na) 3s^1

a. E–pH diagram. The E–pH diagram for $10^{-1.0}$ M Na is presented in Figure 5.2. The figure legend provides an equation for the line that separates

Na^+ and Na. The horizontal line appears at an E value of -2.77 v. Considerably above the Na^+/Na line, the $HOH \equiv H^+/H_2$ line appears, which indicates that Na metal is unstable in HOH, reacting with it to produce H_2 and Na^+. Note further that Na^+ dominates the diagram reflecting that the aqueous chemistry of Na is largely that of the ion Na^+. In the region above the Na^+/Na line, no oxide or hydroxide precipitates as the pH is increased, denoting that neither is insoluble in water. The region of Na^+ extends all the way to the top of the diagram, pointing out that further oxidation is not possible. And the reaction of Na with HOH indicates that the metal cannot be prepared in water solution.

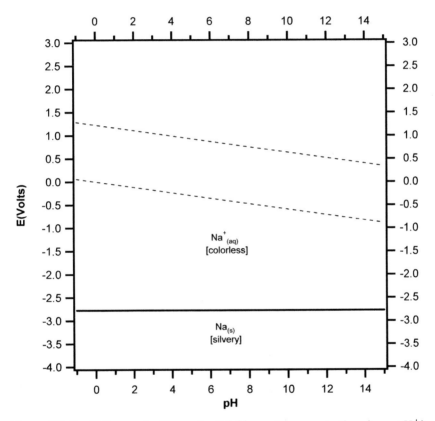

Figure 5.2 E–pH diagram for Na species. Soluble species concentrations (except H^+) $= 10^{-1.0}$ M. Soluble species and most solids are hydrated. No agents producing complexes or insoluble compounds are present other than HOH and OH^-.

Species ($\Delta G°$ in kJ/mol): Na (0.0), Na^+ (-261.9), HOH (-237.2), H^+ (0.0), and OH^- (-157.3)

Equation for the line:

$$Na^+/Na \qquad E = -2.71 + 0.059 \log[Na^+]$$

b. Discovery.   The element Na in its mineral forms has been known since ancient times. The name is derived from the Italian word soda and the symbol from the German natrium, which probably goes back to the Egyptian word neter which was a designation of sodium carbonate. The metal was isolated first in 1807 by Humphry Davy who obtained it by the electrolysis of molten NaOH.

c. Extraction.   Sodium occurs naturally in large deposits of a number of its compounds: rock salt or halite ($NaCl$), saltpeter ($NaNO_3$), thenardite (Na_2SO_4), trona ($Na_2CO_3 \cdot NaHCO_3 \cdot 2HOH$), borax ($Na_2B_4O_7 \cdot 10HOH$), and others. There are also vast supplies of $NaCl$ in oceanic and some lake waters. The most commonly employed source of sodium is the mineral $NaCl$, most of the other minerals being used chiefly as sources of their anions. The purity of $NaCl$ in rock salt deposits ranges between 90.0 and 99.9%. Further purification is attained by evaporation and recrystallization. Aqueous $NaCl$ is electrolyzed to produce $NaOH$ which is generally the basis of Na chemistry.

d. The element.   Molten $NaCl$ is electrolyzed to produce Na metal and Cl_2. Industrially, other chlorides ($CaCl_2$, $BaCl_2$) are added to the melt to lower the temperature at which the process can be carried out. The soft, silvery Na is exceptionally active as indicated by the E–pH diagram. It reacts violently with HOH to give Na^+, H_2, and OH^-, and it oxidizes upon heating in O_2 to produce the peroxide Na_2O_2. The reaction with water may be written two ways:

$$2Na + 2HOH \rightarrow 2Na^+ + 2OH^- + H_2, \text{ or}$$

$$2Na + 2H^+ \rightarrow 2Na^+ + H_2.$$

The first equation indicates an increase in $[OH^-]$ and thus a decrease in $[H^+]$, whereas the second equation indicates a decrease in $[H^+]$ and thus an increase in $[OH^-]$. Hence they are signifying essentially the same thing. The first equation can be obtained by addition of the second equation to this equation

$$2HOH \rightarrow 2H^+ + 2OH^-,$$

this being a reaction that always occurs in HOH solution.

e. Oxides and hydroxides.   The pale yellow solid Na_2O_2 is formed upon heating Na metal in dry air or dry O_2. The white solid $NaOH$ can be prepared by adding Na_2O_2 to water and then evaporating the solution. The reaction upon heating the solution is $2Na_2O_2 + 4HOH \rightarrow O_2 + 4Na^+ + 4OH^-$, and when the solution is evaporated, hydrated $NaOH$ remains. Heating of the hydrated $NaOH$ drives the HOH off to yield the anhydrous compound. $NaOH$ is very soluble, is completely dissociated in HOH, and acts as a strong base.

The monoxide compound Na_2O can be prepared by fusing the peroxide with elemental Na.

f. Compounds and solubilities. Table 5.2 presents the names, formulas, states, solubilities, and standard free energies of a number of the more important inorganic species of Na. A few organic salts are also shown. The species Na, Na_2O, Na_2O_2, NaOH, $NaHCO_3$, and Na_2CO_3 can be treated with appropriate acids to produce the corresponding salts. Note that almost all Na^+ inorganic salts are soluble with exceptions being sodium zinc uranyl

Table 5.2
Sodium Species

Name	Formula	State	Color	Solubility	$\Delta G°$ (kJ/mole)
Sodium	Na	s	Silvery	Decomp	0.0
Sodium ion	Na^+	aq	Colorless		−261.9
Sodium(I)					
acetate	$NaC_2H_3O_2$	s	White	Sol	
arsenate	Na_3AsO_4	s	White	Sol	−1425.9
azide	NaN_3	s	White	Sol	
borate	Na_3BO_3	s	White	Sol	
borohydride	$NaBH_4$	s	White	Sol	
bromate	$NaBrO_3$	s	White	Sol	−252.7
bromide	NaBr	s	White	Sol	−349.4
carbonate	Na_2CO_3	s	White	Sol	−1048.1
chlorate	$NaClO_3$	s	White	Sol	−274.9
chloride	NaCl	s	White	Sol	−384.1
chlorite	$NaClO_2$	s	White	Sol	−228.0
chromate	Na_2CrO_4	s	Yellow	Sol	−1232.2
citrate	$Na_3C_6H_5O_7$	s	White	Sol	
cyanate	NaOCN	s	White	Sol	−359.4
cyanide	NaCN	s	White	Sol	
dichromate	$Na_2Cr_2O_7$	s	Red	Sol	−1800.4
dihydrogen arsenate	NaH_2AsO_4	s	White	Sol	
dihydrogen phosphate	NaH_2PO_4	s	White	Sol	−1385.7
fluoride	NaF	s	White	Sol	−543.5
formate	NaOOCH	s	White	Sol	
germanate	Na_2GeO_3	s	White	Sol	
hexafluorosilicate	Na_2SiF_6	s	White	Insol	−2553.9
hydride	NaH	s	White	Decomp	
hypochlorite	NaClO	s	White	Sol	
hydrogen arsenate	Na_2HAsO_4	s	White	Sol	
hydrogen carbonate	$NaHCO_3$	s	White	Sol	−851.9
hydrogen oxalate	$NaHC_2O_4$	s	White	Sol	
hydrogen phosphate	Na_2HPO_4	s	White	Sol	−1623.8
hydrogen selenate	$NaHSeO_4$	s	White	Sol	

Table 5.2
(Continued)

Name	Formula	State	Color	Solubility	$\Delta G°$ (kJ/mole)
Sodium(I)—*cont'd*					
hydrogen sulfate	$NaHSO_4$	s	White	Sol	-1020.9
hydrogen sulfide	$NaHS$	s	White	Sol	-231.4
hydroxide	$NaOH$	s	White	Sol	-381.6
hypoiodite	$NaIO$	s	White	Sol	
hypophosphite	$Na(H_2PO_2)$	s	White	Sol	
iodate	$NaIO_3$	s	White	Sol	-397.9
iodide	NaI	s	White	Sol	-284.5
laurate	$NaC_{12}H_{23}O_2$	s	White	Sl sol	
metaborate	$Na_3B_3O_6$	s	White	Sol	-2767.8
metasilicate	Na_2SiO_3	s	White	Sol	-1466.9
metavanadate	$NaVO_3$	s	White	Sol	-1065.7
molybdate	Na_2MoO_4	s	White	Sol	-1353.9
nitrate	$NaNO_3$	s	White	Sol	-366.1
nitrite	$NaNO_2$	s	White	Sol	-283.7
oxalate	$Na_2C_2O_4$	s	White	Sol	-1289.1
oxide	Na_2O	s	White	Decomp	-377.0
perbromate	$NaBrO_4$	s	White	Sol	
perchlorate	$NaClO_4$	s	White	Sol	-254.4
periodate	$NaIO_4$	s	White	Sol	-323.0
peroxydisulfate	$Na_2S_2O_8$	s	White	Sol	
permanganate	$NaMnO_4$	s	Purplish	Sol	-1582.8
peroxide	Na_2O_2	s	White	Sol	-449.8
perrhenate	$NaReO_4$	s	White	Sol	-948.5
phosphate	Na_3PO_4	s	White	Sol	-1802.0
phosphite	Na_2HPO_3	s	White	Sol	
pyrophosphate	$Na_4P_2O_7$	s	White	Sol	
selenate	Na_2SeO_4	s	White	Sol	-970.3
selenite	Na_2SeO_3	s	White	Sol	
silicate	Na_4SiO_4	s	White	Sol	
sulfate	Na_2SO_4	s	White	Sol	-1269.4
sulfide	Na_2S	s	White-yellow	Sol	-361.5
sulfite	Na_2SO_3	s	White	Sol	-1002.1
tartrate	$NaC_4H_4O_6$	s	White	Sol	
telluride	Na_2Te	s	White	Sol	-345.2
tetraborate	$Na_2B_4O_5(OH)_4$	s	White	Sol	
tetrafluoroborate	$NaBF_4$	s	White	Sol	
thiocarbonate	Na_2CS_3	s	Yellow	Sol	
thiocyanate	$NaSCN$	s	White	Sol	-153.1
thiosulfate	$Na_2S_2O_3$	s	White	Sol	-1021.7
tungstate	Na_2WO_4	s	White	Sol	-1429.7
vanadate	Na_3VO_4	s	White	Sol	-1643.1

acetate $NaZn(UO_2)_3(C_2H_3O_2)_9 \cdot 6HOH$, Na_2SiF_6, and $NaSb(OH)_6$. Further, most Na compounds are white, except for those which have a colored anion such as Na_2CrO_4 and $NaMnO_4$. When Na salts of weak-acid anions are dissolved they give basic solutions by freeing OH^- when the anions take H^+ away from HOH. A typical reaction of this type is

$$Na_2S(solid)\text{---dissolve in water} \rightarrow Na^+ + S^{-2}, \text{then}$$

$$S^{-2} + HOH \rightarrow HS^- + OH^-$$

The most common strong-acid anions are $Cl^-, Br^-, I^-, ClO_3^-, BrO_3^-, IO_3^-$, $ClO_4^-, BrO_4^-, IO_4^-, HSO_4^-, HSeO_4^-, NO_3^-, H_2PO_4^-$, and $H_2AsO_4^-$. Almost all other anions are weak-acid anions, and their salts will give basic solutions when they are dissolved. The extent to which reactions of the type $S^{-2} + HOH \rightarrow HS^- + OH^-$ occur can be determined by a free-energy calculation of the reaction energy of this equation, followed by conversion of the reaction free energy to a K value, followed by solving for the pH.

g. Redox reactions. As can be seen from the E–pH diagram, Na ordinarily shows only one valence +1 other than 0, a seeming exception being Na_2O_2, a highly unstable compound in which the O oxidation number is –I. The metal Na is a powerful reducing agent, reducing water readily by virtue of its E° value of -2.71 v. The E–pH diagram also indicates that Na^+ is highly stable, resisting both oxidation and reduction.

h. Complexes. The Na^+ ion probably exists in HOH solution in the form of an exceptionally stable complex with six HOH molecules attached, $Na(HOH)_6^+$, the oxygens on the water molecules being the points of attachment. The ion is said to have a coordination number of 6, this value referring to the number of points of attachment that Na^+ provides to molecules or ions (ligands) which can attach. This water complex with attachments to oxygen is so very stable that displacement of water by other ligands (attaching molecules or ions) is difficult. As a result, Na^+ forms very few other stable complexes, and when it does, the complexes are weak. Some complexation constants (log β_1) for ligands with Na^+ are as follows: $C_2H_3O_2^-(-0.1)$, $B(OH)_4^-(1.9)$, $HPO_4^{-2}(0.2)$, $P_2O_7^{-4}(2.3)$, $P_3O_{10}^{-5}(2.7)$, $SO_4^{-2}(0.7)$, $S_2O_3^{-2}(0.5)$, maleate$^{-2}(0.7)$, malic$^{-2}(0.3)$, Htartrate$^-(0.8)$, citrate$^{-3}(0.6)$, phthalate^{-2} (0.7), iminodiacetate$^{-2}(0.4)$, nitrilotriacetate^{-3} (1.2), edta$^{-4}(1.6)$.

i. Analysis. Na or a Na compound in the flame gives a bright yellow color due to its emission of 589.0 and 589.6 nm photons. This is the property that allows for the spectrophotometric determination of Na by emission or absorption flame or plasma spectroscopy. AAS and ETAAS are capable down to 10 ppb. ICPAES extends this to 1 ppb, and a limit of detection of 0.1 ppb can be obtained by ICPMS and IC. A spot test for 250 ppm or more can be carried out using zinc uranyl acetate. Unfortunately, there are numerous interfering ions.

j. Health aspects. Sodium is an essential element in the human body, the daily requirement being about 1.0 g. The toxicity of chemicals is usually measured by the LD50 in rats fed orally, the quantities being expressed in mg/kg of body weight. The LD50 is the median lethal dose, that is, the dose that is estimated to be fatal to 50% of the rats. It is generally assumed that the rat values can be approximately applied to humans, although safety factors of 100 or so are often applied. It is to be carefully noted that symptoms of poisoning can show up at values considerably less than the LD50 value. Highly toxic sodium salts (LD50 less than 50 mg/kg) include the arsenate, arsenite, azide, cyanide, selenite, and tellurate. Somewhat toxic sodium salts (LD50 between 50 and 500 mg/kg) include the fluoride, hydroxide, iodate, iodide, nitrite, peroxydisulfate, sulfite, and thiosulfate. Low toxicity sodium salts (LD50 over 500 mg/kg) include the acetate, borate, bromide, chlorate, chloride, nitrate, phosphate, and sulfate. In each of the above categories, Li, K, Rb, or Cs may be substituted for Na, since the values probably will be about the same. Material Safety Data Sheets (MSDS), which may be accessed on the internet, provide details on the safety aspects of these and other alkali metal compounds.

4. Potassium (K) 4s^1

a. E–pH diagram. The E–pH diagram for $10^{-1.0}$ M K is presented in Figure 5.3. The figure legend provides an equation for the line that separates K^+ and K. The horizontal line appears at an E value of -2.99 v. Considerably above the K^+/K line, the HOH $\equiv$ H$^+$/H$_2$ line appears, which indicates that K metal is unstable in HOH, reacting with it to produce H$_2$ and K^+. Note further that K^+ dominates the diagram reflecting that the aqueous chemistry of K is largely that of the ion K^+. In the region above the K^+/K line, no oxide nor hydroxide precipitates as the pH is increased, denoting that neither is insoluble in water. The region of K^+ extends all the way to the top of the diagram, pointing out that further oxidation is not possible. And the reaction of K with HOH indicates that the metal cannot be prepared in water solution.

b. Discovery. The element K in its carbonate form has been known since ancient times when it was produced by leaching wood ashes. The compound was known as potash and the English name is derived from that. The symbol originates from the Latin kalium. The metal was first isolated in 1807 by Humphry Davy who obtained it by the electrolysis of molten KOH.

c. Extraction. Potassium occurs naturally in large deposits of a number of its compounds: sylvite (KCl), carnallite (KMgCl$_3$·6HOH), and langbeinite

($K_2Mg_2(SO_4)_3$), and others. The most commonly employed source of potassium is the mineral KCl, which is usually fairly pure. Further purification is attained by evaporation and recrystallization. Aqueous KCl is electrolyzed to produce KOH which is generally the basis of K chemistry.

d. The element. Metallic K is obtained by treating molten KCl with sodium vapor. The soft, silvery K is exceptionally active as indicated by the E–pH diagram. It reacts violently with HOH to give K^+, H_2, and OH^-, and it oxidizes upon heating in air or O_2 to produce the orange superoxide KO_2. This latter compound contains the ion O_2^-.

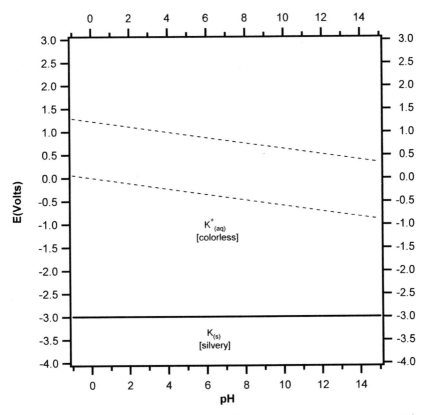

Figure 5.3 E–pH diagram for K species. Soluble species concentrations (except H^+) $= 10^{-1.0}$ M. Soluble species and most solids are hydrated. No agents producing complexes or insoluble compounds are present other than HOH and OH^-.

Species ($\Delta G°$ in kJ/mol): K (0.0), K^+ (−282.4), HOH (−237.2), H^+ (0.0), and OH^- (−157.3).

Equation for the line:

$$K^+/K \qquad E = -2.93 + 0.059 \log[K^+]$$

e. *Oxides and hydroxides.* The orange KO_2 is formed upon heating K metal in dry air or dry O_2. The white solid KOH can be made by adding KO_2 to water and then evaporating the solution. The reaction upon adding the KO_2 to the water is

$$2KO_2 + 2HOH \rightarrow O_2 + 2K^+ + 2OH^- + H_2O_2.$$

Heating the solution to evaporate it decomposes H_2O_2 as follows

$$2H_2O_2 \rightarrow O_2 + 2HOH.$$

KOH is very soluble, is completely dissociated in HOH, and acts as a strong base. Other oxides of K (K_2O and K_2O_2) may be prepared by careful regulation of the amount of O_2 allowed to react with metallic K. These oxides also are readily hydrolyzed by water to produce KOH.

f. *Compounds and solubilities.* Table 5.3 presents the names, formulas, states, solubilities, and standard free energies of a number of the more important inorganic species of K. A few organic salts are also shown. The species K, K_2O, K_2O_2, KO_2, KOH, $KHCO_3$, and K_2CO_3 can be treated with appropriate acids to produce the corresponding salts. Note that almost all K^+ inorganic salts are soluble, notable exceptions being potassium tetraphenylborate ($KB(C_6H_5)_4$), potassium hexachloroplatinate (K_2PtCl_6), potassium hydrogen tartrate ($KHC_4H_5O_6$), potassium picrate ($KC_6H_2N_3O_7$), potassium perchlorate ($KClO_4$), and dipotassium sodium hexanitritocobaltate(III) ($K_2NaCo(NO_2)_6$). Further, most K compounds are white, except for those which have a colored anion such as K_2CrO_4 and $KMnO_4$, and the several oxides. When K salts of weak-acid anions are dissolved they give basic solutions by freeing OH^- when the anions take H^+ away from HOH. The most common strong-acid anions are Cl^-, Br^-, I^-, ClO_3^-, BrO_3^-, IO_3^-, ClO_4^-, BrO_4^-, IO_4^-, HSO_4^-, $HSeO_4^-$, NO_3^-, $H_2PO_4^-$, and $H_2AsO_4^-$. Almost all other anions are weak-acid anions, and their K salts will give basic solutions when they are dissolved. The extent to which reactions of this type occur can be determined by a free-energy calculation of the reaction energy of the appropriate equation, followed by conversion of the reaction free energy to an equilibrium constant K value, followed by solving for the pH.

g. *Redox reactions.* As can be seen from the E–pH diagram, K ordinarily shows only one valence +1 other than 0, seeming exceptions being K_2O_2 and KO_2, compounds in which the O oxidation number is $-I$ or $-1/2$, since O_2^{-2} has a charge of -2 and the superoxide ion carries a charge of -1. The metal K is a powerful reducing agent, reducing water readily by virtue of its $E°$ value of -2.93 v. The E–pH diagram also indicates that K^+ is highly stable, resisting both oxidation and reduction.

h. *Complexes.* The K^+ ion probably exists in HOH solution in the form of an exceptionally stable complex with six HOH molecules attached,

Table 5.3
Potassium species

Name	Formula	State	Color	Solubility	$\Delta G°$ (kJ/mole)
Potassium	K	s	Silvery	Decomp	0.0
Potassium ion	K^+	aq	Colorless		−282.4
Potassium(I)					
acetate	$KC_2H_3O_2$	s	White	Sol	
arsenate	K_3AsO_4	s	White	Sol	−1548.9
azide	KN_3	s	White	Sol	
borate	K_3BO_3	s	White	Sol	
borohydride	KBH_4	s	White	Sol	
bromate	$KBrO_3$	s	White	Sol	−243.9
bromide	KBr	s	White	Sol	−380.3
carbonate	K_2CO_3	s	White	Sol	−1064.4
chlorate	$KClO_3$	s	White	Sol	−290.0
chloride	KCl	s	White	Sol	−408.8
chlorite	$KClO_2$	s	White	Sol	
chromate	K_2CrO_4	s	Yellow	Sol	−1299.1
citrate	$K_3C_6H_5O_7$	s	White	Sol	
cyanate	$KOCN$	s	White	Sol	−374.0
cyanide	KCN	s	White	Sol	−102.1
dihydrogen arsenate	KH_2AsO_4	s	White	Sol	−991.6
dihydrogen phosphate	KH_2PO_4	s	White	Sol	−1418.8
dichromate	$K_2Cr_2O_7$	s	Red	Sol	−1882.8
fluoride	KF	s	White	Sol	−537.6
formate	$KOOCH$	s	White	Sol	−591.2
germanate	K_2GeO_3	s	White	Sol	
hexafluorosilicate	K_2SiF_6	s	White	Insol	
hydride	KH	s	White	Decomp	
hydrogen arsenate	K_2HAsO_4	s	White	Sol	
hydrogen carbonate	$KHCO_3$	s	White	Sol	−860.6
hydrogen oxalate	KHC_2O_4	s	White	Sol	
hydrogen phosphate	K_2HPO_4	s	White	Sol	−1636.4
hydrogen selenate	$KHSeO_4$	s	White	Sol	
hydrogen sulfate	$KHSO_4$	s	White	Sol	−1043.1
hydrogen sulfide	KHS	s	White	Sol	
hydroxide	KOH	s	White	Sol	−379.9
hypochlorite	$KClO$	s	White	Sol	
hypoiodite	KIO	s	White	Sol	
hypophosphite	$K(H_2PO_2)$	s	White	Sol	
iodate	KIO_3	s	White	Sol	−425.5
iodide	KI	s	White	Sol	−324.3
laurate	$KC_{12}H_{23}O_2$	s	White	Sl sol	
metaborate	$K_3B_3O_6$	s	White	Sol	−2810.4
metasilicate	K_2SiO_3	s	White	Sol	
metavanadate	KVO_3	s	White	Sol	

Table 5.3
(Continued)

Name	Formula	State	Color	Solubility	$\Delta G°$ (kJ/mole)
Potassium(I)—*con'd*					
molybdate	K_2MoO_4	s	White	Sol	
nitrate	KNO_3	s	White	Sol	−393.3
nitrite	KNO_2	s	White	Sol	−282.0
oxalate	$K_2C_2O_4$	s	White	Sol	−1241.0
oxide	K_2O	s	Yellow	Decomp	−322.2
perbromate	$KBrO_4$	s	White	Sol	−174.5
perchlorate	$KClO_4$	s	White	Sol	−300.0
periodate	KIO_4	s	White	Sol	−360.7
permanganate	$KMnO_4$	s	Purplish	Sol	−713.8
peroxide	K_2O_2	s	Yellow	Sol	−429.7
peroxydisulfate	$K_2S_2O_8$	s	White	Sol	−1692.8
perrhenate	$KReO_4$	s	White	Sol	−997.9
pertechnetate	$KTcO_4$	s	White	Sol	−927.6
phosphate	K_3PO_4	s	White	Sol	−1859.0
phosphite	K_2HPO_3	s	White	Sol	
pyrophosphate	$K_4P_2O_7$	s	White	Sol	
selenate	K_2SeO_4	s	White	Sol	−1002.9
selenite	K_2SeO_3	s	White	Sol	
silicate	K_4SiO_4	s	White	Sol	
sulfate	K_2SO_4	s	White	Sol	−1316.3
sulfide	K_2S	s	White-yellow	Sol	−404.2
sulfite	K_2SO_3	s	White	Sol	
superoxide	KO_2	s	Orange	Decomp	
tartrate	$KC_4H_4O_6$	s	White	Sol	
tetraphenylborate	$KB(C_6H_5)_4$	s	White	Insol	
telluride	K_2Te	s	White	Sol	
tetraborate	$K_2B_4O_5(OH)_4$	s	White	Sol	
tetrafluoroborate	KBF_4	s	White	Sol	
thiocarbonate	K_2CS_3	s	Yellow	Sol	
thiocyanate	$KSCN$	s	White	Sol	−178.2
thiosulfate	$K_2S_2O_3$	s	White	Sol	
tungstate	K_2WO_4	s	White	Sol	
vanadate	K_3VO_4	s	White	Sol	−1025.9

$K(HOH)_6{}^+$, the oxygens on the water molecules being the points of attachment. The ion is said to have a coordination number of 6, this value referring to the number of points of attachment that K^+ provides to molecules or ions (ligands) which can attach. This water complex with attachments to oxygen is so very stable that displacement of the water molecules by other ligands (attaching molecules or ions) is difficult. As a result, K^+

forms very few other stable complexes, and when it does the complexes are weak. Some complexation constants ($\log \beta_1$) for ligands with K^+ are as follows: $NO_3^-(-0.1), HPO_4^{-2}(0.2), P_2O_7^{-4}(2.1), P_3O_{10}^{-5}(2.8), SO_4^{-2}(0.9)$ $S_2O_3^{-2}(1.0), Cl^-(-0.5), Br^-(-0.6), I^-(-0.2), edta^{-4}(0.8)$.

i. Analysis.   K or a K compound in the flame gives a reddish-violet (lilac) color. This is due to its emission of 404.4, 404.7, 766.5, and 769.9 nm photons. These photon lines allow for the determination of K by AAS (limit of detection = 20 ppb), ETAAS (1 ppb), and ICPAES (0.1 ppb). This latter limit is also attainable by ICPMS and IC. Spots tests with sodium hexanitritocobaltate(III) and dipicrylamine both detect 100 ppm of K.

j. Health aspects.   Potassium is an essential element in the human body, the daily requirement being about 3 g. Toxic compounds of K parallel those of Na, which are treated in Section 3j of this chapter.

5. Rubidium (Rb) $5s^1$

a. E–pH diagram.   The E–pH diagram for $10^{-1.0}$ M Rb is presented in Figure 5.4. The figure legend provides an equation for the line that separates Rb^+ and Rb. The horizontal line appears at an E value of -2.98 v. Considerably above the Rb^+/Rb line, the $HOH \equiv H^+/H_2$ line appears, which indicates that Rb metal is unstable in HOH, reacting with it to produce H_2 and Rb^+. Note further that Rb^+ dominates the diagram reflecting that the aqueous chemistry of Rb is largely that of the ion Rb^+. In the region above the Rb^+/Rb line, no oxide or hydroxide precipitates as the pH is increased, denoting that neither is insoluble in water. The region of Rb^+ extends all the way to the top of the diagram, pointing out that further oxidation is not possible. And the reaction of Rb with HOH indicates that the metal cannot be prepared in water solution.

b. Discovery.   The element Rb was discovered in 1861 by Robert W. Bunsen and Gustav R. Kirchoff in the mineral lepidolite by observation of its red flame color in their recently-developed spectroscope. The name is derived from the Latin rubidus which means deep red. Bunsen then proceeded shortly thereafter to prepare the metal by reduction of rubidium hydrogen tartrate with carbon.

c. Extraction.   Rubidium does not occur as the major constituent in any mineral. The chief source of it is the mineral lepidolite ($KRbLi(OH,F)$ $Al_2Si_3O_{10}$) which may contain up to 3.2% Rb. The ore is treated with H_2SO_4 to produce a solution containing K^+, Rb^+, Li^+, Al^{+3}, and SO_4^{-2}, leaving a residue of SiO_2. The solution is evaporated, with rubidium alum $RbAl(SO_4)_2$ precipitating first. This compound is then put into solution and brought to neutrality to precipitate $Al(OH)_3$. After filtration of the insoluble $Al(OH)_3$, the solution is treated with $Ba(OH)_2$ which precipitates $BaSO_4$ and leaves a solution of RbOH. Evaporation gives the solid.

d. The element. Rb metal is produced by reducing lepidolite ore with Ca or Mg, followed by vacuum distillation. Alternatively pure RbOH, RbCl, or Rb_2CO_3 can be reduced by Ca or Mg. The soft, silvery Rb is exceptionally active as indicated by the E–pH diagram. It reacts violently with HOH to give Rb^+, H_2, and OH^-, and it burns in air or O_2 to dark brown RbO_2.

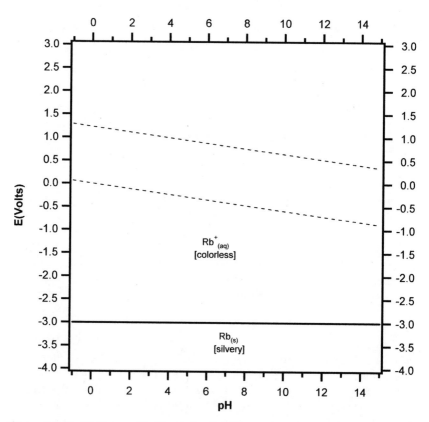

Figure 5.4 E–pH diagram for Rb species. Soluble species concentrations (except H^+) $= 10^{-1.0}$ M. Soluble species and most solids are hydrated. No agents producing complexes or insoluble compounds are present other than HOH and OH^-.

Species ($\Delta G°$ in kJ/mol): Rb (0.0), Rb^+ (−282.0), HOH (−237.2), H^+ (0.0) and OH^- (−157.3).

Equation for the line:

$$Rb^+/Rb \qquad E = -2.92 + 0.059 \log[Rb^+]$$

e. Oxides and hydroxides.   As indicated above, the dark brown RbO_2 is formed from Rb metal in dry air or dry O_2. The white solid RbOH can be made by adding RbO_2 to water and then evaporating the solution. Equations similar to those for K (see Section 4e above) apply. RbOH is very soluble, is completely dissociated in HOH, and acts as a strong base. Other oxides of Rb (Rb_2O and Rb_2O_2) may be made by careful regulation of the amount of O_2 allowed to react with metallic Rb. These oxides also are readily hydrolyzed by water to produce RbOH.

f. Compounds and solubilities.   Table 5.4 presents the names, formulas, states, solubilities, and standard free energies of a number of the more important inorganic species of Rb. A few organic salts are also shown. The species Rb, Rb_2O, Rb_2O_2, RbO_2, RbOH, $RbHCO_3$, and Rb_2CO_3 can be treated with appropriate acids to produce the corresponding salts. Note that almost all Rb^+ inorganic salts are soluble, exceptions are rubidium tetraphenylborate ($RbB(C_6H_5)_4$), rubidium hexachloroplatinate (Rb_2PtCl_6), rubidium hexafluorosilicate (Rb_2SiF_6), rubidium hydrogen tartrate ($RbHC_4H_5O_6$), rubidium picrate ($RbC_6H_2N_3O_7$), rubidium perchlorate ($RbClO_4$), rubidium periodate ($RbIO_4$), and dirubidium sodium hexanitritocobaltate(III) ($Rb_2NaCo(NO_2)_6$) . Further, most Rb compounds are white, except for those which have a colored anion such as Rb_2CrO_4 and $RbMnO_4$, and the several oxides. When Rb salts of weak-acid anions are dissolved they give basic solutions by freeing OH^- when the anions take H^+ away from HOH. The extent to which reactions of this type occur can be determined by a free-energy calculation of the reaction energy of the appropriate equation, followed by conversion of the reaction free energy to an equilibrium constant K value, followed by solving for the pH.

g. Redox reactions.   As can be seen from the E–pH diagram, Rb ordinarily shows only one valence +1 other than 0, seeming exceptions being Rb_2O_2 and RbO_2, compounds in which the O oxidation number is $-I$ or $-1/2$, since the peroxide ion O_2^{-2} has a charge of -2 and the superoxide ion carries a charge of -1. The metal Rb is a powerful reducing agent, reducing water readily by virtue of its $E°$ value of -2.92 v. The E–pH diagram also indicates that Rb^+ is highly stable, resisting both oxidation and reduction.

h. Complexes.   The Rb^+ ion probably exists in HOH solution in the form of an exceptionally stable complex with six or eight HOH molecules attached, $Rb(HOH)_6^+$ or $Rb(HOH)_8^+$, the oxygens on the water molecules being the points of attachment. The ion is said to have a coordination number of 6 or 8, this value referring to the number of points of attachment that Rb^+ provides to molecules or ions (ligands) which can attach. This water complex with attachments to oxygen is so very stable that displacement of

Table 5.4
Rubidium Species

Name	Formula	State	Color	Solubility	$\Delta G°$ (kJ/mole)
Rubidium	Rb	s	Silvery	Decomp	0.0
Rubidium Ion	Rb^+	aq	Colorless		−282.0
Rubidium(I)					
acetate	$RbC_2H_3O_2$	s	White	Sol	
arsenate	Rb_3AsO_4	s	White	Sol	−1546.8
azide	RbN_3	s	White	Sol	
borate	Rb_3BO_3	s	White	Sol	
borohydride	$RbBH_4$	s	White	Sol	
bromate	$RbBrO_3$	s	White	Sol	−282.0
bromide	RbBr	s	White	Sol	−378.2
carbonate	Rb_2CO_3	s	White	Sol	−1043.1
chlorate	$RbClO_3$	s	White	Sol	−292.0
chloride	RbCl	s	White	Sol	−405.0
chlorite	$RbClO_2$	s	White	Sol	
chromate	Rb_2CrO_4	s	Yellow	Sol	−1300.0
citrate	$Rb_3C_6H_5O_7$	s	White	Sol	
cyanate	RbOCN	s	White	Sol	−68.2
cyanide	RbCN	s	White	Sol	
dichromate	$Rb_2Cr_2O_7$	s	Red	Sol	
dihydrogen arsenate	RbH_2AsO_4	s	White	Sol	
dihydrogen phosphate	RbH_2PO_4	s	White	Sol	
fluoride	RbF	s	White	Sol	−520.1
formate	RbOOCH	s	White	Sol	
germanate	Rb_2GeO_3	s	White	Sol	
hexafluorosilicate	Rb_2SiF_6	s	White	Sl sol	
hydride	RbH	s	White	Decomp	
hydrogen arsenate	Rb_2HAsO_4	s	White	Sol	
hydrogen carbonate	$RbHCO_3$	s	White	Sol	−855.2
hydrogen oxalate	$RbHC_2O_4$	s	White	Sol	
hydrogen phosphate	Rb_2HPO_4	s	White	Sol	
hydrogen selenate	$RbHSeO_4$	s	White	Sol	
hydrogen sulfate	$RbHSO_4$	s	White	Sol	
hydrogen sulfide	RbHS	s	White	Sol	
hydroxide	RbOH	s	White	Sol	−372.4
hypochlorite	RbClO	s	White	Sol	
hypoiodite	RbIO	s	White	Sol	
hypophosphite	$Rb(H_2PO_2)$	s	White	Sol	
iodate	$RbIO_3$	s	White	Sol	
iodide	RbI	s	White	Sol	−325.9
laurate	$RbC_{12}H_{23}O_2$	s	White	Sl sol	
metaborate	$Rb_3B_3O_6$	s	White	Sol	−2751.3
metasilicate	Rb_2SiO_3	s	White	Sol	

Continued

Table 5.4
(Continued)

Name	Formula	State	Color	Solubility	$\Delta G°$ (kJ/mole)
Rubidium(I)—*cont'd*					
metavanadate	$RbVO_3$	s	White	Sol	
molybdate	Rb_2MoO_4	s	White	Sol	
nitrate	$RbNO_3$	s	White	Sol	−389.9
nitrite	$RbNO_2$	s	White	Sol	−302.5
oxalate	$Rb_2C_2O_4$	s	White	Sol	
oxide	Rb_2O	s	Yellow	Decomp	−291.2
perbromate	$RbBrO_4$	s	White	Sol	
perchlorate	$RbClO_4$	s	White	Sol	−306.3
periodate	$RbIO_4$	s	White	Sol	
permanganate	$RbMnO_4$	s	Purplish	Sol	
peroxide	Rb_2O_2	s	Yellow	Sol	−352.7
peroxydisulfate	$Rb_2S_2O_8$	s	White	Sol	
perrhenate	$RbReO_4$	s	White	Sol	
phosphate	Rb_3PO_4	s	White	Sol	
phosphite	Rb_2HPO_3	s	White	Sol	
pyrophosphate	$Rb_4P_2O_7$	s	White	Sol	
selenate	Rb_2SeO_4	s	White	Sol	
selenite	Rb_2SeO_3	s	White	Sol	
silicate	Rb_4SiO_4	s	White	Sol	
sulfate	Rb_2SO_4	s	White	Sol	−1309.2
sulfide	Rb_2S	s	White-yellow	Sol	−336.8
sulfite	Rb_2SO_3	s	White	Sol	
superoxide	RbO_2	s	Brown	Decomp	−220.1
tartrate	$RbC_4H_4O_6$	s	White	Sol	
telluride	Rb_2Te	s	White	Sol	
tetraborate	$Rb_2B_4O_5(OH)_4$	s	White	Sol	
tetrafluoroborate	$RbBF_4$	s	White	Sol	
tetraphenylborate	$RbB(C_6H_5)_4$	s	White	Insol	
thiocarbonate	Rb_2CS_3	s	Yellow	Sol	
thiocyanate	$RbSCN$	s	White	Sol	
thiosulfate	$Rb_2S_2O_3$	s	White	Sol	
tungstate	Rb_2WO_4	s	White	Sol	
vanadate	Rb_3VO_4	s	White	Sol	

the water molecules by other ligands (attaching molecules or ions) is difficult. As a result, Rb^+ forms very few other stable complexes, and when it does the complexes are weak. Most complexation constants for ligands with Rb^+ are generally slightly less than those for K^+.

i. Analysis.   Rb or a Rb compound in the flame gives a red color due to its emission of 420.2, 422.6, 780.0, and 794.8 nm photons. This is the property that allows for the analysis by AAS (down to 20 ppb), ETAES (1 ppb), and ICPAES (10 ppb). A spot test which goes down to 10 ppm is provided by a mixture of $AuBr_3$ and AgBr.

j. Health aspects.   Toxic compounds of Rb parallel those of Na, which are treated in Section 3j of this chapter.

6. Cesium (Cs) 6s^1

a. E–pH diagram.   The E–pH diagram for $10^{-1.0}$ M Cs is presented in Figure 5.5. The figure legend provides an equation for the line that separates Cs^+ and Cs. The horizontal line appears at an E value of -2.98 v. Considerably above the Cs^+/Cs line, the HOH $\equiv$ H$^+$/H$_2$ line appears, which indicates that the Cs metal is unstable in HOH, reacting with it to produce H_2 and Cs^+. Note further that Cs^+ dominates the diagram reflecting that the aqueous chemistry of Cs is largely that of the ion Cs^+. In the region above the Cs^+/Cs line, no oxide or hydroxide precipitates as the pH is increased, denoting that neither is insoluble in water. The region of Cs^+ extends all the way to the top of the diagram, pointing out that further oxidation is not possible. And the reaction of Cs with HOH indicates that the metal cannot be prepared in water solution.

b. Discovery.   The element Cs was discovered in 1860 by Robert W. Bunsen and Gustav R. Kirchoff in spring water by observation of its blue flame color in their recently-developed spectroscope. The name is derived from the Latin caesius which means pale blue. The metal was not prepared until 1881 by C. Setterberg who employed electrolysis of a molten mixture of CsCN and $Ba(CN)_2$.

c. Extraction.   The major source of Cs is the mineral pollucite, $CsAlSi_2O_6$, which is usually found with considerable amounts of quartz SiO_2. The material is treated by either base or acid digestion. In the former, the pollucite–silica is mixed with lime and $CaCl_2$ and heated to 900°C, or is mixed with NaCl and Na_2CO_3 and heated to 800°C. Leaching then produces a solution of CsCl. Alternatively, acid digestion can be carried out with H_2SO_4 to produce insoluble $CsAl(SO_4)_2$, which, when roasted with carbon, converts the Al to insoluble Al_2O_3, and leaves the soluble Cs_2SO_4. Or HBr can be used to treat the mineral to put bromides of Cs and Al into solution; the addition of ethanol then precipitates CsBr.

d. The element.   Cs metal is produced by reducing CsCl with Ca or Ba, followed by vacuum distillation. The soft Cs metal, which is light golden yellow, is exceptionally active as indicated by the E–pH diagram. It reacts violently with HOH to give Cs^+, H_2, and OH^-, and it burns in air or O_2 to orange CsO_2.

e. Oxides and hydroxides.   As indicated above, the orange CsO_2 is formed from Cs metal in dry air or dry O_2. The white solid CsOH can be made by adding CsO_2 to water and then evaporating the solution. Equations similar to those for K (see Section 4e above) apply. CsOH is very soluble, is completely dissociated in HOH, and acts as a strong base. Other oxides of Cs (Cs_2O and Cs_2O_2), both of which are orange, may be prepared by careful regulation of

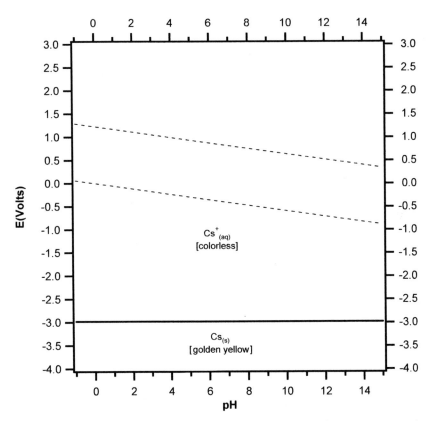

Figure 5.5 E–pH diagram for Cs species. Soluble species concentrations (except H^+) $= 10^{-1.0}$ M. Soluble species and most solids are hydrated. No agents producing complexes or insoluble compounds are present other than HOH and OH^-.

Species (ΔG° in kJ/mol): Cs (0.0), Cs^+ (−282.0), HOH (−237.2), H^+ (0.0), and OH^- (−157.3)

Equation for the line:

$$Cs^+/Cs \qquad E = -2.92 + 0.059 \log [Cs^+]$$

the amount of O_2 allowed to react with metallic Cs. These oxides also are readily hydrolyzed by water to produce CsOH.

f. Compounds and solubilities.

Table 5.5 presents the names, formulas, states, solubilities, and standard free energies of a number of the more important inorganic species of Cs. A few organic salts are also shown. The species Cs, Cs_2O, Cs_2O_2, CsO_2, CsOH, CsHCO3, and Cs_2CO_3 can be treated with appropriate acids to produce the corresponding salts.

Table 5.5
Cesium Species

Name	Formula	State	Color	Solubility	$\Delta G°$ (kJ/mole)
Cesium	Cs	s	Golden yellow	Decomp	0.0
Cesium Ion	Cs^+	aq	Colorless		-282.0
Cesium(I)					
acetate	$CsC_2H_3O_2$	s	White	Sol	
arsenate	Cs_3AsO_4	s	White	Sol	-1543.9
azide	CsN_3	s	White	Sol	
borate	Cs_3BO_3	s	White	Sol	
borohydride	$CsBH_4$	s	White	Sol	
bromate	$CsBrO_3$	s	White	Sol	-284.9
bromide	CsBr	s	White	Sol	-382.8
carbonate	Cs_2CO_3	s	White	Sol	-1018.8
chlorate	$CsClO_3$	s	White	Sol	-283.7
chloride	CsCl	s	White	Sol	-404.2
chlorite	$CsClO_2$	s	White	Sol	
chromate	Cs_2CrO_4	s	Yellow	Sol	
citrate	$Cs_3C_6H_5O_7$	s	White	Sol	
cyanate	CsOCN	s	White	Sol	
cyanide	CsCN	s	White	Sol	-99.2
dichromate	$Cs_2Cr_2O_7$	s	Red	Sol	
dihydrogen arsenate	CsH_2AsO_4	s	White	Sol	
dihydrogen phosphate	CsH_2PO_4	s	White	Sol	
fluoride	CsF	s	White	Sol	-525.5
formate	CsOOCH	s	White	Sol	
germanate	Cs_2GeO_3	s	White	Sol	
hexafluorosilicate	Cs_2SiF_6	s	White	Sl sol	
hydride	CsH	s	White	Decomp	
hydrogen arsenate	Cs_2HAsO_4	s	White	Sol	
hydrogen carbonate	$CsHCO_3$	s	White	Sol	-831.8
hydrogen oxalate	$CsHC_2O_4$	s	White	Sol	
hydrogen phosphate	Cs_2HPO_4	s	White	Sol	
hydrogen selenate	$CsHSeO_4$	s	White	Sol	
hydrogen sulfate	$CsHSO_4$	s	White	Sol	

Continued

Table 5.5
(Continued)

Name	Formula	State	Color	Solubility	$\Delta G°$ (kJ/mole)
Cesium Ion—*cont'd*					
hydrogen sulfide	$CsHS$	s	White	Sol	
hydroxide	$CsOH$	s	White	Sol	−354.8
hypochlorite	$CsClO$	s	White	Sol	
hypoiodite	$CsIO$	s	White	Sol	
hypophosphite	$Cs(H_2PO_2)$	s	White	Sol	
iodate	$CsIO_3$	s	White	Sol	
iodide	CsI	s	White	Sl sol	−333.0
laurate	$CsC_{12}H_{23}O_2$	s	White	Sol	
metaborate	$Cs_3B_3O_6$	s	White	Sol	−2715.0
metasilicate	Cs_2SiO_3	s	White	Sol	
metavanadate	$CsVO_3$	s	White	Sol	
molybdate	Cs_2MoO_4	s	White	Sol	
nitrate	$CsNO_3$	s	White	Sol	−392.9
nitrite	$CsNO_2$	s	White	Sol	−316.7
oxalate	$Cs_2C_2O_4$	s	White	Sol	
oxide	Cs_2O	s	Orange	Decomp	−274.1
perbromate	$CsBrO_4$	s	White	Sol	
perchlorate	$CsClO_4$	s	White	Sl sol	−306.7
periodate	$CsIO_4$	s	White	Sl sol	
permanganate	$CsMnO_4$	s	Purplish	Sol	
peroxide	Cs_2O_2	s	Orange	Sol	−340.6
peroxydisulfate	$Cs_2S_2O_8$	s	White	Sol	
perrhenate	$CsReO_4$	s	White	Sol	
phosphate	Cs_3PO_4	s	White	Sol	
phosphite	Cs_2HPO_3	s	White	Sol	
pyrophosphate	$Cs_4P_2O_7$	s	White	Sol	
selenate	Cs_2SeO_4	s	White	Sol	−1010.9
selenite	Cs_2SeO_3	s	White	Sol	
silicate	Cs_4SiO_4	s	White	Sol	
sulfate	Cs_2SO_4	s	White	Sol	−1300.0
sulfide	Cs_2S	s	White-yellow	Sol	
sulfite	Cs_2SO_3	s	White	Sol	
superoxide	CsO_2	s	Orange	Decomp	−210.9
tartrate	$CsC_4H_4O_6$	s	White	Sol	
telluride	Cs_2Te	s	White	Sol	
tetraborate	$Cs_2B_4O_5(OH)_4$	s	White	Sol	
tetrafluoroborate	$CsBF_4$	s	White	Sol	
tetraphenylborate	$CsB(C_6H_5)_4$	s	White	Insol	
thiocarbonate	Cs_2CS_3	s	Yellow	Sol	
thiocyanate	$CsSCN$	s	White	Sol	
thiosulfate	$Cs_2S_2O_3$	s	White	Sol	−325.1
tungstate	Cs_2WO_4	s	White	Sol	
vanadate	Cs_3VO_4	s	White	Sol	

Note that almost all Cs^+ inorganic salts are soluble with exceptions being cesium tetraphenylborate $(CsB(C_6H_5)_4)$, cesium hexachloroplatinate (Cs_2PtCl_6), cesium hexafluorosilicate (Cs_2SiF_6), cesium hydrogen tartrate $(CsHC_4H_5O_6)$, cesium picrate $(CsC_6H_2N_3O_7)$, cesium perchlorate $(CsClO_4)$, cesium periodate $(CsIO_4)$, and dicesium sodium hexanitritocobaltate(III) $(Cs_2NaCo(NO_2)_6)$. Further, most Cs compounds are white, except for those which have a colored anion such as Cs_2CrO_4 and $CsMnO_4$, and the several oxides. When Cs salts of weak-acid anions are dissolved they give basic solutions by freeing OH^- when the anions take H^+ away from HOH. The extent to which reactions of this type occur can be determined by a free-energy calculation of the reaction energy of the appropriate equation, followed by conversion of the reaction free energy to an equilibrium constant K value, followed by solving for the pH.

g. Redox reactions. As can be seen from the E–pH diagram, Cs ordinarily shows only one valence +1 other than 0, seeming exceptions being Cs_2O_2 and CsO_2, compounds in which the O oxidation number is −1 or −1/2, since the peroxide ion O_2^{-2} has a charge of −2 and the superoxide ion carries a charge of −1. The metal Cs is a powerful reducing agent, reducing water readily by virtue of its E° value of −2.92 v. The E–pH diagram also indicates that Cs^+ is highly stable, occupying most of the area, and thus resisting both oxidation and reduction.

h. Complexes. The Cs^+ ion probably exists in HOH solution in the form of an exceptionally stable complex with eight HOH molecules attached, or $Cs(HOH)_8^+$, the oxygens on the water molecules being the points of attachment. The ion is said to have a coordination number of 8, this value referring to the number of points of attachment that Cs^+ provides to molecules or ions (ligands) which can attach. This water complex with attachments to oxygen is so very stable that displacement of the water molecules by other ligands (attaching molecules or ions) is difficult. As a result, Cs^+ forms very few other stable complexes, and when it does the complexes are very weak. Most complexation constants for ligands with Cs^+ are generally less than those for Rb^+.

i. Analysis. Cs or a Cs compound in the flame gives a blue-violet color due to its emission of 455.4, 459.3, 852.1, and 894.4 nm photons. AAS determines Cs above a level of 200 ppb. IC is about the same, but ICPMS can go as low as 0.1 ppb. The reagent $KBiI_4$ permits a spot test which can detect about 20 ppm.

j. Health aspects. Toxic compounds of Cs parallel those of Na, which are treated in Section 3j of this Chapter. The Cs^+ ion is somewhat more toxic than the Na^+ ion, but less toxic than the corresponding ions of Li, K, and Rb. The oral LD50 for CsCl in mice is 2.3 g/kg, and that for CsF is 0.5 g/kg.

7. Francium (Fr) 7s^1

The element Fr was discovered by Marguerite Perey in 1939 in the decay products of U-235, and it also can be prepared artificially by proper nuclear reactions, particularly the bombardment of thorium with protons. The longest-lived isotope is Fr-223, which has a half-life of 21.8 min. This property of the element obviously makes it very difficult to work with, and only very dilute solutions can be investigated because of the energy liberated in the radioactive decay. Practically all of the experimental information regarding the chemistry of Fr comes from radiochemical tracer investigations. No weighable amount of the element or its compounds has been produced. The experimental information fits well into the concept that the behavior of Fr resembles that of Cs quite closely. In line with this, Figure 5.6 presents the

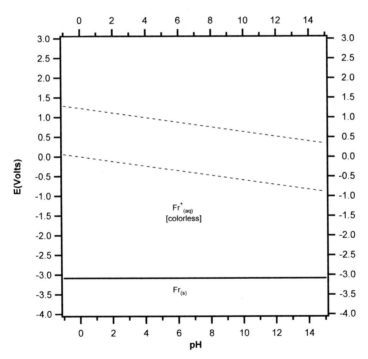

Figure 5.6 E–pH diagram for Fr species. Soluble species concentrations (except H$^+$) = $10^{-1.0}$ M. Soluble species and most solids are hydrated. No agents producing complexes or insoluble compounds are present other than HOH and OH$^-$.

Species (ΔG$^\circ$ in kJ/mol): Fr (0.0), Fr$^+$ (−292.0), HOH (−237.2), H$^+$ (0.0) and OH$^-$ (−157.3).

Equation for the line:

$$\text{Fr}^+/\text{Fr} \qquad E = -3.03 + 0.059 \log [\text{Fr}^+]$$

E–pH diagram of Fr at a concentration of $10^{-1.0}$ M using an estimated $E°$ value of -3.03 v. In tracer quantities, the concentration of Fr would be much less, making the E value very negative, which would favor the Fr^+ species even more.

8. The Ammonium Ion (NH_4^+)

a. E–pH diagram. The E–pH diagram for $10^{-1.0}$ M NH_4^+ is presented in Figure 5.7. Note that this ion is related to other N-containing species: NH_4OH, HNO_3, NO_3^-, HNO_2, and NO_2^-. In the figure legend are equations for lines which separate these various species. Both NH_4^+ and NH_4OH are stable in water solution as is indicated by their presence within the water lines. And oxidation of them will produce the nitrite and nitrate species. The regions of NH_4^+ and NH_4OH extend to the bottom of the diagram, indicating that further reduction is not possible. At a pH of 9.2 the predominant species changes from NH_4^+ to NH_4OH, the designation NH_4OH being a shorthand for hydrated NH_3. In actuality, the undissociated NH_4OH molecule probably does not exist, but it is convenient to write NH_4OH for the hydrated NH_3 ($NH_3 \cdot HOH = NH_4OH$).

b. Discovery. The parent compound of NH_4^+, ammonia NH_3, has been known from antiquity as the sharp smelling gas emitted from some decomposing nitrogen-containing substances. The gas was first isolated and characterized in 1774 by Joseph Priestly, who collected it over Hg. Its very high solubility in water precluded its collection in that fashion.

c. Preparation. As indicated by the E–pH diagram, the ammonium ion NH_4^+ can be prepared by acidifying a basic solution of NH_4OH (actually hydrated NH_3) until a pH of 9.2 or below is attained. The counterion (anion) associated with the NH_4^+ is determined by the anion of the acid used in the acidification. The precursor of NH_4^+, namely NH_3, is produced by the reaction of N_2 with H_2 at an elevated temperature in the presence of a catalyst. The H_2 for this reaction is generated from various substances including coke and natural gas (CH_4).

d. Amalgam. The elemental analog of the NH_4^+ ion would be the neutral NH_4, which has not been isolated. However, an unusual amalgam is formed when an ammonium sulfate solution is electrolyzed using Hg as a cathode, or when an ammonium ion solution is treated with Na amalgam. The form of the compound in the amalgam could be the radical NH_4. The radical in the amalgam decomposes at room temperature, but at low temperatures is stable. It is thought that the situation in the amalgam consists of NH_4^+ ions with negative Hg clusters. The $E°$ of the couple NH_4^+/NH_4 amalgam is approximately -1.70 v.

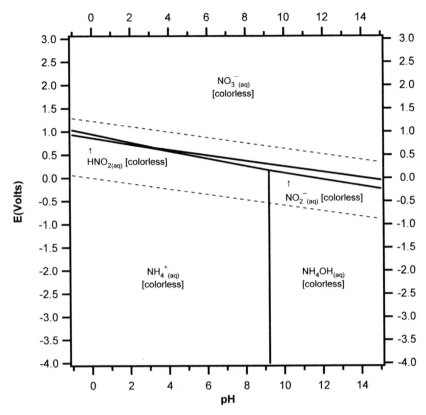

Figure 5.7 E–pH diagram for N species. Soluble species concentrations (except H^+) $= 10^{-1.0}$ M. Soluble species and most solids are hydrated. No agents producing complexes or insoluble compounds are present other than HOH and OH^-.

Species ($\Delta G°$ in kJ/mol): N_2 (0.0), HNO_3 (−103.4), NO_3^- (−110.9), HNO_2 (−55.7), NO_2^- (−36.8), NH_4^+ (−79.5), NH_4OH (−264.0), HOH (−237.2), H^+ (0.0), and OH^- (−157.3).

Equations for the lines:

$$HNO_2/NH_4^+ \qquad E = 0.86 - 0.069\ pH$$

$$NO_2^-/NH_4^+ \qquad E = 0.89 - 0.079\ pH$$

$$NO_2^-/NH_4OH \qquad E = 0.80 - 0.069\ pH$$

$$NO_3^-/HNO_2 \qquad E = 0.94 - 0.089\ pH$$

$$NO_3^-/NO_2^- \qquad E = 0.85 - 0.059\ pH$$

$$NO_2^-/HNO_2 \qquad pH = 3.2$$

$$NH_4OH/NH_4^+ \qquad pH = 9.2$$

e. Compounds and solubilities. Table 5.6 presents the names, formulas, states, solubilities, and standard free energies of a number of the more important inorganic species of NH_4^+. A few organic salts are also shown. The species NH_3 (NH_4OH), NH_4HCO_3 and $(NH_4)_2CO_3$, and the NH_4 amalgam can be treated with appropriate acids to produce the corresponding salts. Note that almost all NH_4^+ inorganic salts are soluble with exceptions being ammonium tetraphenylborate ($NH_4B(C_6H_5)_4$), ammonium hexachloroplatinate ($(NH_4)_2PtCl_6$), ammonium hexafluorosilicate ($(NH_4)_2SiF_6$), ammonium hydrogen tartrate ($NH_4HC_4H_5O_6$), ammonium picrate ($NH_4C_6H_2N_3O_7$), ammonium periodate (NH_4IO_4), and diammonium sodium hexanitritocobaltate(III) ($(NH_4)_2NaCo(NO_2)_6$). Further, most NH_4^+ compounds are white, except for those which have a colored anion such as $(NH_4)_2CrO_4$

Table 5.6
Ammonium Species

Name	Formula	State	Color	Solubility	$\Delta G°$ (kJ/mole)
Ammonia	NH_3	aq	Colorless	Sol	−26.8
Ammonium Ion	NH_4^+	aq	Colorless		−79.5
acetate	$(NH_4)C_2H_3O_2$	s	White	Sol	
arsenate	$(NH_4)_3AsO_4$	s	White	Sol	
azide	NH_4N_3	s	White	Sol	274.2
borate	$(NH_4)_3BO_3$	s	White	Sol	
borohydride	NH_4BH_4	s	White	Sol	
bromate	NH_4BrO_3	s	White	Sol	
bromide	NH_4Br	s	White	Sol	−176.1
carbonate	$(NH_4)_2CO_3$	s	White	Sol	
chlorate	NH_4ClO_3	s	White	Sol	
chloride	NH_4Cl	s	White	Sol	−202.9
chlorite	NH_4ClO_2	s	White	Sol	
chromate	$(NH_4)_2CrO_4$	s	Yellow	Sol	−889.9
citrate	$(NH_4)_3C_6H_5O_7$	s	White	Sol	
cyanate	NH_4OCN	s	White	Sol	
cyanide	NH_4CN	s	White	Sol	
dichromate	$(NH_4)_2Cr_2O_7$	s	Orange	Sol	
dihydrogen arsenate	$NH_4H_2AsO_4$	s	White	Sol	−833.0
dihydrogen phosphate	$NH_4H_2PO_4$	s	White	Sol	−1210.4
fluoride	NH_4F	s	White	Sol	−348.5
formate	NH_4OOCH	s	White	Sol	
hexafluorosilicate	$(NH_4)_2SiF_6$	s	White	Sl sol	
hydrogen arsenate	$(NH_4)_2HAsO_4$	s	White	Sol	
hydrogen carbonate	NH_4HCO_3	s	White	Sol	−669.0
hydrogen oxalate	$NH_4HC_2O_4$	s	White	Sol	
hydrogen phosphate	$(NH_4)_2HPO_4$	s	White	Sol	

Continued

Table 5.6
(Continued)

Name	Formula	State	Color	Solubility	$\Delta G°$ (kJ/mole)
Ammonium Ion—*cont'd*					
hydrogen selenate	NH_4HSeO_4	s	White	Sol	
hydrogen sulfate	NH_4HSO_4	s	White	Sol	−785.3
hydrogen sulfide	NH_4HS	s	White	Sol	−52.7
hydroxide	NH_4OH (actually NH_3+HOH)	aq	Colorless	Sol	−264.0
hypochlorite	NH_4ClO	s	White	Sol	
hypophosphite	$NH_4(H_2PO_2)$	s	White	Sol	
iodate	NH_4IO_3	s	White	Sol	
iodide	NH_4I	s	White	Sol	−112.1
laurate	$NH_4C_{12}H_{23}O_2$	s	White	Sl sol	
metaborate	$(NH_4)_3B_3O_6$	s	White	Sol	
metasilicate	$(NH_4)_2SiO_3$	s	White	Sol	
metavanadate	NH_4VO_3	s	White	Sol	−888.3
molybdate	$(NH_4)_2MoO_4$	s	White	Sol	
nitrate	NH_4NO_3	s	White	Sol	−183.7
nitrite	NH_4NO_2	s	White	Sol	−111.7
oxalate	$(NH_4)_2C_2O_4$	s	White	Sol	
perchlorate	NH_4ClO_4	s	White	Sol	−88.7
periodate	NH_4IO_4	s	White	Sol	
permanganate	NH_4MnO_4	s	Purplish	Sol	
peroxydisulfate	$(NH_4)_2S_2O_8$	s	White	Sol	
perrhenate	NH_4ReO_4	s	White	Sol	−775.3
phosphate	$(NH_4)_3PO_4$	s	White	Sol	
phosphite	$(NH_4)_2HPO_3$	s	White	Sol	
pyrophosphate	$(NH_4)_4P_2O_7$	s	White	Sol	
selenate	$(NH_4)_2SeO_4$	s	White	Sol	
selenide	$(NH_4)_2Se$	s	White	Sol	
silicate	$(NH_4)_4SiO_4$	s	White	Sol	
sulfate	$(NH_4)_2SO_4$	s	White	Sol	−901.7
sulfide	$(NH_4)_2S$	s	White-yellow	Sol	
sulfite	$(NH_4)_2SO_3$	s	White	Sol	
tellurate	$(NH_4)_2TeO_4$	s	White	Sol	
tartrate	$NH_4C_4H_4O_6$	s	White	Sol	
tetraborate	$(NH_4)_2B_4O_5(OH)_4$	s	White	Sol	
tetrafluoroborate	NH_4BF_4	s	White	Sol	
tetraphenylborate	$NH_4B(C_6H_5)_4$	s	White	Insol	
thiocarbonate	$(NH_4)_2CS_3$	S	Yellow	Sol	
thiocyanate	NH_4SCN	s	White	Sol	
thiosulfate	$(NH_4)_2S_2O_3$	s	White	Sol	
tungstate	$(NH_4)_2WO_4$	s	White	Sol	
vanadate	$(NH_4)_3VO_4$	s	White	Sol	

and NH_4MnO_4. The resemblance of NH_4^+ to Rb^+ is evident in all these characteristics. When NH_4^+ salts of a strong acid anion are dissolved, they give acidic solutions by freeing H^+ when the NH_4^+ cation takes OH^- away from HOH. The extent to which this type of reaction occurs can be determined by a calculation of the reaction free energy of the appropriate equation, followed by conversion of the reaction free energy to an equilibrium constant K value, and then by solving for the pH. When the NH_4^+ salt of the anion of a weak acid is dissolved, both the hydrolysis of the NH_4^+ and that of the weak acid anion must be taken into account to determine the resulting pH value.

f. Redox reactions. The E–pH diagram for N as shown in Figure 5.7 indicates that the NH_4^+ ion at a pH from strongly acid up into the low basicity region can be oxidized to HNO_2/NO_2^- and HNO_3/NO_3^-, the exact product depending on the pH and E value of the oxidant. At pH values in the basic region, NH_4OH (or more accurately, aqueous NH_3) can be oxidized to NO_2^- or NO_3^-. No reduction of NH_4^+ or NH_3 can be effected.

g. Complexes. The NH_4^+ ion does not undergo complexation, but the molecule NH_3 is an agent which forms moderately strong or strong complexes with a wide variety of cations. Among the most well-characterized of these cations are Cr^{+3}, Mn^{+2}, Fe^{+2}, Co^{+2}, Co^{+3}, Ni^{+2}, Pd^{+2}, Pt^{+2}, Cu^+, Cu^{+2}, Ag^+, Au^+, Zn^{+2}, Cd^{+2}, Hg^{+2}, and Tl^{+3}. In some instances, a maximum of six NH_3 can attach, in others four, and in a few cases two. Complexation constants β_n for many ammonia (also called ammine) complexes are given under the sections on the various elements.

h. Analysis. The most widely used analytical procedure for NH_4^+ is color development by indophenol followed by spectrophotometric determination. The limit of detection is in the range of 20–100 ppb. A spot test capable of detecting 20 ppm is based upon p-nitrobenzenediazonium chloride. Less than that can be detected by adding NaOH, then heating for several minutes at 40°C in a closed vessel containing a wet piece of red litmus paper.

i. Health aspects. Most NH_4^+ salts are about as toxic as the corresponding K^+ or Rb^+ salts. The main danger they pose is the conversion to gaseous NH_3 and aqueous NH_3 under basic conditions. These substances are corrosive and are basic enough to attack human tissue, producing pulmonary edema, dyspnea, bronchospasm, chest pain, and skin burns. Typical LD50 values for ingestion of NH_4Cl or $(NH_4)_2SO_4$ are 1.6 g/kg and 3 g/kg.

6

The Beryllium Group

1. Introduction

The elements which constitute the Be Group of the Periodic Table are known as the alkaline earths. They are beryllium Be, magnesium Mg, calcium Ca, strontium Sr, barium Ba, and radium Ra. All six of the elements have atoms characterized by an outer electron structure of ns^2 with n representing the principal quantum number. The elements exhibit marked resemblances to each other with Be differing considerably. This deviation is assignable to the small size of Be which causes the positive charge of Be^{+2} to be concentrated, that is, the charge density is high. The higher charge-density ions attack HOH to attach OH^- and to liberate H^+, that is, they hydrolyze readily. All of the elements exhibit oxidation numbers of 0 and II, with their chemistries being dominated by the oxidation state II. The six metals are exceptionally reactive, being strong reductants, reacting with acids, HOH, and bases at all pH values to give H_2. The other product of such reactions are M^{+2} ions and $M(OH)_2$, the ions being present at lower pH values and the hydroxides being present at higher pH values. The transition from M^{+2} to $M(OH)_2$ occurs at increasing pH values from Be to Ra, such that the hydroxide of Sr is slightly soluble and those of Ba and Ra are soluble. These soluble hydroxides are strong bases. Ionic sizes in pm for the members of the group are as follows: Be (59), Mg (72), Ca (100), Sr (132), Ba (135), and Ra (148). The E° values for the M^{+2}/M couples are as follows: Be (-1.97 v), Mg (-2.36 v), Ca (-2.87 v),

Sr (-2.89 v), Ba (-2.91 v), and Ra (-2.91 v), indicating that they are very reactive metals.

2. Beryllium (Be) $2s^2$

a. E–pH diagram. The E–pH diagram for $10^{-1.0}$ M Be is presented in Figure 6.1. In the figure legend are equations which describe the lines separating the species. It can be seen that Be is thermodynamically unstable with respect to H^+, HOH, and OH^-. However, Be is relatively inactive in the middle pH range due to a protective oxide coat. To dissolve Be, dilute acids such as HCl or H_2SO_4, or bases such as NaOH or KOH are required. Concentrated oxidizing acids, HNO_3 for example, secure the oxide coat. Above the Be region, the diagram is dominated by $Be(OH)_2$, this compound being amphoteric (soluble in both acid and base). The Be^{+2} designation is simplified in that the species is actually $Be(HOH)_4^{+2}$. In the region of transformation from Be^{+2} to $Be(OH)_2$ there exist partially hydrolyzed and polymeric species such as $Be_3(OH)_3^{+3}$ and Be_2OH^{+3}. These species illustrate the strong tendency that Be^{+2} has to hydrolyze. The three divalent forms of Be extend to the top of the diagram indicating that they are not oxidized.

b. Discovery. In 1798 Nicolas-Louis Vauquelin extracted BeO from the mineral beryl. He named the element glucinium (Greek glykys which means sweet) because the salts of BeO tasted sweet. The name berylerde (beryl earth) was used for the oxide in Germany. Elemental Be was first isolated independently by Friedrich Wöhler and Antoine Bussy in 1828 by reduction of $BeCl_2$ with K. Wöhler then introduced the name beryllium (Greek word beryllos, which means beryl), which superseded glucinium.

c. Extraction. The main minerals of Be are beryl $3BeO\cdot Al_2O_3\cdot 6SiO_2$ and bertrandite $4BeO\cdot SiO_2\cdot HOH$. Be is recovered from from beryl by roasting the mineral with Na_2SiF_6. HOH is added to leach out the resulting Na_2BeF_4, and the solution is made alkaline to precipitate $Be(OH)_2$. The process for bertrandite is more complicated but the end product is again $Be(OH)_2$.

d. The element. There are two main methods for the production of Be. One is the reduction of BeF_2 with Mg, and the other is the electrolysis of melts consisting of BeF_2, BeO, or $BeCl_2$ with alkali or alkaline earth halides. Elemental Be is a brittle, light-gray metal which is thermodynamically active, but which forms an inert oxide coat in air making it inactive in some reactions.

e. Oxides and hydroxides. BeO can be formed by heating Be in air, by heating of the hydroxide, basic carbonate, acetate, or sulfate. The oxide does not dissolve in HOH, but is soluble in acids and bases, dissolving very slowly if it has been fired at high temperatures. White gelatinous hydrated $Be(OH)_2$ can be precipitated from solutions of Be^{+2} by addition of OH^-. Freshly precipitated $Be(OH)_2$ dissolves readily in acids and bases, but as it ages, dissolution becomes slower.

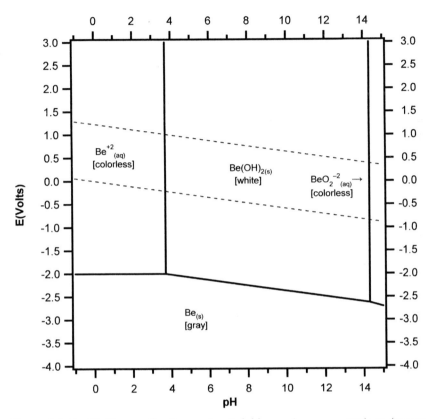

Figure 6.1 E–pH diagram for Be species. Soluble species concentrations (except H^+) = $10^{-1.0}$ M. Soluble species and most solids are hydrated. No agents producing complexes or insoluble compounds are present other than HOH and OH^-.

Species ($\Delta G°$ in kJ/mol): Be (0.0), Be^{+2} (−380.3), $Be(OH)_2$ (−818.0), BeO_2^{-2} (−648.9), HOH (−237.2), H^+ (0.0), and OH^- (−157.3).

Equations for the lines:

Be^{+2}/Be	$E = -1.97 + 0.030 \log [Be^{+2}]$
$Be(OH)_2/Be$	$E = -1.79 - 0.059 \ pH$
BeO_2^{-2}/Be	$E = -0.90 - 0.119 \ pH + 0.030 \log [BeO_2^{-2}]$
$Be(OH)_2/Be^{+2}$	$2 \ pH = 6.4 - \log [Be^{+2}]$
$BeO_2^{-2}/Be(OH)_2$	$2 \ pH = 29.7 + \log [BeO_2^{-2}]$

Table 6.1
Beryllium Species

Name	Formula	State	Color	Solubility	$\Delta G°$ (kJ/mole)
Beryllium	Be	s	Gray	Insol	0.0
Beryllium ion	Be^{+2}	aq	Colorless		-380.3
Hydroxoberyllium ion	$Be(OH)^+$	aq	Colorless		-600.5
Dioxoberyllate ion	BeO_2^{-2}	aq	Colorless		-648.9
Beryllium (II)					
acetate	$Be(C_2H_3O_2)_2$	s	White	Insol	
basic carbonates	various	s	White	Insol	
bromide	$BeBr_2$	s	White	Sol	
carbonate	$BeCO_3$	s	White	Decomp	
chloride	$BeCl_2$	s	White	Sol	-445.6
cyanide	$Be(CN)_2$	s	White	Decomp	
fluoride	BeF_2	s	White	Sol	-979.5
formate	$Be(OOCH)_2$	s	White	Decomp	
hydroxide	$Be(OH)_2$	s	White	Insol	-818.0
hydroxide hydrated	$Be(OH)_2$	s	White	Insol	-810.4
iodide	BeI_2	s	White	Decomp	
nitrate	$Be(NO_3)_2$	s	White	Sol	
oxalate	BeC_2O_4	s	White	Sol	
oxide	BeO	s	White	Insol	-580.3
oxoacetate	$Be_4O(C_2H_3O_2)_6$	s	White	Insol	
perchlorate	$Be(ClO_4)_2$	s	White	Sol	
sulfate	$BeSO_4$	s	White	Sol	-1093.9
sulfide	BeS	s	White	Insol	
sulfite	$BeSO_3$	s	White	Decomp	

f. Compounds and solubilities. Table 6.1 presents details on the more important Be compounds for aqueous chemistry. Be, BeO, $Be(OH)_2$, and $BeCO_3$ can be dissolved in strong acids to produce the corresponding Be^{+2} salts. Dissolution in weaker acids often produces basic salts because of the tendency for Be^{+2} to hydrolyze. Most Be^{+2} salts are white soluble substances, the main exception to the solubility rules being the acetate. The high charge to radius ratio of Be^{+2} gives rise to $Be(OH)^+$ and a variety of hydrolyzed polymeric cations even at low pH values, especially at higher Be^{+2} concentrations.

g. Redox reactions. As seen from the E–pH diagram, Be shows only one oxidation state II other than 0. The metal is a strong reducing agent, but this action is limited in many cases by its non-reactivity. The divalent species of Be are highly stable, resisting both oxidation and reduction.

h. Complexes. A number of complexes are formed by Be^{+2}. Among those that have been characterized in detail are the following, along with their

β_n values (log β_1, log β_2, log β_3, log β_4) OH$^-$(8.6, 14.3, 18.7, 18.6), SO$_4^{-2}$(2.0, 1.8), F$^-$(4.7, 8.3, 11.1, 13.1), HCOO$^-$(1.4), C$_2$H$_3$O$_2^-$(1.6, 2.4), C$_2$O$_4^{-2}$(4.1, 5.4), tartrate^{-2}(1.7), citrate^{-3}(4.5), salicylate^{-2}(12.4, 22.0), phthalate^{-2}(4.0, 5.7), nta^{-3}(7.1), edta^{-4}(9.2), acac$^-$(7.9, 14.6), 1,2-dihydroxybenzene^{-2}(13.5, 23.4). The acetylacetone complex Be(acac)$_2$ is a neutral compound with an organic ligand, and hence is extractable from HOH into immiscible non-aqueous solvents such as benzene, carbon tetrachloride, chloroform, and diethylether.

i. Analysis. Be can be quantitatively determined by colorimetry down to 40 ppb using eriochrome cyanine R or acetylacetone. The sensitivity may be improved by electrothermal absorption spectroscopy (ETAS) to 1 ppb and to 0.1 ppb by inductively-coupled plasma emission spectroscopy (ICPES) or inductively-coupled plasma mass spectroscopy (ICPMS). A simple spot test for qualitative detection of Be is one with quinalizarin in alcoholic NaOH which can detect 3 ppm. The color is produced by both Be and Mg. If the color persists after the addition of Br$_2$ water, Be is present. If the color is bleached, Mg is indicated.

j. Health aspects. LD50 values (rat) for ingestion of Be salts include BeCl$_2$ (86 mg/kg), BeF$_2$ (89 mg/kg), and BeSO$_4$ (82 mg/kg). The inhalation of air-borne Be in any form may result in severe pulmonary disease. Also, the introduction of Be compounds into a wound can produce slow-healing ulceration.

3. Magnesium (Mg) 3s^2

a. E–pH diagram. Figure 6.2 sets out the E–pH diagram for $10^{-1.0}$ M Mg. In the figure legend are equations which describe the lines separating the species. Mg metal is thermodynamically unstable toward acids, HOH, and bases; however, in practice, the metal shows some degree of non-reactivity, especially toward cold HOH. This is due to the formation of an oxide coat from exposure to air. Mg is readily attacked by most acids and by hot HOH, especially if the metal is powdered. Bases do not readily react. Above the Mg region, the diagram is dominated by the Mg^{+2} ion, which converts to Mg(OH)$_2$ at a pH value of 8.9. This is considerably higher than the corresponding value for Be (3.7). Mg(OH)$_2$ is not amphoteric in that it will not dissolve in excess OH$^-$. The upper O$_2$/HOH line in the diagram points out that Mg metal is subject to attack by O$_2$. No further oxidation of the Mg(II) species can occur since they extend to the top of the diagram, and reduction is not possible in HOH because of the HOH/H$_2$ line.

b. Discovery. During antiquity, the term magnesia (a district in Thessaly) was applied to the Mg mineral steatite, also called soapstone or talc. In 1755,

Joseph Black recognized Mg as an element, and in 1808, Humphry Davy electrolytically made Mg amalgam. The metal was isolated by Antoine Bussy in 1828 by reduction of $MgCl_2$ with K.

c. *Extraction.*   Mg occurs widely in nature in seawater and in minerals, the main ones being dolomite $CaCO_3 \cdot MgCO_3$, magnesite $MgCO_3$, brucite $Mg(OH)_2$, carnallite $MgCl_2 \cdot KCl \cdot 6HOH$, and various silicates such as talc $3MgO \cdot 2SiO_2 \cdot HOH$. The first three are thermally decomposed to produce

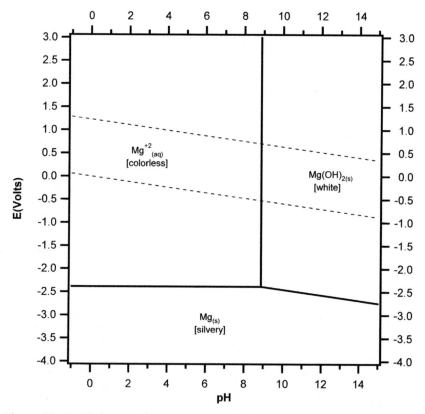

Figure 6.2 E–pH diagram for Mg species. Soluble species concentrations (except H^+) = $10^{-1.0}$ M. Soluble species and most solids are hydrated. No agents producing complexes or insoluble compounds are present other than HOH and OH^-.

Species (ΔG° in kJ/mol): Mg (0.0), Mg^{+2} (−454.8), $Mg(OH)_2$ (−833.5), HOH (−237.2), H^+ (0.0), and OH^- (−157.3).

Equations for the lines:

$$Mg^{+2}/Mg \qquad E = -2.36 + 0.030\log [Mg^{+2}]$$

$$Mg(OH)_2/Mg \qquad E = -1.86 - 0.059\, pH$$

$$Mg(OH)_2/Mg^{+2} \qquad 2\, pH = 16.8 - \log [Mg^{+2}]$$

MgO which provides the basis for the preparation of Mg and other Mg compounds. Mg is recovered from seawater by precipitation of $Mg(OH)_2$, which is dehydrated to MgO.

d. The element. Elemental Mg is produced commercially by either molten-salt reduction or solid thermal reduction. The first process is usually carried out with molten $MgCl_2$. The second involves the use of C or ferrosilicon FeSi as a reductant. Elemental Mg is a light, silvery white metal which oxidizes in air to give an adherent coat. Mg dissolves in acids and hot HOH to produce Mg^{+2} and H_2.

e. Oxides and hydroxides. MgO can be formed by heating Mg metal in air, or by heating the hydroxide, carbonate, or sulfate. The oxide is insoluble in HOH and bases, although it reacts slowly with HOH to give $Mg(OH)_2$. (The reaction $MgO + HOH \rightarrow Mg(OH)_2$ has a ΔG° value of about -30 kJ.) If the oxide has been fired, it is more resistant to dissolution in acid and transformation to the hydroxide. White, gelatinous, hydrated $Mg(OH)_2$ can be precipitated from Mg^{+2} solutions by addition of OH^-. It is readily redissolved in acid. The hydroxide can be dissolved by the addition of an ammonium salt according to the equation

$$Mg(OH)_2 + 2NH_4{}^+ \rightarrow 2NH_4OH + Mg^{+2}.$$

This dissolution is brought about by the weakness of the base NH_4OH ($NH_3 \cdot HOH$).

f. Compounds and solubilities. Table 6.2 presents details on inorganic compounds of Mg which are important in aqueous solution chemistry. Mg, MgO, $Mg(OH)_2$, and $MgCO_3$ may be dissolved in acids to yield the corresponding Mg^{+2} salts. Most Mg^{+2} salts are white unless the anion is colored, and many of them are soluble. As one moves down from Be to Ra, the solubilities of fluorides and hydroxides increase while the solubilities of chromates and sulfates decrease. The Mg^{+2} is written more properly as $Mg(HOH)_6{}^{+2}$, thereby reflecting the coordination number of 6.

g. Redox reactions. Mg shows only one oxidation state II other than 0. The metal is thermodynamically a good reducing agent, but this capability is somewhat damped by its non-reactivity, although this is less than with Be. The E–pH diagram indicates that oxidation and reduction are both resisted by Mg(II) species.

h. Complexes. Complexes of Mg^{+2} which have been investigated in detail are represented by the following list of β_n values: (log β_1, log β_2) $CO_3{}^{-2}$ (2.9), NH_3(0.2, 0.0), $PO_4{}^{-3}$(3.4), OH^-(2.6), $SO_4{}^{-2}$(2.2), F^-(1.8), en(0.4), oxine$^-$ (4.7), 1,10-phenanthroline (1.2), $C_2H_3O_2{}^-$(1.3), $C_2O_4{}^{-2}$(2.8, 4.2), malonate^{-2}(2.1), tartrate^{-2}(1.6), citrate^{-3}(3.2), 3,4-dihydroxybenzoate^{-3} (5.7, 9.8), iminodiacetate^{-2}(3.7), picolinate$^-$(2.6, 4.0), nta^{-3}(6.5), edta^{-4}

Table 6.2
Magnesium Species

Name	Formula	State	Color	Solubility	$\Delta G°$ (kJ/mole)
Magnesium	Mg	s	Silvery	insol	0.0
Magnesium ion	Mg^{+2}	aq	Colorless		−454.8
Hydroxomagnesium ion	$MgOH^+$	aq	Colorless		−626.5
Magnesium(II)					
acetate	$Mg(C_2H_3O_2)_2$	s	White	Sol	
basic carbonate	$Mg_2CO_3(OH)_2$	s	White	Insol	
borate	$Mg_3(BO_3)_2$	s	White	Insol	
bromate	$Mg(BrO_3)_2$	s	White	Sol	
bromide	$MgBr_2$	s	White	Sol	−468.5
carbonate	$MgCO_3$	s	White	Insol	−1011.6
chlorate	$Mg(ClO_3)_2$	s	White	Sol	
chloride	$MgCl_2$	s	White	Sol	−591.5
chromate	$MgCrO_4$	s	Yellow	Sol	
cyanide	$Mg(CN)_2$	s	White	Sol	
fluoride	MgF_2	s	White	Insol	−1069.7
formate	$Mg(OOCH)_2$	s	White	Sol	
hydroxide	$Mg(OH)_2$	s	White	Insol	−833.5
hydroxochloride	$Mg(OH)Cl$	s	White	Insol	−731.4
iodate	$Mg(IO_3)_2$	s	White	Sol	
iodide	MgI_2	s	White	Sol	−358.0
metaborate	$Mg(BO_2)_2$	s	White	Insol	
metavanadate	$Mg(VO_3)_2$	s	White	Insol	−2038.4
molybdate	$MgMoO_4$	s	White	Sol	−1295.1
nitrate	$Mg(NO_3)_2$	s	White	Sol	−589.2
nitrite	$Mg(NO_2)_2$	s	White	Sol	
oxalate	MgC_2O_4	s	White	Insol	−1158.2
oxide	MgO	s	White	Insol	−569.2
oxide(unfired)	MgO	s	White	Insol	−565.7
perchlorate	$Mg(ClO_4)_2$	s	White	Sol	
periodate	$Mg(IO_4)_2$	s	White	Sol	−439.0
permanganate	$Mg(MnO_4)_2$	s	Purple	Sol	
phosphate	$Mg_3(PO_4)_2$	s	White	Insol	−3537.0
pyrophosphate	$Mg_2P_2O_7$	s	White	Insol	
sulfate	$MgSO_4$	s	White	Sol	−1170.1
sulfide	MgS	s	White	Decomp	−341.7
sulfite	$MgSO_3$	s	White	Sol	
thiosulfate	MgS_2O_3	s	White	Sol	
tungstate	$MgWO_4$	s	White	Insol	−1420.2

(9.1), acac$^-$(3.7, 6.3), 1,2-dihydroxybenzene^{-2}(5.7). This list reveals that Mg^{+2} complexes are only moderately strong except for those involving chelation.

i. Analysis. Mg is quantitatively determined by colorimetry down to 30 ppb, by atomic absorption spectroscopy (AAS) to 10 ppb, and to 0.1 ppb by electrothermal absorption spectroscopy (ETAS), inductively-coupled plasma emission spectroscopy (ICPES), and inductively-coupled plasma mass spectroscopy (ICPMS). A spot test for Mg which extends to 3 ppm is provided by quinalizarin in alcoholic NaOH. If Mg is present the color is bleached by Br_2 water. If not, Be is indicated.

j. Health aspects. Mg is generally considered to be relatively non-toxic. Mg in powdered form can be a significant fire hazard, particularly if it is exposed to oxidizing agents or flames. The LD50 (oral rat) for $MgCl_2$ is 2.8 g/kg, that for $Mg(OH)_2$ is 8.5 g/kg, and that for $Mg(NO_3)_2$ is 5.4 g/kg.

4. Calcium (Ca) $4s^2$

a. E–pH diagram. Figure 6.3 is the E–pH diagram for $10^{-1.0}$ M Ca with equations in the legend which describe the lines separating the species. Ca metal is seen to react with aqueous solutions to produce H_2 and either soluble Ca^{+2} or insoluble $Ca(OH)_2$ depending upon the pH. The diagram shows Ca^{+2} as the only soluble species involved, this going over to insoluble $Ca(OH)_2$ at a pH value of about 12.0. This pH value occurs farther to the right than is the case for Be (3.7) or Mg (8.9) indicating the trend in basicity in the series. $Ca(OH)_2$ is not amphoteric, and it is more stable than CaO. The divalent state Ca(II) resists both oxidation and reduction. The upper rung on the HOH electron ladder and the Ca^{+2}/Ca and $Ca(OH)_2$/Ca rungs point out that Ca is subject to attack by O_2, which it is. The metal rapidly forms CaO in air.

b. Discovery. Limestone ($CaCO_3$) and its thermal decomposition product CaO date far back. The name calcium is derived from the Latin word for lime which is calx. Amalgams of Ca were first prepared by Humphry Davy in 1808 by the electrolysis of a mixture of mercuric chloride and moistened lime. Later that year, Davy isolated Ca metal by distilling off the mercury.

c. Extraction. Ca occurs widely in seawater and in several mineral forms, chief of which are limestone, marble, and chalk, all $CaCO_3$, dolomite $CaCO_3 \cdot MgCO_3$, gypsum $CaSO_4 \cdot 2HOH$, fluorite CaF_2, and apatite $Ca_5(PO_4)_3F$. Limestone is the principal source for Ca, with recovery from the carbonate either by thermal decomposition to CaO or by leaching with HCl to give $CaCl_2$.

d. The element. The main process for the production of elemental Ca is reduction of CaO by Al at very high temperature. The resulting Ca is distilled from the residue. Ca is a lustrous, silvery-white metal which readily forms an oxide coat, a hydroxide coat, and a carbonate coat when exposed to air. This coat does not protect the Ca sufficiently so that it dissolves readily in acids and reacts with HOH.

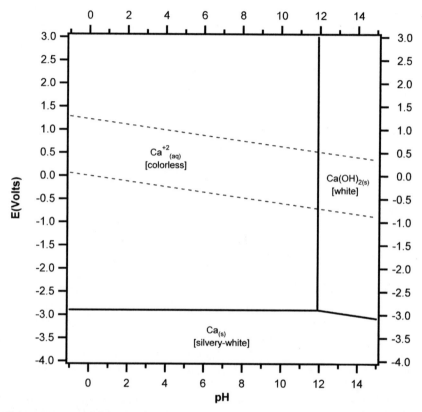

Figure 6.3 E–pH diagram for Ca species. Soluble species concentrations (except H^+) = $10^{-1.0}$ M. Soluble species and most solids are hydrated. No agents producing complexes or insoluble compounds are present other than HOH and OH^-.

Species ($\Delta G°$ in kJ/mol): Ca (0.0), Ca^{+2} (−553.5), $Ca(OH)_2$ (−897.5), HOH (−237.2), H^+ (0.0), and OH^- (−157.3)

Equations for the lines:

$$Ca^{+2}/Ca \qquad E = -2.87 + 0.030 \log [Ca^{+2}]$$

$$Ca(OH)_2/Ca \qquad E = -2.19 - 0.059 \, pH$$

$$Ca(OH)_2/Ca^{+2} \qquad 2 \, pH = 22.9 - \log [Ca^{+2}]$$

e. Oxides and hydroxides.   In aqueous systems Ca forms one oxide CaO and one hydroxide $Ca(OH)_2$, the former being less stable than the latter (CaO + HOH → $Ca(OH)_2$ $\Delta G° = -57.4$ kJ). The oxide is soluble in acids, but in HOH and bases it goes over to $Ca(OH)_2$. Treatment of Ca^{+2} solutions with OH^- precipitates $Ca(OH)_2$, which redissolves in acid, but does not dissolve in excess base. The hydroxide in itself is a moderately strong base, its basicity being due to the OH^- produced by its slight solubility ($K_{sp} = 10^{-5.3}$).

f. Compounds and solubilities.   Details regarding the main compounds of Ca which are pertinent to its aqueous chemistry are presented in Table 6.3. Ca, CaO, $Ca(OH)_2$, and $CaCO_3$ can be dissolved in acids to produce the corresponding Ca^{+2} salts. Most Ca^{+2} salts are white, except those with colored anions, and their solubilities follow the solubility rules of Chapter 4 quite well. These rules incorporate the insoluble CaF_2 and $CaSO_4$ as exceptions to the major generalizations. In solution the Ca^{+2} cation is hydrated with 6 to 9 water molecules depending upon the counter anion and the concentration.

g. Redox reactions.   Ca exhibits oxidation states of II and 0, with the II state predominating in the E–pH diagram. Ca is a very good reducing agent and has the potential to reduce many metal ions. The divalent Ca species are resistant to both oxidation and reduction, and the preparation of Ca in HOH solution is difficult because of the metal's relation to the HOH/H_2 line.

h. Complexes.   Ca forms moderately strong complexes with a number of complexing agents, particularly those that attach through O and those that form chelate rings. Among the complexes that have been thoroughly investigated are the following, with log β_n values attached: (log β_1, log β_2) CO_3^{-2} (3.2), NO_3^-(0.7, 0.6), PO_4^{-3}(6.5), OH^-(1.2), SO_4^{-2}(2.3), $S_2O_3^{-2}$(2.0), F^- (1.1), oxine$^-$(3.3), 1,10-phenanthroline(0.7), $C_2H_3O_2^-$(1.2), $C_2O_4^{-2}$(1.7, 2.7), malonate^{-2}(1.3), tartrate^{-2}(1.0), citrate^{-3}(2.8), 3,4-dihydroxybenzoate^{-3} (3.7, 6.4), phthalate^{-2}(2.4, 1.1), mandelate$^-$(1.5), iminodiacetate^{-2} (3.4), picolinate$^-$ (1.9, 3.5), nta^{-3} (7.6, 8.8), edta^{-4} (11.0). Note that the log β_1 value for Ca^{+2} with edta^{-4}(11.0) is greater than that of the cation above it Mg^{+2}(9.1) and of the cation below it Sr^{+2}(8.8). This reflects the fact that the edta^{-4} fits the Ca^{+2} better than either of the others.

i. Analysis.   Ca gives a brick-red flame coloration, indicating that various optical spectroscopies will be effective in its determination. Ca is quantitatively determined by colorimetry down to 100 ppb using murexide or o-cresolphthalein, by atomic absorption spectroscopy (AAS) to 20 ppb, to 1 ppb by electrothermal absorption spectroscopy (ETAS), to 0.01 ppb by inductively-coupled plasma emission spectroscopy (ICPES), and to 10 ppb by inductively-coupled plasma mass spectroscopy (ICPMS). A spot test for Ca which extends to 3 ppm is provided by glyoxal bis(2-hydroxyanil).

Table 6.3
Calcium Species

Name	Formula	State	Color	Solubility	$\Delta G°$ (kJ/mole)
Calcium	Ca	s	Silvery-white	Decomp	0.0
Calcium ion	Ca^{+2}	aq	Colorless		−553.5
Hydroxocalcium ion	$CaOH^+$	aq	Colorless		−718.4
Calcium (II)					
acetate	$Ca(C_2H_3O_2)_2$	s	White	Sol	
bromate	$Ca(BrO_3)_2$	s	White	Sol	
bromide	$CaBr_2$	s	White	Sol	−663.6
carbonate	$CaCO_3$	s	White	Insol	−1128.8
chlorate	$Ca(ClO_3)_2$	s	White	Sol	
chloride	$CaCl_2$	s	White	Sol	−748.1
chlorite	$Ca(ClO_2)_2$	s	White	Decomp	
chromate	$CaCrO_4$	s	Yellow	Sol	
cyanide	$Ca(CN)_2$	s	White	Decomp	
fluoride	CaF_2	s	White	Insol	−1167.3
formate	$Ca(OOCH)_2$	s	White	Sol	
hydroxide	$Ca(OH)_2$	s	White	Insol	−897.5
hypochlorite	$Ca(ClO)_2$	s	White	Decomp	
iodate	$Ca(IO_3)_2$	s	White	Insol	−839.3
iodide	CaI_2	s	White	Sol	−528.9
metaborate	$Ca(BO_2)_2$	s	White	Sl sol	−1924.1
metavanadate	$Ca(VO_3)_2$	s	White	Insol	−2169.7
molybdate	$CaMoO_4$	s	White	Insol	−1434.7
nitrate	$Ca(NO_3)_2$	s	White	Sol	−743.2
nitrite	$Ca(NO_2)_2$	s	White	Sol	
oxalate	CaC_2O_4	s	White	Insol	−1276.8
oxide	CaO	s	White	Decomp	
perchlorate	$Ca(ClO_4)_2$	s	White	Sol	−1476.8
phosphate	$Ca_3(PO_4)_2$	s	White	Insol	−3884.1
pyrophosphate	$Ca_2P_2O_7$	s	White	Insol	−3132.1
sulfate	$CaSO_4$	s	White	Insol	−1321.9
tetraborate	CaB_4O_7	s	White	Insol	
thiosulfate	CaS_2O_3	s	White	Sol	
sulfide	CaS	s	White	Decomp	−477.4
sulfite	$CaSO_3$	s	White	Insol	
tungstate	$CaWO_4$	s	White	Insol	−1538.5

j. Health aspects. Ca compounds are relatively non-toxic, except for those with poisonous anions. Care must be exercised with regard to the corrosive character of $Ca(OH)_2$ and the fire hazard presented by finely-divided Ca. The LD50 value (oral rat) for $CaCl_2$ is 1 g/kg, but for the poisonous anion compound $Ca(CN)_2$, the value is 39 mg/kg.

5. Strontium (Sr) $5s^2$

a. E–pH diagram. The E–pH relationships for Sr at a soluble species concentration level of $10^{-1.0}$ M are shown in Figure 6.4. In the figure legend are equations describing the lines which separate the species. Sr is seen to react with acids or HOH to give H_2 and either Sr^{+2} or $Sr(OH)_2$. The aqueous

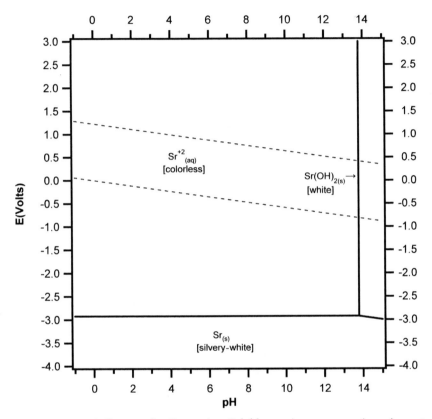

Figure 6.4 E–pH diagram for Sr species. Soluble species concentrations (except H^+) $= 10^{-1.0}$ M. Soluble species and most solids are hydrated. No agents producing complexes or insoluble compounds are present other than HOH and OH^-.

Species ($\Delta G°$ in kJ/mol): Sr (0.0), Sr^{+2} (−558.6), $Sr(OH)_2$ (−881.2), HOH (−237.2), H^+ (0.0), and OH^- (−157.3).

Equations for the lines:

$$Sr^{+2}/Sr \qquad E = -2.89 + 0.030\log [Sr^{+2}]$$

$$Sr(OH)_2/Sr \qquad E = -2.11 - 0.059\,pH$$

$$Sr(OH)_2/Sr^{+2} \qquad 2\,pH = 26.6 - \log [Sr^{+2}]$$

chemistry of Sr is dominated by the Sr^{+2} ion which is not oxidizable, cannot be reduced to Sr in aqueous solution, and precipitates as $Sr(OH)_2$ only at high pH values. Similar to the hydroxides of Mg and Ca, excess base does not dissolve $Sr(OH)_2$, that is, it is not amphoteric.

b. Discovery.   In 1790, Adair Crawford and William Cruickshank heated the mineral strontianite ($SrCO_3$) to give SrO. The name derives from a mine at Strontian in Scotland, which was the source of the mineral. Humphry Davy in 1808 electrolyzed a mixture of moistened SrO and HgO to obtain Sr amalgam. He then distilled the Hg from the amalgam to obtain Sr metal.

c. Extraction.   Strontianite ($SrCO_3$) and celestite ($SrSO_4$) are the two major minerals of Sr. The former mineral is dissolved in HNO_3 which gives $Sr(NO_3)_2$. The celestite is treated with hot Na_2CO_3 solution to give $SrCO_3$ which is then dissolved in HNO_3. The nitrate and the carbonate function as the basis for the preparation of Sr and other Sr salts.

d. The element.   As was the case with Ca, the primary method for producing Sr is thermal reduction of SrO with Al. This is followed by distillation of Sr from the Al_2O_3 residue. Sr is a hard, silvery-white metal which reacts with air to produce SrO, $Sr(OH)_2$, and $SrCO_3$. The metal reacts with HOH and acids to give Sr^{+2} and H_2, or $Sr(OH)_2$ and H_2 at high pH values.

e. Oxides and hydroxides.   Sr has one oxide—white SrO, one hydroxide—white $Sr(OH)_2$, and one peroxide—white SrO_2. SrO is primarily made by heating $SrCO_3$, or by reaction of Sr with pure O_2. It is soluble in acids, and in HOH or base readily goes over to the insoluble $Sr(OH)_2$. The white, gelatinous $Sr(OH)_2$ also results from the treatment of Sr^{+2} salts with base, but excess base does not put it back into solution. The hydroxide is a strong base. The peroxide SrO_2 can be made by the addition of H_2O_2 to $Sr(OH)_2$. This compound does not represent an oxidation state of IV for the Sr, since it is the combination of Sr^{+2} with the peroxide anion O_2^{-2}.

f. Compounds and solubilities.   A listing of Sr compounds along with some of their properties is given in Table 6.4. Sr, SrO, $Sr(OH)_2$, and $SrCO_3$ dissolve in acids to give the corresponding Sr^{+2} salts. Most of these salts are white, except for those with a colored anion, and their solubilities follow the solubility rules (SrF_2 and $SrSO_4$ being noted in the rules as insoluble).

g. Redox reactions.   Sr exhibits oxidation states of II and 0, with the II state predominating in the E–pH diagram. Sr is a very good reducing agent and has the potential to reduce many metal ions. The divalent Sr species are resistant to both oxidation and reduction, and the preparation of Sr in HOH solution is difficult because of the metal's relation to the HOH/H_2 line.

Table 6.4
Strontium Species

Name	Formula	State	Color	Solubility	$\Delta G°$ (kJ/mole)
Strontium	Sr	s	Silvery-white	Decomp	0.0
Strontium ion	Sr^{+2}	aq	Colorless		−558.6
Hydroxostrontium ion	$SrOH^+$	aq	Colorless		−721.3
Strontium (II)					
acetate	$Sr(C_2H_3O_2)_2$	s	White	Sol	
bromate	$Sr(BrO_3)_2$	s	White	Sol	
bromide	$SrBr_2$	s	White	Sol	−697.1
carbonate	$SrCO_3$	s	White	Insol	−1140.1
chlorate	$Sr(ClO_3)_2$	s	White	Sol	
perchlorate	$Sr(ClO_4)_2$	s	White	Sol	
chloride	$SrCl_2$	s	White	Sol	−781.2
chromate	$SrCrO_4$	s	Yellow	Insol	
cyanide	$Sr(CN)_2$	s	White	Decomp	
fluoride	SrF_2	s	White	Insol	−1164.8
formate	$Sr(OOCH)_2$	s	White	Sol	
hydrogen sulfide	$Sr(HS)_2$	s	White	Decomp	
hydroxide	$Sr(OH)_2$	s	White	Sl sol	−881.2
iodate	$Sr(IO_3)_2$	s	White	Insol	−855.2
iodide	SrI_2	s	White	Sol	
metavanadate	$Sr(VO_3)_2$	s	White	Insol	
molybdate	$SrMoO_4$	s	White	Insol	
nitrate	$Sr(NO_3)_2$	s	White	Sol	−780.2
nitrite	$Sr(NO_2)_2$	s	White	Sol	
oxalate	SrC_2O_4	s	White	Insol	
oxide	SrO	s	White	Decomp	−561.9
permanganate	$Sr(MnO_4)_2$	s	Purple	Sol	
phosphate	$Sr_3(PO_4)_2$	s	White	Insol	
sulfate	$SrSO_4$	s	White	Insol	−1341.0
tetraborate	SrB_4O_7	s	White	Sol	
thiosulfate	SrS_2O_3	s	White	Sol	
sulfide	SrS	s	White	Decomp	−467.8
sulfite	$SrSO_3$	s	White	Insol	
tungstate	$SrWO_4$	s	White	Insol	−1531.0

h. Complexes. Sr resembles Ca in its complexation properties, with the stabilities of the complexes being somewhat less. Among the complexes that have been quantitatively characterized are the following. The values of log β_n have been appended to each of the complexing agents: (log β_1, log β_2) CO_3^{-2}(2.8), NO_3^-(0.6, 0.5), OH^-(0.6), SO_4^{-2}(2.7, 1.4), $S_2O_3^{-2}$(2.3), oxine$^-$(2.1), $C_2H_3O_2^-$(1.0), $C_2O_4^{-2}$(2.3), phthalate^{-2}(2.3), mandelate$^-$(0.8), iminodiacetate^{-2}(1.7), nta^{-3}(5.9), edta^{-4}(8.8).

i. Analysis. The bright scarlet flame color of Sr indicates that atomic emission and absorption methods will be good for its analysis. Sr is quantitatively determined by colorimetry down to 200 ppm using chloranilic acid, by atomic absorption spectroscopy (AAS) to 100 ppb, to 1 ppb by electrothermal absorption spectroscopy (ETAS), and to 0.1 ppb by inductively-coupled plasma emission spectroscopy (ICPES) and inductively-coupled plasma mass spectroscopy (ICPMS). A spot test for Sr which extends to 40 ppm is provided by K_2CrO_4 and sodium rhodizonate.

j. Health aspects. Sr and its compounds are only slightly toxic. The element tends to concentrate in the bones. Care must be exercised with regard to the corrosive character of $Sr(OH)_2$ and the fire hazard presented by finely divided Sr. The LD50 (oral rat) for $SrCl_2$ is 2.2 g/kg, that for $Sr(NO_3)_2$ is 2.7 g/kg, but that for the insoluble SrF_2 is 10.6 g/kg.

6. Barium (Ba) $6s^2$

a. E–pH diagram. The E–pH diagram for $10^{-1.0}$ M Ba in Figure 6.5 shows a number of variations from those for Ca and Sr. Chief among these are the presence of the white peroxide and the octahydrate $Ba(OH)_2 \cdot 8HOH$. Ba metal reacts with acids or HOH to give H_2 and either Ba^{+2} or $Ba(OH)_2 \cdot 8HOH$ depending upon the pH. The upper region of the diagram shows the peroxide BaO_2. (The reason this is included is that the compound is produced upon burning Ba metal. Such is not the case with Sr, where SrO_2 was omitted from the E–pH diagram, but where its preparation was described.) The dominant species in the diagram is Ba^{+2} which can be converted to the peroxide in the region where the O_2^{-2} ion is stable, and which converts to $Ba(OH)_2 \cdot 8HOH$ at a pH of 12.7. The base $Ba(OH)_2$ is very strong, resembling the alkali metal hydroxides.

b. Discovery. Several minerals of Ba were known in the seventeenth century, but the composition of them was not worked out until 1774–79 by Carl Wilhelm Scheele and Johann Gottlieb Gahn. They recognized barite or barytes or heavy spar as $BaSO_4$. Scheele isolated some other salts and baryta BaO, which Humphry Davy in 1808 used to prepare Ba amalgam. Upon distillation of the Hg from the amalgam, metallic Ba was obtained. The name barium comes from the Greek word barys which means heavy.

c. Extraction. Ba occurs in nature chiefly in two minerals, barytes $BaSO_4$ and witherite $BaCO_3$. Ba is recovered from barytes by reduction of the sulfate with carbon to give BaS. The resulting solid is leached with HOH to dissolve the BaS, and the solution is employed to prepare Ba salts.

d. The element. The main method for preparing Ba is by the thermal reduction of BaO with Al. The Ba metal is then distilled off. Ba is a malleable, silvery-white metal which reacts rapidly when exposed to air to form BaO,

Ba(OH)$_2$, and BaCO$_3$. The metal reacts with HOH and acids to produce H$_2$ and Ba^{+2} and at high pH values Ba(OH)$_2$•8HOH.

e. Oxides and hydroxides. Ba has one oxide BaO, one hydroxide Ba(OH)$_2$•8HOH, and one peroxide BaO$_2$. BaO can be obtained by exposing

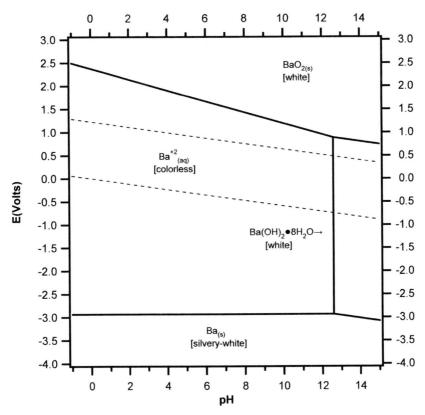

Figure 6.5 E–pH diagram for Ba species. Soluble species concentrations (except H$^+$) = 10$^{-1.0}$ M. Soluble species and most solids are hydrated. No agents producing complexes or insoluble compounds are present other than HOH and OH$^-$.

Species (ΔG^0 in kJ/mol): Ba (0.0), Ba^{+2} (−561.1), Ba(OH)$_2$•8H$_2$O (−2794.4), BaO$_2$ (−582.4), HOH (−237.2), H$^+$ (0.0), and OH$^-$ (−157.3).

Equations for the lines:

Ba^{+2}/Ba	E = −2.91 + 0.030 log [Ba^{+2}]
Ba(OH)$_2$•8H$_2$O/Ba	E = −2.19 − 0.059 pH
BaO$_2$/Ba^{+2}	E = 2.35 − 0.118 pH − 0.030 log [Ba^{+2}]
BaO$_2$/Ba(OH)$_2$•8H$_2$O	E = 1.63 − 0.059 pH
Ba(OH)$_2$•8H$_2$O/Ba^{+2}	2 pH = 24.3 − log [Ba^{+2}]

Ba to O_2; if Ba is burned, BaO_2 forms. BaO can also be made by thermal decomposition of $BaCO_3$, which is the most common method. The oxide dissolves in acids to give Ba^{+2} salts. The hydroxide can be precipitated from Ba^{+2} solutions by adding an excess of strong base. This precipitated hydroxide is a strong base. BaO_2 is conventionally prepared by precipitation upon the addition of H_2O_2 to an alkaline solution of Ba^{+2}.

f. Compounds and solubilities. Table 6.5 sets out the Ba compounds which are of chief significance to the aqueous chemistry of the element. Ba, BaO, $Ba(OH)_2 \cdot 8HOH$, and $BaCO_3$ may all be treated with acids to produce the corresponding salts. Ba salts are usually white, unless the anion is colored.

Table 6.5
Barium Species

Name	Formula	State	Color	Solubility	$\Delta G°$ (kJ/mole)
Barium	Ba	s	Silvery-white	Decomp	0.0
Barium ion	Ba^{+2}	aq	Colorless		−561.1
Hydroxobarium	$BaOH^+$	aq	Colorless		−730.5
Barium (II)					
acetate	$Ba(C_2H_3O_2)_2$	s	White	Sol	
bromate	$Ba(BrO_3)_2$	s	White	Sl sol	−577.4
bromide	$BaBr_2$	s	White	Sol	−736.8
carbonate	$BaCO_3$	s	White	Insol	−1137.6
chlorate	$Ba(ClO_3)_2$	s	White	Sol	
chloride	$BaCl_2$	s	White	Sol	−810.4
chlorite	$Ba(ClO_2)_2$	s	White	Sol	
chromate	$BaCrO_4$	s	Yellow	Insol	−1345.3
cyanide	$Ba(CN)_2$	s	White	Sol	
fluoride	BaF_2	s	White	Insol	−1156.9
formate	$Ba(OOCH)_2$	s	White	Sol	
hydrogen sulfide	$Ba(HS)_2$	s	Yellow	Decomp	
hydroxide	$Ba(OH)_2 \cdot 8HOH$	s	White	Insol	−2794.4
iodate	$Ba(IO_3)_2$	s	White	Insol	−864.8
iodide	BaI_2	s	White	Sol	
molybdate	$BaMoO_4$	s	White	Insol	−1439.7
nitrate	$Ba(NO_3)_2$	s	White	Sol	−796.7
nitrite	$Ba(NO_2)_2$	s	White	Sol	
oxalate	BaC_2O_4	s	White	Insol	
oxide	BaO	s	White	Decomp	−525.1
perchlorate	$Ba(ClO_4)_2$	s	White	Sol	−796.3
permanganate	$Ba(MnO_4)_2$	s	Purple	Sol	−1119.2
peroxide	BaO_2	s	White	Sl sol	−582.4
perrhenate	$Ba(ReO_4)_2$	s	White	Sol	−2443.9
phosphate	$Ba_3(PO_4)_2$	s	White	Insol	

Table 6.5
(*Continued*)

Name	Formula	State	Color	Solubility	$\Delta G°$ (kJ/mole)
Barium (II) (Continued)					
pyrovanadate	$Ba_2V_2O_7$	s	White	Insol	
sulfate	$BaSO_4$	s	White	Insol	−1362.3
sulfide	BaS	s	White	Decomp	−456.0
sulfite	$BaSO_3$	s	White	Insol	
thiocarbonate	$BaCS_3$	s	Yellow	Sl sol	
thiocyanate	$Ba(CNS)_2$	s	White	Sol	
thiosulfate	BaS_2O_3	s	White	Insol	
tungstate	$BaWO_4$	s	White	Sl sol	

Ba salt solubilities follow the general solubility rules, especially when the exceptions BaF_2 and $BaSO_4$ are recognized.

g. Redox reactions. Other than 0, Ba shows only the oxidation state II. This latter oxidation state is resistant toward both oxidation and reduction. For reasons similar to the other alkaline earth metals, Ba cannot be produced from its salts in an aqueous solution.

h. Complexes. The trend toward weaker complexes as one goes from Be to Ba is further exemplified by the following values for Ba^{+2} complexes: (log β_1, log β_2) CO_3^{-2}(2.8), NO_3^-(0.6, 0.5), OH^-(0.6), SO_4^{-2}(2.7, 1.4), $S_2O_3^{-2}$(2.3), oxine$^-$(2.1), $C_2H_3O_2^-$(1.0), $C_2O_4^{-2}$(2.3), phthalate^{-2}(2.3), mandelate$^-$(0.8), iminodiacetate^{-2}(1.7), nta^{-3}(5.9), edta^{-4}(7.8).

i. Analysis. The green flame color of Ba is an indicator that it may be determined readily by atomic emission or absorption spectroscopy. Ba is quantitatively determined by colorimetry down to 1 ppm using o-cresolphthalein at a pH of 11, by atomic absorption spectroscopy (AAS) to 200 ppb, to 10 ppb by electrothermal absorption spectroscopy (ETAS), and to 0.1 ppb by inductively-coupled plasma emission spectroscopy (ICPES) and inductively-coupled plasma mass spectroscopy (ICPMS). A spot test for Ba which extends to 30 ppm is provided by a controlled combination of $KMnO_4$, H_2SO_4, and H_2SO_3.

j. Health aspects. The soluble salt $BaCl_2$ is poisonous, as is evidenced by the LD50 value of 110 mg/kg. If this were extrapolated to humans, it means that ingestion of a small amount could be life threatening. $BaSO_4$ is used as a contrasting agent in X-ray investigations, but it has no serious effect because of its very small solubility product.

7. Radium (Ra) 7s^2

The E–pH diagram for $10^{-7.0}$ M Ra is presented in Figure 6.6. This concentration is used because such a solution of the most long-lived radium isotope Ra-226 (half-life 1620 years) would be decaying at the rate of about 3 billion atoms per minute per liter. Such a radioactivity could be worked with given special apparatus and precautions, but more concentrated solutions would require more demanding measures. The discovery of the element Ra was in 1898 by Marie Sklodowska Curie, Pierre Curie, and M. G. Bemont who isolated its salts from large quantities of pitchblende,

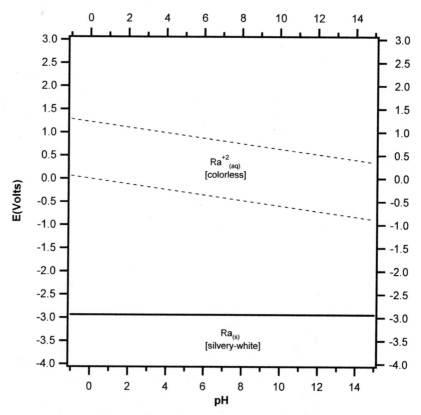

Figure 6.6 E–pH diagram for Ra species. Soluble species concentrations (except H$^+$) = $10^{-1.0}$ M. Soluble species and most solids are hydrated. No agents producing complexes or insoluble compounds are present other than HOH and OH$^-$.

Species ($\Delta G°$ in kJ/mol): Ra (0.0), Ra^{+2} (−561.5), Ra(OH)$_2$•8H$_2$O (−2794.4), HOH (−237.2), H$^+$ (0.0), and OH$^-$ (−157.3).

Equations for the lines:

$$Ra^{+2}/Ra \quad E = -2.91 + 0.030 \log [Ra^{+2}]$$

Table 6.6
Radium Species

Name	Formula	State	Color	Solubility	$\Delta G°$ (kJ/mole)
Radium	Ra	s	Silvery-white	Decomp	0.0
Radium ion	Ra^{+2}	aq	Colorless		−561.5
Radium(II)					
bromate	$Ra(BrO_3)_2$	s	White	Sl sol	−537.2
bromide	$RaBr_2$	s	White	Sol	−748.9
carbonate	$RaCO_3$	s	White	Insol	−1143.9
chlorate	$Ra(ClO_3)_2$	s	White	Sol	−569.4
chloride	$RaCl_2$	s	White	Sol	−839.3
fluoride	RaF_2	s	White	Insol	−1148.9
hydroxide	$Ra(OH)_2·8HOH$	s	White	Insol	−2794.4
iodate	$Ra(IO_3)_2$	s	White	Insol	−868.6
iodide	RaI_2	s	White	Sol	−636.8
molybdate	$RaMoO_4$	s	White	Insol	−1464.8
nitrate	$Ra(NO_3)_2$	s	White	Sol	−796.2
oxide	RaO	s	White	Decomp	−492.5
peroxide	RaO_2	s	White	Sl sol	−516.7
sulfate	$RaSO_4$	s	White	Insol	−1365.2
sulfide	RaS	s	White	Decomp	−440.6
sulfite	$RaSO_3$	s	White	Insol	−1097.0
tungstate	$RaWO_4$	s	White	Sl sol	−1589.1

a U ore. Ra-226 results as a product of the natural decay of U-238, the most abundant isotope of U, an element that occurs in nature. The extraction was performed by separating $BaCl_2$ from a solution of the mineral and discovering that it was highly radioactive due to the coprecipitation of $RaCl_2$. Many successive recrystallizations of the $BaCl_2$ were necessary to obtain weighable amounts of Ra compounds.

The element was isolated in 1910, when Marie Curie and André Debierne electrolyzed a solution of $RaCl_2$ using a Hg cathode to obtain Ra-amalgam. The Hg was then separated by distillation. The oxides and hydroxides of Ra parallel those of Ba, both in formulas and properties, except that the octahydrate of the hydroxide does not precipitate at the $10^{-7.0}$ M level. Compounds and solubilities are likewise very similar to those of Ba as is shown by Table 6.6. Many of the $\Delta G°$ values in the table have been estimated and are subject to error. The redox reactions also resemble Ba closely, the $E°$ for the M^{+2}/M couple being the same for both elements. The ready reaction with air and HOH for both metals further emphasizes their similarity. Complexes for Ra would be expected to be slightly weaker. Analysis of Ra is generally carried out by radioactivity measurements, these permitting detection down to very low levels. The flame color of Ra is carmine as compared to the green of Ba. The radioactivity of Ra presents a major health hazard. Ingestion is to be avoided since it is a bone-seeking alpha-emitter.

7

The Boron Group

1. Introduction

The elements which constitute the Boron Group of the Periodic Table are boron B, aluminum Al, gallium Ga, indium In, and thallium Tl. All five of the elements have atoms characterized by an outer electron structure of ns^2np^1 with n representing the principal quantum number. There are marked similarities in the elements, except for B whose small size and high charge density make it a non-metal. B evidences an oxidation state of III but shows no aqueous cation chemistry. The other elements all show cation chemistries involving an oxidation state of III, but the I oxidation state becomes progressively more stable until at Tl it is the predominant state. All ions in the group are colorless. Ionic sizes in pm are $B^{+3}(27)$, $Al^{+3}(53)$, $Ga^{+3}(62)$, $In^{+3}(80)$, $Tl^{+3}(89)$, and $Tl^+(150)$, with the B^{+3} value being hypothetical since B bonds only covalently. In line with the increasing sizes, the basicity of the oxides and hydroxides increases: H_3BO_3 or $B(OH)_3$ is weakly acidic, $M(OH)_3$ for Al, Ga, and In are amphoteric, and $Tl(OH)_3$ or Tl_2O_3 is basic. The $E°$ values in volts for the M(III)/M couples are as follows: H_3BO_3/B (−0.89), Al^{+3}/Al (−1.68), Ga^{+3}/Ga (−0.55), and In^{+3}/In (−0.35). The $E°$ value for the Tl^+/Tl couple is −0.33 v.

2. Boron (B) 2s²2p¹

a. E–pH diagram. The E–pH diagram for $10^{-1.0}$ M B is presented in Figure 7.1. In the figure legend are equations which describe the lines which separate species. Considerably above the H_3BO_3/B line, the $HOH \equiv H^+/H_2$

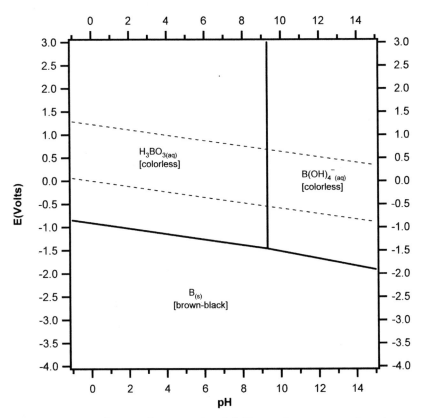

Figure 7.1 E–pH diagram for B species. Soluble species concentrations (except H^+) = $10^{-1.0}$ M. Soluble species and most solids are hydrated. No agents producing complexes or insoluble compounds are present other than HOH and OH^-.

Species ($\Delta G°$ in kJ/mol): B (0.0), H_3BO_3 (−969.0), $B(OH)_4^-$ (−1153.1), HOH (−237.2), H^+ (0.0), and OH^- (−157.3).

Equations for the lines:

H_3BO_3/B $E = -0.89 - 0.059\,pH + 0.020\,\log[H_3BO_3]$

$B(OH)_4^-/B$ $E = -0.71 - 0.079\,pH + 0.020\,\log[B(OH)_4^-]$

$B(OH)_4^-/H_3BO_3$ $pH = 9.3$

line appears, which indicates that elemental B is thermodynamically unstable in HOH, but in practice B has a strong tendency to be non-reactive, vigorous treatment usually being required to oxidize it. Note that the two soluble species H_3BO_3 and $B(OH)_4^-$ dominate the diagram, with the transformation between the two species occurring at pH $= 9.3$. These two species extend all the way to the top of the diagram, pointing out that further oxidation is not possible. It should also be noted that the upper O_2/HOH line predicts that B will react with O_2, but again this is retarded by the recalcitrant character of B. As is the case with most E–pH diagrams, Figure 7.1 is simplified. This is particularly so because in the region of the H_3BO_3 to $B(OH)_4^-$ transition, some polymeric species occur, these being products of condensation of H_3BO_3 units and $B(OH)_4^-$ units. Chief among them are $B_3O_3(OH)_4^-$, $B_4O_5(OH)_4^{-2}$, and $B_5O_6(OH)_4^-$. In dilute solutions, they are not significant, but as the B concentration goes up they increase.

b. Discovery. The element boron B (Arabic buraq, Persian burah) was discovered in 1808 by Humphry Davy and simultaneously by Joseph Louis Gay-Lussac and Louis Jacques Thenard. Davy produced the impure element by electrolysis of H_3BO_3 and by reduction of H_3BO_3 with K, this latter being the process also used by Gay-Lussac and Thenard. High-purity B was not produced until 1892 by Ferdinand Frederick Henri Moissan, and very pure B in 1909 by W. Weintraub.

c. Extraction. Boron occurs in the form of several minerals, chief of which are borax (tincal) $Na_2B_4O_5(OH)_4 \cdot 8HOH$ and kernite $Na_2B_4O_6$ $(OH)_2 \cdot 3HOH$. When these finely divided minerals are treated with hot sulfuric acid, they are decomposed, and when cooled, boric acid H_3BO_3 precipitates out. For borax, the equation is

$$Na_2B_4O_5(OH)_4 \cdot 8HOH + H^+ + HSO_4^- \rightarrow 4H_3BO_3 + 2Na^+$$
$$+ 5HOH + SO_4^{-2}.$$

The cooling to induce precipitation is effective because H_3BO_3 is much less soluble in cold water (47 g/L) than in boiling water (275 g/L). Recrystallizations are employed to increase the purity of H_3BO_3. This compound is the starting point for all other B compounds.

d. The element. Commercially, B is prepared in two different forms: a brown-black amorphous solid and a black crystalline solid. The less-expensive amorphous material is made by heating H_3BO_3 to the oxide B_2O_3, and then reducing this oxide with Mg. The crystalline solid is prepared by reduction of gaseous BCl_3 with H_2 on a hot filament. As mentioned above, elemental B is quite inert, the crystalline form being especially so. The amorphous form reacts very slowly with boiling water to give H_3BO_3, reacts slowly with non-oxidizing acids, reacts with HNO_3 and concentrated H_2O_2, and is attacked by hot, concentrated NaOH solution. The speed of the reactions can be increased by using very finely divided B.

e. Oxides and hydroxides. White solid boron oxide B_2O_3 is formed by heating amorphous B in air or oxygen, or preferably by heating H_3BO_3 to drive off HOH. Conversely, the B_2O_3 reacts with HOH to produce H_3BO_3. The oxide is soluble in strong bases to produce $B(OH)_4^-$ at dilute and moderate concentrations and to produce polymeric B anions at high concentrations. Boric acid may be viewed as a hydroxide $B(OH)_3$ which permits its comparison with the other hydroxides of this group. It is moderately soluble in water, where it acts as a weak monoprotic acid. Dissolution of H_3BO_3 or $B(OH)_3$ in OH^- converts it to the tetrahydroxoborate ion $B(OH)_4^-$ as shown by the E–pH diagram. It is important to recognize that this aqueous ion is written as BO_2^- [$B(OH)_4^-$ minus 2HOH] in some of the literature.

f. Compounds and solubilities. Metal salts of $B(OH)_4^-$ can be produced by treating oxides or soluble metal compounds with H_3BO_3 or $B(OH)_4^-$. The composition of these salts can vary widely depending upon the conditions of preparation because in the process of separation of the salts from HOH, polymeric forms of the $B(OH)_4^-$ anion are created. Hence the borate precipitates may have widely varying compositions involving such anions as $B_3O_6^{-3}$ (often written simply as BO_2^- and called metaborate), BO_3^{-3} (called orthoborate), $B_2O_5^{-4}$, $(BO_2)_n^{-n}$, $B_4O_5(OH)_4^{-2}$, and $B_5O_6(OH)_4^-$. In general, most alkali borates are soluble, whereas most other borates are insoluble. Table 7.1 presents the names, formulas, states, solubilities, and standard free energies of a number of the more important simple inorganic species of B. All borate salts are white unless they contain a colored cation.

g. Redox reactions. As can be seen from the E–pH diagram, B ordinarily shows only one valence +3 other than 0, an apparent exception being $Na_2B_2(O_2)_2(OH)_4 \cdot 6HOH$, which is a peroxy compound called sodium perborate (usually written as $NaBO_3 \cdot 4HOH$). Elemental B is not a good reducing agent because of its inertness, even though the $E°$ of the H_3BO_3/B couple is -0.89 v.

h. Complexes. H_3BO_3 and $B(OH)_4^-$ combine with some organic polyhydroxy compounds (more than one –OH group) which function as bidentate agents to form stable complexes. Examples of such compounds are glycerol and mannitol. The complexes enhance the acid strength of H_3BO_3 remarkably. If H_3BO_3 is treated with HF, tetrafluoroboric acid HBF_4 is slowly produced in solution. This strong acid has not been isolated from solution, but numerous salts, tetraflouroborates, have been prepared. These salts are similar in solubilities to perchlorates (salts ClO_4^-).

i. Analysis. If a borate is treated with H_2SO_4 and methanol H_3COH, methyl borate is produced. If a small amount of the mixture is introduced to a flame, a green flame color is produced. The test is sensitive to 20 ppm of B. Ba and Cu interfere, both of these also giving a green color. A spot test with curcumin will detect 5 ppm of B. Titration of H_3BO_3 in a mannitol solution

Table 7.1
Boron Species

Name	Formula	State	Color	Solubility	$\Delta G°$ (kJ/mole)
Boron (amorphous)	B	s	Brown-black	Insol	0.0
Boron (crystalline)	B	s	Black	Insol	3.5
Boric acid	H_3BO_3	s	White	Sol	−969.0
Tetrahydroxoborate	$B(OH)_4{}^-$	aq	Colorless		−1153.1
Tetrafluoroborate	$BF_4{}^-$	aq	Colorless		−1486.6
Borates (simple)*					
aluminum	$AlBO_3$	s	White	Insol	
ammonium	NH_4BO_2	s	White	Sol	
barium	$Ba(BO_2)_2$	s	White	Insol	
cadmium	$Cd(BO_2)_2$	s	White	Insol	
calcium	$Ca(BO_2)_2$	s	White	Insol	
cesium	$CsBO_2$	s	White	Sol	
cobalt	$Co(BO_2)_2$	s	Pink	Insol	
copper	$Cu(BO_2)_2$	s	Blue	Insol	
lanthanoids	$LnBO_3$	s	Various	Insol	
lead	$Pb(BO_2)_2$	s	White	Insol	
lithium	$LiBO_2$	s	White	Sol	
magnesium	$Mg(BO_2)_2$	s	White	Insol	
manganese	$Mn(BO_2)_2$	s	Pale pink	Insol	
nickel	$Ni(BO_2)_2$	s	Green	Insol	
potassium	KBO_2	s	White	Sol	
rubidium	$RbBO_2$	s	White	Sol	
scandium	$ScBO_3$	s	White	Insol	
silver	$AgBO_2$	s	White	Insol	
sodium	$NaBO_2$	s	White	Sol	
strontium	$Sr(BO_2)_2$	s	White	Insol	
thallium	$TlBO_2$	s	White	Insol	
yttrium	YBO_3	s	White	Insol	
zinc	$Zn(BO_2)_2$	s	White	Insol	
Boron (III)					
tribromide	BBr_3	l	Colorless	Decomp	−236.8
trichloride	BCl_3	g	Colorless	Decomp	−387.9
trifluoride	BF_3	g	Colorless	Decomp	−1120.1
triiodide	BI_3	s	Colorless	Decomp	20.8

* These are the simplest borates. The $BO_2{}^-$ ion in some cases is actually the $B_3O_6{}^{-3}$ ring structure and in other cases the $(BO_2)_n{}^{-n}$ chain structure. Carefully controlled conditions must often be employed to obtain the above simple compounds. As discussed in the text, other conditions give more complex borates. However, in most cases the solubilities of the more complex borates resemble those of their simpler forms.

provides a quantitative method for macro amounts that is widely employed. Colorimetric methods afford a 10-ppb detection limit. ICPAES (inductively-coupled plasma atomic emission spectroscopy) will detect down to 1 ppb and ICPMS (inductively-coupled plasma mass spectroscopy) is 10 times better.

j. Health and environmental aspects. B is an essential trace element in plants. The LD50 (oral rat) for H_3BO_3 is 2.6 g/kg and that for $Na_2B_4O_7$ is the same. The major symptoms of poisoning are central nervous depression and gastro-intestinal irritation.

3. Aluminum (Al) $3s^2 3p^1$

a. E–pH diagram. The E–pH diagram for $10^{-1.0}$ M Al is presented in Figure 7.2. In the figure legend are equations for the lines that separate the species. The horizontal line in the lower pH region between Al^{+3} and Al appears at an E value of -1.70 v. Considerably above this Al^{+3}/Al line, the $HOH \equiv H^+/H_2$ line appears, which indicates that Al metal is unstable in acid reacting with it to produce H_2 and Al^{+3}. However, Al metal generally has a highly adherent, refractory, non-reactive oxide coat which resists attack. It will dissolve slowly in dilute mineral acids such as HCl, more rapidly if the acid is concentrated, but a concentrated oxidizing acid such as HNO_3 secures the oxide coat. HOH will not attack Al unless it is very finely divided. The sloped line which separates $Al(OH)_4^-$ and Al indicates that Al is soluble in strong base, which it is. Particular attention should be paid to the horizontal scan from pH -1.0 to 15.0 at an E value of 0.00 v. The transformations of Al^{+3} to $Al(OH)_3$ to $Al(OH)_4^-$ indicate that $Al(OH)_3$ is amphoteric, that is, it is soluble in both acid and base. In the region of the transformation from Al^{+3} to $Al(OH)_3$, there exist partially hydrolyzed and polymeric species, the identities and amounts being dependent upon both concentration and pH. Some of these species include $AlOH^{+2}$, $Al(OH)_2^+$, $Al_2(OH)_2^{+4}$, and $Al_3(OH)_{11}^{-2}$. The complex $Al(OH)_4^-$ often appears in the older literature in its dehydrated form as AlO_2^-. The three trivalent forms of Al extend to the top of the E–pH diagram indicating that they are not oxidized.

b. Discovery. The element Al (Latin alumen which means bitter salt) was first prepared in an impure form by Hans Christian Oersted in 1825 by reacting potassium amalgam (K dissolved in Hg) with $AlCl_3$. Later preparations by Friedrich Wöhler, Henri Sainte-Claire Deville, and Robert Wilhelm Bunsen employed K, Na, and the electrolysis of molten $NaAlCl_4$.

c. Extraction. Al occurs in nature as silicates, but the chief source used for isolation of the element is the mineral bauxite, which occurs as various hydrates of aluminum oxide $Al_2O_3 \cdot nHOH$ (n = 1, 2, 3). The mineral is treated

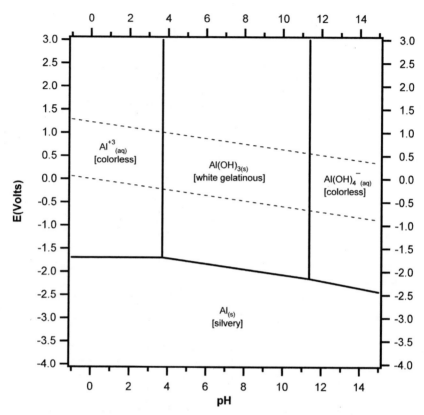

Figure 7.2 E–pH diagram for Al species. Soluble species concentrations (except H^+) = $10^{-1.0}$ M. Soluble species and most solids are hydrated. No agents producing complexes or insoluble compounds are present other than HOH and OH^-.

Species ($\Delta G°$ in kJ/mol): Al (0.0), Al^{+3} (−485.3), $Al(OH)_3$ (−1138.9), $Al(OH)_4^-$ (−1305.4), HOH (−237.2), H^+ (0.0), and OH^- (−157.3).

Equations for the lines:

Al^{+3}/Al	$E = -1.68 + 0.020 \log [Al^{+3}]$
$Al(OH)_3/Al$	$E = -1.47 - 0.059 \, pH$
$Al(OH)_4^-/Al$	$E = -1.23 - 0.079 \, pH + 0.020 \log [Al(OH)_4^-]$
$Al(OH)_3/Al^{+3}$	$3 \, pH = 10.2 - \log [Al^{+3}]$
$Al(OH)_4^-/Al(OH)_3$	$pH = 12.4 + \log [Al(OH)_4^-]$

with NaOH solution to form $Al(OH)_4{}^-$, insoluble impurity hydroxides like $Fe(OH)_3$ are filtered out, the pH of the solution is reduced, and $Al(OH)_3$ precipitates. This product provides the basis for the preparation of Al compounds.

d. The element.   The $Al(OH)_3$ from the above process is dehydrated at a high temperature, the resulting Al_2O_3 is dissolved in molten cryolite Na_3AlF_6, and electrolyzed. Al is a hard, strong, silvery metal which is thermodynamically very active, but whose oxide coat makes it inactive in some reactions.

e. Oxides and hydroxides.   Al is oxidized in air or oxygen to produce a thin coat of resistant oxide on the surface which inactivates it. If very finely divided and heated, it can go completely to white Al_2O_3. This oxide is insoluble in HOH, and soluble in strong acids and strong bases. If the Al_2O_3 is fired (heated very hot), it dissolves in the above reagents very slowly, the higher the temperature of firing, the slower its dissolution. The white $Al(OH)_3$ can be precipitated from Al^{+3} solutions by strong base, or by dissolution of Al_2O_3 in strong base to give $Al(OH)_4{}^-$ then lowering the pH. The hydroxide is insoluble in HOH, soluble in strong bases to give $Al(OH)_4{}^-$, and soluble in strong acids to give Al^{+3}. There is uncertainty with regard to the transformation of $Al(OH)_3$ into Al_2O_3. This uncertainty centers around the question of whether $Al(OH)_3$ is actually a hydroxide or a hydrated oxide ($Al_2O_3 \cdot 3HOH$). If $\Delta G°$ values for $Al(OH)_3$ as precipitated in the laboratory (amorphous) and Al_2O_3 as represented by mineral forms can be trusted, then Al_2O_3 is the more stable compound. However, it is assumed here that the non-equilibrium amorphous, gelatinous $Al(OH)_3$ that is prepared from solution is truly $Al(OH)_3$, but upon aging and moving toward equilibrium it approaches Al_2O_3.

f. Compounds and solubilities.   Table 7.2 presents the names, formulas, states, solubilities, and standard free energies of a number of the more important inorganic species of Al. The metal, Al_2O_3, and $Al(OH)_3$ are soluble in strong acids to give the corresponding salts, and are also soluble in strong base to give $Al(OH)_4{}^-$. Notice that most of the compounds are salts of strong acids. Most Al compounds with weak acid anions are not stable in HOH solution. This is because the anions react with water, become protonated, and generate OH^- which precipitates the Al^{+3} as $Al(OH)_3$. Consider the addition of NaCN to a solution of $Al(NO_3)_3$. The hydrolysis of the CN^- is described by a combination of the protonation constant Kp of CN^- and the ion product K_w of HOH, thus:

$$CN^- + H^+ \rightarrow HCN \qquad K_p = 10^{9.2}$$

$$HOH \rightarrow H^+ + OH^- \qquad K_w = 10^{-14.0}$$

$$CN^- + HOH \rightarrow HCN + OH^- \qquad K_pK_w = 10^{-4.8}$$

Table 7.2
Aluminum Species

Name	Formula	State	Color	Solubility	$\Delta G°$ (kJ/mole)
Aluminum	Al	s	Silvery	Insol	0.0
Aluminum ion	Al^{+3}	aq	Colorless		−485.3
Hexafluoroaluminate	AlF_6^{-3}	aq	Colorless		−2291.2
Tetrahydroxoaluminate	$Al(OH)_4^-$	aq	Colorless		−1305.4
Aluminum (III)					
arsenate	$AlAsO_4$	s	White	Insol	−1224.1
borate	$AlBO_3$	s	White	Insol	
bromate	$Al(BrO_3)_3$	s	White	Sol	
bromide	$AlBr_3$	s	White	Sol	
chlorate	$Al(ClO_3)_3$	s	White	Sol	
chloride	$AlCl_3$	s	White	Sol	−630.1
fluoride	AlF_3	s	White	Insol	−1430.9
hydroxide	$Al(OH)_3$	s	White	Insol	−1138.9
hydroxyacetate	$AlOH(C_2H_3O_2)_2$	s	White	Sl sol	
iodide	AlI_3	s	White	Sol	−299.2
metaphosphate	$Al(PO_3)_3$	s	White	Sol	
nitrate	$Al(NO_3)_3$	s	White	Sol	
oxalate	$Al_2(C_2O_4)_3$	s	White	Insol	
oxide	Al_2O_3	s	White	Insol	−1582.0
perchlorate	$Al(ClO_4)_3$	s	White	Sol	
phosphate	$AlPO_4$	s	White	Insol	−1617.5
selenide	Al_2Se_3	s	Tan	Decomp	
silicate	Al_2OSiO_4	s	White	Insol	
sulfate	$Al_2(SO_4)_3$	s	White	Sol	−3099.5
sulfide	Al_2S_3	s	Yellow	Decomp	−713.4

If the NaCN concentration is $10^{-1.0}$ M (0.10 M), the approximate $[OH^-]$ can be calculated as follows:

$$K_p K_w = 10^{-4.8} = [HCN][OH^-]/[CN^-][HOH] = [OH^-]^2/[0.10][1]$$

$$[OH^-] = 10^{-2.9} \text{ M}.$$

The solublity product of $Al(OH)_3$ is $10^{-32.3}$, and if the $[OH^-]$ is generated to be $10^{-2.9}$ M and the $[Al^{+3}]$ is $10^{-1.0}$ M (0.10 M), then $Al(OH)_3$ will precipitate.

$$K_{sp} = 10^{-32.3} = [Al^{+3}][OH^-]^3$$

$$[OH^-] = (10^{-32.3}/[Al^{+3}])^{1/3} = (10^{-32.3}/10^{-1.0})^{1/3} = 10^{-10.4}$$

This calculation shows that any $[OH^-]$ greater than $10^{-10.4}$ will produce the precipitate. The most common strong-acid anions are Cl^-, Br^-, I^-, ClO_3^-,

BrO_3^-, IO_3^-, ClO_4^-, BrO_4^-, IO_4^-, HSO_4^-, $HSeO_4^-$, NO_3^-, $H_2PO_4^-$, and $H_2AsO_4^-$. Almost all other anions are weak-acid anions, and their salts will give basic solutions when they are dissolved. The extent to which weak acid anions produce OH^- can be determined as illustrated above. K_p values may be found in the literature or they may be calculated from $\Delta G°$ values of the weak acid anion and the weak acid. Salts can be formed between cations and the $Al(OH)_4^-$ anion, these being called tetrahydroxoaluminates. Those of the alkali metals are soluble, but most others are insoluble.

g. Redox reactions. As can be seen from the E–pH diagram, Al ordinarily shows only one valence +3 other than 0. The metal Al is thermodynamically a strong reducing agent, but as mentioned above, its non-reactivity under certain conditions must be taken into account. The E–pH diagram also indicates that the trivalent species of Al are highly stable, resisting both oxidation and reduction.

h. Complexes. The Al^{+3} ion probably exists in HOH solution in the form of an exceptionally stable complex with six HOH molecules attached, $Al(HOH)_6^{+3}$, the oxygens on the water molecules being the points of attachment. Al appears to retain this coordination number of 6 in many of its compounds and complex ions. For example, the species $Al(OH)_4^-$ is more properly written as $Al(HOH)_2(OH)_4^-$. Al^{+3} has a strong attraction for O-containing ligands and little attraction for N-containing ligands. Some complexation constants (log β_1 through log β_6) for ligands with Al^{+3} are as follows: OH^- (9.0, 18.7, 27.0, 33.0), F^- (7.0, 12.6, 16.7, 19.1, 19.4, 19.6), $HCOO^-$ (1.4), $C_2O_4^{-2}$ (6.1, 11.1, 15.1), $C_2H_3O_2^-$ (1.5), tartrate^{-2} (5.3, 9.8), salicylate^{-2} (12.9, 23.2, 29.8), phthalate^{-2} (3.2, 6.3), iminodiacetate^{-2} (8.1, 15.1), nta^{-3} (11.4), edta^{-4} (16.3). Notice that many of these are chelate structures. Al^{+3} forms neutral complexes by attachment of three bidentate groups such as the 2,4-pentanedione anion and other β-diketone anions, and the 8-hydroxyquinoline anion. These complexes are insoluble in HOH, but dissolve in organic media.

i. Analysis. Colorimetric methods provide for the determination of Al down to 10 ppb. AAS will analyze for Al down to 50 ppb, ETAAS to 10 ppb, and ICPAES to 1 ppb. ICPMS improves upon this by going down to 0.1 ppb. Alizarin is a reagent which provides for a spot test of Al down to 20 ppm. Al^{+3} can also be detected in a spot test by morin in methanol down to 10 ppm.

j. Health and environmental aspects. The LD50 (oral rat) of anhydrous $Al(NO_3)_3$ is 4.3 g/kg and that for anhydrous $AlCl_3$ is 3.7 g/kg. These relatively high values are probably due to the fact that not much absorption of Al occurs when it is ingested.

4. Gallium (Ga) 4s²4p¹

a. E–pH diagram.   The E–pH diagram for $10^{-1.0}$ M Ga is presented in Figure 7.3. In the figure legend are equations for the lines that separate the major aqueous species. The diagram strongly resembles the Al diagram except that $Ga(OH)_3$ occupies a larger area and that the lines above Ga occur about a volt higher. The amphoterism of $Ga(OH)_3$ is shown and the indications are that Ga metal is dissolved by acids and by strong bases. Ga metal shows some of the inertness of Al, in that it dissolves very slowly in cold HOH, but dissolves easily in boiling HOH. In the region of the transformation from Ga^{+3} to $Ga(OH)_3$, there exist partially hydrolyzed and polymeric species. Again, as with Al, the oxidation states of Ga are 0 and III, and no further oxidation can be worked in HOH solution.

b. Discovery.   In 1875 Paul Émile Lecoq de Boisbaudran detected Ga spectroscopically at 414 nm wave length (violet), then later in the year isolated it by electrolysis of an ammoniacal solution of the sulfate. He named it gallium after the old Latin name for France (Gallia).

c. Extraction.   Ga occurs naturally in trace quantities in sulfide, Zn, and Al minerals. Its industrial source is as a by-product of Al extraction from bauxite. When bauxite is treated with NaOH solution, Ga goes into solution as $Ga(OH)_4^-$.

d. The element.   The solution from the above process is placed in contact with kerosene containing 8-hydroxyquinoline. Both Al^{+3} and Ga^{+3} are extracted as the neutral chelate complexes M(8-hydroxyquinolinate)$_3$. The kerosene containing these complexes is then removed and contacted with dilute acid which removes the Al^{+3} into the aqueous acid solution. Again the kerosene phase is isolated and contacted with concentrated acid which takes out the Ga^{+3}. The metal can then be plated out by electrolysis. This electrolytic process is successful because of the sizable overvoltage of Ga for the evolution of H_2. The element dissolves in strong acids, bases, and boiling HOH, but only slowly in cold HOH. Ga is only superficially oxidized in air or O_2, even at elevated temperatures. An interesting property of the element is its melting point, the solid becoming a liquid at 30°C (86°F).

e. Oxides and hydroxides.   White Ga_2O_3 is usually produced by the thermal decomposition of the nitrate, sulfate, or hydroxide. It is soluble in strong acids and strong bases, but if it has been fired, it dissolves very slowly. White $Ga(OH)_3$ can be produced by precipitation of Ga^{+3} with OH^- using a soluble salt of Ga and a base such as NaOH. The freshly prepared hydroxide is amorphous, but as it ages it goes over to GaO(OH). The hydroxide and the oxide are both amphoteric in that they dissolve in both acids and bases.

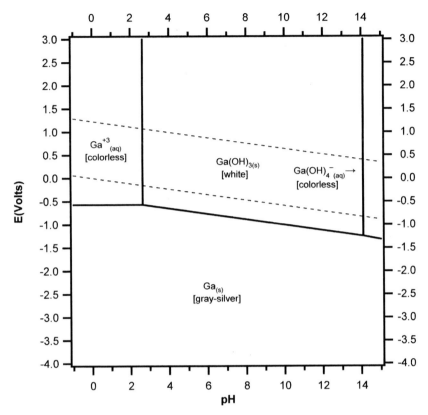

Figure 7.3 E–pH diagram for Ga species. Soluble species concentrations (except H^+) = $10^{-1.0}$ M. Soluble species and most solids are hydrated. No agents producing complexes or insoluble compounds are present other than HOH and OH^-.

Species ($\Delta G°$ in kJ/mol): Ga (0.0), Ga^{+3} (−159.4), $Ga(OH)_3$ (−831.4), $Ga(OH)_4^-$ (−982.4), HOH (−237.2), H^+ (0.0), and OH^- (−157.3).

Equations for the lines:

Ga^{+3}/Ga $E = -0.55 + 0.020 \log [Ga^{+3}]$

$Ga(OH)_3/Ga$ $E = -0.41 - 0.059 \, pH$

$Ga(OH)_4^-/Ga$ $E = -0.12 - 0.079 \, pH + 0.020 \log [Ga(OH)_4^-]$

$Ga(OH)_3/Ga^{+3}$ $3 \, pH = 6.9 - \log [Ga^{+3}]$

$Ga(OH)_4^-/Ga(OH)_3$ $pH = 15.1 + \log [Ga(OH)_4^-]$

Table 7.3
Gallium Species

Name	Formula	State	Color	Solubility	$\Delta G°$ (kJ/mole)
Gallium	Ga	s	Gray-silver	Decomp	0.0
Gallium ion	Ga^{+3}	aq	Colorless		−159.4
Hexafluorogallate	GaF_6^{-3}	aq	Colorless		
Tetrahydroxogallate	$Ga(OH)_4^-$	aq	Colorless		−982.4
Gallium (III)					
arsenate	$GaAsO_4$	s	White	Insol	
borate	$GaBO_3$	s	White	Insol	
bromate	$Ga(BrO_3)_3$	s	White	Sol	
bromide	$GaBr_3$	s	White	Sol	−359.8
chlorate	$Ga(ClO_3)_3$	s	White	Sol	
chloride	$GaCl_3$	s	White	Sol	−453.1
fluoride	GaF_3	s	White	Insol	−1092.0
hydroxide	$Ga(OH)_3$	s	White	Insol	−831.4
hydroxyacetate	$GaOH(C_2H_3O_2)_2$	s	White	Sl sol	
iodate	$Ga(IO_3)_3$	s	White	Sl sol	
iodide	GaI_3	s	White	Sol	−299.2
metaphosphate	$Ga(PO_3)_3$	s	White	Sol	
nitrate	$Ga(NO_3)_3$	s	White	Sol	
oxalate	$Ga_2(C_2O_4)_3$	s	White	Insol	
oxide	Ga_2O_3	s	White	Insol	−998.3
perchlorate	$Ga(ClO_4)_3$	s	White	Sol	
phosphate	$GaPO_4$	s	White	Insol	
selenide	Ga_2Se_3	s	Tan	Decomp	
silicate	Ga_2OSiO_4	s	White	Insol	
sulfate	$Ga_2(SO_4)_3$	s	White	Sol	
sulfide	Ga_2S_3	s	Yellow	Decomp	−505.9
thiocyanate	$Ga(NCS)_3$	s	White	Sol	

f. Compounds and solubilities.   Table 7.3 presents the names, formulas, states, solubilities, and standard free energies of a number of the more important inorganic species of Ga. Salts of Ga^{+3} may be prepared by treatment of Ga, Ga_2O_3, or $Ga(OH)_3$ with strong acids. There is a strong resemblance to Al in the cases of Ga salts of weak-acid anions, in the formation of $Ga(OH)_4^-$ and its salts, and in the occurrence of complex species around the transition from $Ga(OH)_3$ to $Ga(OH)_4^-$. Salts of this tetrahydroxogallate ion are usually insoluble, except for those of the alkali metals.

g. Redox reactions.   In its redox behavior, Ga again shows strong parallels with Al. These relate to its oxidation states and the reducing power of the metal. Ga is not quite as inert as Al, but in some cases, the position of the metal on the E–pH diagram is deceptive unless this inertness is taken into account.

h. Complexes. The Ga^{+3} ion exists in solution as $Ga(HOH)_6^{+3}$ and the $Ga(OH)_4^-$ ion is more properly written as $Ga(HOH)_2(OH)_4^-$, both reflecting a coordination number of 6. Some complexation constants (log β_1 through log β_4) for ligands with Ga^{+3} are as follows: SCN^- (2.1), OH^- (11.4, 22.1, 31.6, 39.4), F^- (5.9, 8.0, 10.5, 11.5), 8-hydroxyquinolinate$^-$ (14.5, 28.0, 40.5), $C_2O_4^{-2}$ (6.4, 12.4, 17.9), nta^{-3} (13.6), edta^{-4} (20.3). Other complexes probably resemble those of Al^{+3} and likely have similar β values. Ga^{+3} forms neutral complexes by attachment of three bidentate groups such as the 2,4-pentanedione anion and other β-diketone anions, and the 8-hydroxyquinoline anion. These complexes are insoluble in HOH, but dissolve in organic media.

i. Analysis. Colorimetric analysis of Ga permits detection of 20 ppb. Ga or a Ga compound in the flame gives a violet (lilac) color. This is due to its emission of 414.0 nm photons. This photon line allows for the spectrophotometric determination of Ga by AAS, which is sensitive to about 10 ppm. ETAAS increases this sensitivity to 5 ppb, and ICPAES to about 1 ppb, with ICPMS going as low as 0.1 ppb. Rhodamine B is a reagent which facilitates the spottest detection at a level of 10 ppm or above.

j. Health and environmental aspects. The toxicity of Ga salts is very low. The LD50 (oral rat) of $Ga(NO_3)_3$ is about 4 g/kg.

5. Indium (In) $5s^25p^1$

a. E–pH diagram. The E–pH diagram for $10^{-1.0}$ M In is presented in Figure 7.4. In the figure legend are equations for the lines that separate the species. The diagram strongly resembles the Al and Ga diagrams except that the lines above In occur about 0.2 v higher than Ga, indicating that In is slightly less active. The amphoterism of $In(OH)_3$ is somewhat in doubt because there is disagreement about the precise character of what is claimed to be $In(OH)_4^-$. Some indicate that the action of $In(OH)_3$ in concentrated base produces a transparent colloid, but if it is actually the tetrahydroxy ion, it could be added to the far right of the diagram. The diagram indicates that In is dissolved by acids, HOH, and bases, but some degree of inertness is there since it dissolves in hot or concentrated acids, and in hot water and bases only after being finely powdered. Partially hydrolyzed and polymeric species exist in the region of transition of In^{+3} to $In(OH)_3$, their prevalence and character being dependent upon pH and concentration of In. As with the elements that rest above it, the oxidation states of In are 0 and III, and no further oxidation can be worked in HOH solution.

b. Discovery.   Ferdinand Reich and Hieronymous Theodor Richter in 1863 first identified In by its characteristic blue line obtained spectroscopically from a solution of $ZnCl_2$ obtained from sphalerite, a ZnS ore. They named it indium after the Latin word indicum for the indigo blue color. They then proceeded to concentrate the In by dissolving the ore in HNO_3, which brought about the precipitation of sulfides of numerous impurities, leaving Zn^{+2} and In^{+3} in solution. The solution was then treated with NH_4OH, which precipitated

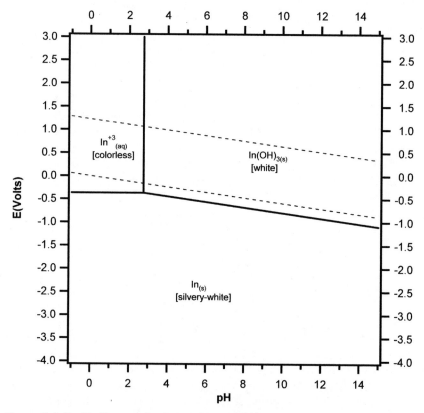

Figure 7.4 E–pH diagram for In species. Soluble species concentrations (except H^+) = $10^{-1.0}$ M. Soluble species and most solids are hydrated. No agents producing complexes or insoluble compounds are present other than HOH and OH^-.

Species ($\Delta G°$ in kJ/mol): In (0.0), In^{+3} (−101.3), $In(OH)_3$ (−771.2), HOH (−237.2), H^+ (0.0) and OH^- (−157.3).

Equations for the lines:

In^{+3}/In	$E = -0.35 + 0.020 \log [In^{+3}]$
$In(OH)_3/In$	$E = -0.21 - 0.059 \, pH$
$In(OH)_3/In^{+3}$	$3 \, pH = 7.3 - \log [In^{+3}]$

$In(OH)_3$. In 1867, they produced the metal by treating the heated oxide with hydrogen.

c. Extraction.   The major industrial source of In is in flue deposits, slags, and residues of the roasting and smelting of Zn and Pb ores. These materials have become concentrated in the element. They are leached with very dilute H_2SO_4 or HCl to put Zn^{+2} in solution, leaving In_2O_3 in the residue. The residue is dissolved in stronger acid, and heavy metals are precipitated out as the sulfides.

d. The element.   The solution from the above process is placed in contact with strips of Zn metal onto which the In plates as a sponge. It is scraped off, dissolved in strong HCl, then electrodeposited. As in the case of Ga, this electrolytic process is successful because of the sizable overvoltage for the evolution of H_2. The silvery-white metal dissolves slowly in cold acids, faster in hot or concentrated acids, but does not readily dissolve in HOH or bases. However, if the In is very finely powdered, it will react with HOH. In is stable at room-temperature in air or O_2, but at elevated temperatures, it burns to In_2O_3.

e. Oxides and hydroxides.   Pale yellow In_2O_3 can be made by the thermal decomposition of $In(OH)_3$, $In(NO_3)_3$, or $In_2(SO_4)_3$, or by heating In in air or O_2. If formed at lower temperatures, it is yellow and dissolves in acids but not bases. If it is heated to high temperatures, it is reddish-brown, and dissolves very slowly. White $In(OH)_3$ can be produced by precipitation of In^{+3} with OH^- using a soluble salt of In and a base such as KOH. The freshly prepared hydroxide is amorphous and dissolves readily in acids. The solubility of the hydroxide in high concentrations of OH^- is questionable as mentioned above.

f. Compounds and solubilities.   Table 7.4 presents the names, formulas, states, solubilities, and standard free energies of a number of the more important inorganic species of In. Salts of In^{+3} may be prepared by treatment of In, In_2O_3, or $In(OH)_3$ with strong acids. Some In salts of weak acids are stable in HOH, especially if they are insoluble [such as yellow-orange In_2S_3 and $In_2(C_2O_4)_3$], but soluble ones are often unstable such as $In(CN)_3$ and $In(NCS)_3$ both of which slowly hydrolyze to $In(OH)_3$. Treatment of $In(OH)_3$ with acetic or other carboxylic acids generally produces basic acetates (containing OH^-) or basic carboxylates.

g. Redox reactions.   In its redox behavior, In again shows strong parallels with Al and Ga. These similarities relate to its oxidation states, and the reducing power of the metal. In is not quite as inert as Al or Ga, but in some cases, the position of the metal on the E–pH diagram is deceptive unless this inertness is taken into account.

Table 7.4
Indium Species

Name	Formula	State	Color	Solubility	$\Delta G°$ (kJ/mole)
Indium	In	s	Silvery-white	Decomp	0.0
Indium ion	In^{+3}	aq	Colorless		−101.3
Tetrafluoroindate	InF_4^-	aq	Colorless		
Tetrahydroxoindate	$In(OH)_4^-$	aq	Colorless		−920.9
Indium(III)					
arsenate	$InAsO_4$	s	White	Insol	
bromide	$InBr_3$	s	Yellow-white	Sol	−397.1
carbonate	$In_2(CO_3)_3$	s	White	Insol	
chlorate	$In(ClO_3)_3$	s	White	Sol	
chloride	$InCl_3$	s	White	Sol	−462.3
chromate	$In_2(CrO_4)_3$	s	Yellow	Insol	
cyanide	$In(CN)_3$	s	White	Decomp	
fluoride	InF_3	s	White	Insol	−1114.6
hydroxide	$In(OH)_3$	s	White	Insol	−771.2
hydroxyacetate	$InOH(C_2H_3O_2)_2$	s	White	Sl sol	
iodide	InI_3	s	Yellow	Sol	−227.6
nitrate	$In(NO_3)_3$	s	White	Sol	
oxalate	$In_2(C_2O_4)_3$	s	White	Insol	
oxide	In_2O_3	s	Yellow	Insol	−830.5
perchlorate	$In(ClO_4)_3$	s	White	Sol	
phosphate	$InPO_4$	s	White	Insol	
silicate	In_2OSiO_4	s	White	Insol	
sulfate	$In_2(SO_4)_3$	s	White	Sol	−1639.7
sulfide	In_2S_3	s	Yellow	Insol	−341.4
thiocyanate	$In(NCS)_3$	s	White	Sol	165.0

h. Complexes. The In^{+3} ion exists in solution as $In(HOH)_6^{+3}$ reflecting a coordination number of 6, which is probably exhibited in most solution species. Complexation constants (log β_1 through log β_4) for ligands with In^{+3} are as follows: SCN^- (3.2, 3.5, 4.6), NO_3^- (0.3), OH^- (10.0, 20.2, 29.6, 33.9), F^- (4.6, 8.1, 10.3, 11.5), Cl^- (2.3, 3.6, 4.0), Br^- (1.9, 3.1, 3.4), I^- (1.6, 2.6), 8-hydroxyquinolinate$^-$ (12.0, 23.9, 35.4), $HCOO^-$ (2.7, 4.7, 5.7, 6.7), $C_2O_4^{-2}$ (5.3, 10.5), $C_2H_3O_2^-$ (3.5, 6.0, 7.9, 9.1), salicylate$^-$ (4.4, 8.5), nta^{-3} (16.9), edta^{-4} (25.0). Other complexes probably resemble those of Al^{+3} and Ga^{+3} and likely have similar β values. In^{+3}, like the two preceding ions, forms neutral complexes by attachment of three bidentate groups such as the 2,4-pentanedione anion and other β-diketone anions, and the 8-hydroxyquinoline anion. These complexes are insoluble in HOH, but dissolve in organic media.

i. Analysis. The colorimetric method for In is capable of a detection limit of 20 ppb. Indium or an In compound in the flame gives an indigo blue color (451.1 nm). This photon line allows for the spectrophotometric determination of In by AAS (atomic absorption flame spectroscopy). The method is sensitive to about 300 ppb. With ETAAS, this limit drops to 10 ppb, as it does with ICPAES. ICPMS drops the limit to 0.01 ppb. Alizarin detects In, as well as Al, but the reaction with Al can be masked by addition of F^- to a spot test. The limit of detection is about 1 ppm.

j. Health and environmental aspects. What little detailed information is available indicates that the toxicity of In salts is not high. The LD50 (oral rat) for the sulfate is 1.2 g/kg.

6. Thallium (Tl) $6s^2 6p^1$

a. E–pH diagrams. The E–pH diagram for $10^{-1.0}$ M Tl is presented in Figure 7.5. In the figure legend are equations for the lines that separate the various species. Tl metal sits beneath the HOH line in the acidic region, but not in the basic region. This indicates that Tl should dissolve in acids, but not in bases. The Tl^+ region occupies a sizable area of the diagram pointing out that TlOH is soluble over most of the pH range. In the upper left corner, the highly unstable Tl^{+3} appears. It persists only in very strong acid and readily goes over to Tl_2O_3 with a slight raise in pH. Figure 7.6 is an E–pH diagram which involves Tl soluble species at $10^{-1.0}$ M and Cl soluble species at $10^{0.0}$ M. Such conditions put the Tl^+ in the form of insoluble TlCl and put much of the Tl^{+3} in the form of the complex $TlCl_4^-$. Notice several things: the stabilization of the Tl(III) in the form of the complex and the stabilization of Tl(I) in the form of TlCl (note the position of the TlCl/Tl line as compared to the Tl^+/Tl). These phenomena illustrate the principle that complexation and insolubilization stabilize oxidation states. Further, pay heed to the fact that the highest portions (above 1.50 v) of both diagrams are identical. This is a result of the oxidation of Cl^- to Cl_2 and/or ClO_4^-, which does not form a complex with Tl^{+3}.

b. Discovery. Tl was first detected by William Crookes in 1861. He observed a bright green line (535.0 nm) in the emission spectrum of a residue from a sulfuric acid plant that was making the acid from sulfide minerals. The name comes from the Latin word thallus which means a new green shoot of a plant. This line was also observed by Claude-Auguste Lamy, and he and Crookes both isolated the element in 1862, Crookes producing a small amount of amorphous Tl, and Lamy making a sizable ingot of the crystalline substance. Lamy also treated residues from an H_2SO_4 plant, isolating the element as TlCl, then converting to the sulfate, and finally obtaining Tl by electrolysis of the Tl_2SO_4 solution.

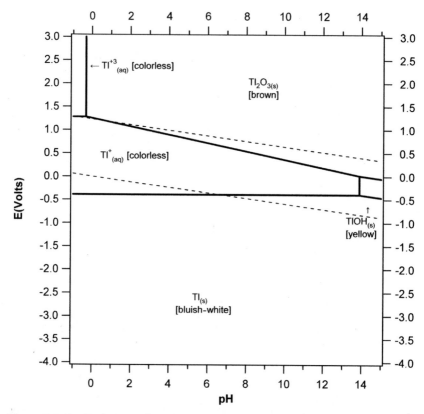

Figure 7.5 E–pH diagram for Tl species. Soluble species concentrations (except H^+) = $10^{-1.0}$ M. Soluble species and most solids are hydrated. No agents producing complexes or insoluble compounds are present other than HOH and OH^-.

Species ($\Delta G°$ in kJ/mol): Tl (0.0), Tl^+ (−32.2), Tl^{+3} (214.6), TlOH (−195.8), Tl_2O_3 (−304.6), HOH (−237.2), H^+ (0.0), and OH^- (−157.3).

Equations for the lines:

Tl^+/Tl	$E = -0.33 + 0.059 \log [Tl^+]$
$TlOH/Tl$	$E = 0.43 - 0.059\ pH$
Tl^{+3}/Tl^+	$E = 1.28$
Tl_2O_3/Tl^+	$E = 1.22 - 0.089\ pH - 0.030 \log [Tl^+]$
$Tl_2O_3/TlOH$	$E = 0.84 - 0.059\ pH$
Tl_2O_3/Tl^{+3}	$6\ pH = -3.9 - 2 \log [Tl^{+3}]$
$TlOH/Tl^+$	$pH = 12.9 - \log [Tl^+]$

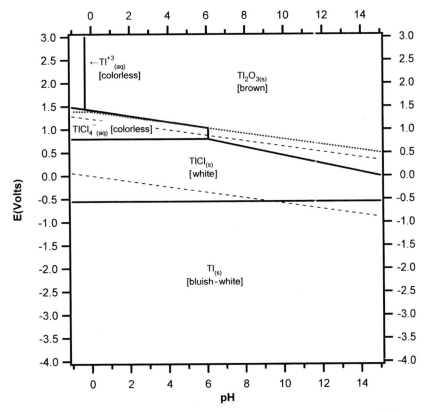

Figure 7.6 E–pH diagram for Tl/Cl species. Tl species concentrations = $10^{-1.0}$ M. Cl species concentrations = $10^{0.0}$ M. Soluble species and most solids are hydrated. No agents producing complexes or insoluble compounds are present other than HOH, OH⁻ and Cl⁻. The finely-dashed line represents the oxidation of Cl⁻ to Cl_2 and/or ClO_4^- and hence the disappearance of all species containing Cl⁻ above the line.

Species ($\Delta G°$ in kJ/mol): Tl (0.0), Tl^+ (−32.2), Tl^{+3} (214.6), TlOH (−195.8), Tl_2O_3 (−304.6), Cl_2 (7.1), Cl⁻ (−131.4), ClO_4^- (−8.4), TlCl (−184.8), $TlCl_3$ (−241.5), $TlCl_4^-$ (−420.1), HOH (−237.2), H^+ (0.0), and OH⁻ (−157.3).

Equations for the lines:

$Tl_2O_3/TlCl_4^-$ $E = 1.42 - 0.065\ pH - 0.002\ \log\ [TlCl_4^-] + 0.007\ \log\ [ClO_4^-]$

$Tl_2O_3/TlCl$ $E = 1.33 - 0.089\ pH + 0.030\ \log\ [Cl^-]$

$TlCl_4^-/TlCl$ $E = 0.82 + 0.030\ \log\ [TlCl_4^-] - 0.089\ \log\ [Cl^-]$

$Tl^{+3}/TlCl_4^-$ $E = 1.42 - 0.059\ pH + 0.007\ \log\ [ClO_4^-]$

$TlCl/Tl$ $E = -0.55 - 0.059\ \log\ [Cl^-]$

$Tl_2O_3/TlCl_4^-$ $6\ pH = 34.4 - 2\ \log\ [TlCl_4^-] + 8\ \log\ [Cl^-]$

Tl_2O_3/Tl^{+3} $6\ pH = -3.9 - 2\ \log\ [Tl^{+3}]$

c. Extraction. The major source of Tl is flue dusts from the roasting of Zn, Cu, and Pb ores. Many Tl compounds are much more volatile that those of other elements, and as a result, they collect in the flues, particularly Tl_2O and Tl_2SO_4. Since both of these compounds are soluble in water, they may be leached out of the dusts to give solutions of TlOH and Tl_2SO_4, leaving behind most impurites as insoluble residue.

d. The element. The Tl solutions from above are converted completely to the sulfate by addition of H_2SO_4 and then electrolysis yields Tl metal. Or the Tl can be deposited on Zn plates. The bluish white Tl metal oxidizes slowly in moist air to black Tl_2O, more rapidly upon heating. In oxygen a mixture of black Tl_2O and red-brown Tl_2O_3 is formed. And with ozone or H_2O_2 or strong heating in O_2, pure Tl_2O_3 is formed. The metal is soluble in acids to give colorless Tl^+ and H_2, but not in HOH or base.

e. Oxides and hydroxides. As indicated above, black Tl_2O and red-brown Tl_2O_3 can be made by proper oxidation of Tl. The Tl_2O is soluble in HOH to give TlOH, which is a strong base. Upon evaporation yellow crystals of TlOH come out. Tl_2O_3 is insoluble in water, base, and dilute acid. In concentrated acid it dissolves to give colorless Tl^{+3}. All present data indicate that $Tl(OH)_3$ does not exist.

f. Compounds and solubilities. Table 7.5 presents the names, formulas, states, solubilities, and standard free energies of a number of the more important inorganic species of Tl. The species Tl, Tl_2O, TlOH, Tl_2CO_3, and Tl_2O_3 can often be treated with appropriate acids to produce the corresponding salts. Note that the Tl^+ salt of F^- is soluble, whereas those of Cl^-, Br^-, and I^- are insoluble (similar to Ag). The salts of Tl^{+3} tend to decompose by hydrolysis to Tl_2O_3 as the E–pH diagram indicates.

g. Redox reactions. The E–pH diagram indicates that Tl shows three oxidation states: 0, I, and III. As mentioned above, the $10^{-1.0}$ M Tl diagram shows that Tl will dissolve in acids but not in bases since the Tl^+/Tl line crosses the H^+/H_2 line in the region of pH $= 7.0$. Based on this relationship only, one might conclude that Tl is not soluble in HOH, but it will dissolve if air or O_2 is present because the O_2/HOH line rests above the Tl^+/Tl line. Such is the case experimentally. The diagram also demonstrates that strong oxidizing agents will oxidize Tl^+ and its compounds to Tl^{+3} in strong acid and to Tl_2O_3 over the rest of the pH range.

h. Complexes. The Tl^+ ion may exist in HOH solution in the form of $Tl(HOH)_6{}^+$, and it is established that the Tl^{+3} ion takes the form $Tl(HOH)_6{}^{+3}$. Tl^+ has little tendency to complex, and when it does the complexes are weak. Some values of complexation constants exemplify this: ($\log \beta_1$, $\log \beta_2$, $\log \beta_3$, $\log \beta_4$), SCN$^-$ (0.6), NO$_3{}^-$ (0.3), P$_2$O$_7{}^{-4}$ (1.7, 1.9),

Table 7.5
Thallium Species

Name	Formula	State	Color	Solubility	$\Delta G°$ (kJ/mole)
Thallium	Tl	s	Bluish-white	Decomp	0.0
Thallium(I) ion	Tl^+	aq	Colorless		−32.2
Thallium(III) ion	Tl^{+3}	aq	Colorless		214.6
Tetrachlorothallate(III)	$TlCl_4^-$	aq	Colorless		−420.1
Thallium(I)					
acetate	$TlC_2H_3O_2$	s	White	Sol	
bromate	$TlBrO_3$	s	White	Sl sol	−53.1
bromide	$TlBr$	s	Yellow	Insol	−167.0
carbonate	Tl_2CO_3	s	White	Sol	−614.6
chlorate	$TlClO_3$	s	White	Sol	
chloride	$TlCl$	s	White	Sol	−184.8
chromate	Tl_2CrO_4	s	Yellow	Insol	
cyanate	$TlCNO$	s	White	Sol	
cyanide	$TlCN$	s	White	Sol	
fluoride	TlF	s	White	Sol	
formate	$TlCHO_2$	s	White	Sol	
hydrogen carbonate	$TlHCO_3$	s	White	Sol	−150.6
hydrogen sulfate	$TlHSO_4$	s	White	Sol	
hydroxide	$TlOH$	s	Yellow	Sol	−195.8
iodate	$TlIO_3$	s	White	Sl sol	−191.9
iodide	TlI	s	Yellow	Insol	−125.4
metavanadate	$TlVO_3$	s	Gray	Sl sol	
molybdate	Tl_2MoO_4	s	White	Insol	
nitrate	$TlNO_3$	s	White	Sol	−152.5
oxalate	$Tl_2C_2O_4$	s	White	Sol	
oxide	Tl_2O	s	Black	Decomp	−147.0
perchlorate	$TlClO_4$	s	White	Sol	
phosphate	Tl_3PO_4	s	White	Sl sol	
sulfate	Tl_2SO_4	s	White	Sol	−830.5
sulfide	Tl_2S	s	Black	Insol	−93.7
thiocyanate	$TlNCS$	s	White	Sl sol	39.0
Thallium(III)					
bromide	$TlBr_3$	s	Yellow	Sol	
chloride	$TlCl_3$	s	White	Sol	−241.5
fluoride	TlF_3	s	Green	Decomp	−505.4
nitrate	$Tl(NO_3)_3$	s	White	Sol	
oxide	Tl_2O_3	s	Brown	Insol	−304.6
sulfate	$Tl_2(SO_4)_3$	s	White	Sol	
sulfide	Tl_2S_3	s	Black	Insol	

OH^- (0.8), SO_4^{-2} (1.4), F^- (0.1), Cl^- (0.5), Br^- (0.9), I^- (0.7, 0.9, 1.1), nta^{-3}(4.7), $edta^{-4}$(6.5). On the contrary, complexes of Tl^{+3} are quite stable. Some values are: (log β_1, log β_2, log β_3, log β_4), CN^- (log β_4 = 35.0), NO_3^- (0.9), OH^- (13.4, 26.4, 38.7, 41.0), SO_4^{-2} (2.3), Cl^- (7.7, 13.5, 16.5, 18.3), Br^- (9.7, 16.6, 21.2, 23.9), I^- (log β_4 = 35.7), $C_2H_3O_2^-$ (6.2, 11.3, 15.1, 18.3), nta^{-3} (20.9, 32.5), $edta^{-4}$ (35.3). As illustrated in the E-pH diagram discussion (Section a), complexation stabilizes Tl^{+3} against both hydrolysis and reduction, the complexes tending to occupy larger areas on the E–pH diagram than the Tl^{+3} cation.

i. Analysis.   Colorimetry for Tl is effective down to 20 ppb. Tl can also be determined by utilization of the 535.0 nm line. AAS affords a detection limit of 200 ppb, and ETAAS and ICPAES a limit of 10 ppb. ICPMS can go down to 0.1 ppb. For a spot test, treatment with KI followed by $Na_2S_2O_3$ gives color for Tl^+ concentrations of 20 ppm and above.

j. Health and environmental aspects.   Tl and its compounds are highly toxic. In this, Tl resembles its neighbors Hg and Pb. The LD50 (oral rat) of Tl_2SO_4 is 16 mg/kg. Tl is excreted from the body quite slowly, the half time sometimes being as long as 30 days.

8

The Carbon Group

1. Introduction

The elements which constitute the Carbon Group of the Periodic Table are carbon C, silicon Si, germanium Ge, tin Sn, and lead Pb. All five of the elements have atoms characterized by an outer electron structure of ns^2np^2 with n representing the principal quantum number. This electron arrangement signals the possibility of oxidation states of IV and II. Such is the case with the II oxidation state becoming more stable from C to Pb. As one descends the group, there is a marked change from non-metallic (C) to metallic character (Pb). Reflecting very high ionization energies, C, Si, and Ge do not form a simple cation, they instead bond covalently. In line with the trends just mentioned, the inorganic aqueous chemistry moves from anionic (C) to cationic (Pb). The inorganic aqueous solution chemistry of C is represented by four acids and their anionic derivatives: carbonic acid H_2CO_3, oxalic acid $H_2C_2O_4$, formic acid HOOCH, and acetic acid $HOOCCH_3$. Note that in all of these the ionizing H^+ ions are not attached to C but to O. The inorganic aqueous chemistry of Si is dominated by anions $SiO(OH)_3^-$ and $SiO_2(OH)_2^{-2}$ and their many polymeric forms and by the hexafluorosilicate anion SiF_6^{-2}. Ge is very similar to Si. Cationic species, largely absent in all three previous elements, are shown in both Sn and Pb. The covalent single bond radii of C, Si, and Ge are 77, 118, and 122 pm, and the ionic radii in pm of the other two elements are $Sn^{+2}(118)$, Sn^{+4} (83), Pb^{+2} (133), Pb^{+4} (92).

2. Carbon (C) $2s^2 2p^2$

a. E–pH diagrams. In order to understand the E–pH relationships of the aqueous species of C, it is important to consider both the thermodynamic and the kinetic relationships. Thermodynamics tells us whether a reaction will occur but it says nothing about how fast. The rate is a kinetic matter. When acetic acid $HC_2H_3O_2$ is entered into a C species E–pH diagram, Figure 8.1 results. This figure shows that at equilibrium $HC_2H_3O_2$ is not stable and disproportionates into H_2CO_3 and CH_4. The same E–pH diagram results when formic acid HOOCH or when oxalic acid $H_2C_2O_4$ is entered. The following equations with their negative $\Delta G°$ values show why this happens.

$$4HOOCH + HOH \rightarrow 3H_2CO_3 + CH_4 \quad \Delta G° = -195.3 \text{ kJ}$$

$$HC_2H_3O_2 + HOH \rightarrow H_2CO_3 + CH_4 \quad \Delta G° = -40.1 \text{ kJ}$$

$$4H_2C_2O_4 + 5HOH \rightarrow 7H_2CO_3 + CH_4 \quad \Delta G° = -406.1 \text{ kJ}$$

However, all three acids can be prepared and persist in HOH solution. The reason is that the above reactions proceed extremely slowly. That is, they are kinetically inert. Some other reactions with these acids or their anions are also very slow, but predictions are difficult and often unreliable.

b. Non-equilibrium E–pH diagrams. Figure 8.1 also represents the E–pH diagram for carbonate species. The diagram shows that carbonate species will not exist in the region labeled CH_4. However, such species do exist there under certain conditions, namely when the intended reduction reaction is kinetically very slow. A theoretical E–pH diagram for HOOCH and $OOCH^-$ is presented as Figure 8.2. At first glance, it appears strange because the lower line represents the carbonate/formate couple and the upper line represents the formate/methane couple. The lines imply that any oxidizing agent above the lower line will oxidize formate species to carbonate species (see up arrowheads), and that any reducing agent below the upper line will reduce formate species to CH_4 (see down arrowheads). This overlap means, of course, that at thermodynamic equilibrium conditions, the formate species will not exist. However, if the reactions between formate and particular oxidizing agents or reducing agents are kinetically very slow, then the formate species can exist. Such existence is referred to as non-equilibrium or rate-retarded existence. For purposes of information, the formic acid to formate transformation for the non-equilibrium conditions is shown in the band between the lines. Figures 8.3 and 8.4 are for acetate species and oxalate species, and similar interpretations apply.

c. Discovery and occurrence. C has been known as a substance since pre-historic times. It occurs as coal, graphite, diamond, soot, charcoal, and in organic compounds and carbonates. It was recognized as an element in the

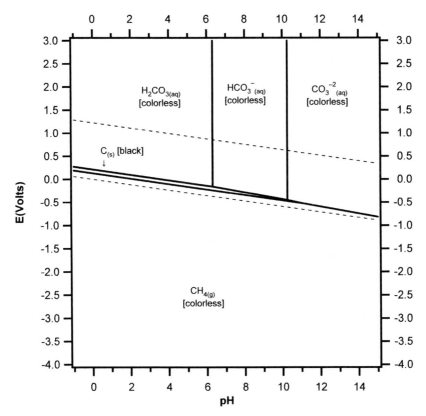

Figure 8.1 E–pH diagram for C species. Soluble species concentrations (except H^+) $= 10^{-1.0}$ M. Gaseous species pressure $= 1.00$ atm. Soluble species and most solids are hydrated. No agents producing complexes or insoluble compounds are present other than HOH and OH^-.

Species ($\Delta G°$ in kJ/mol): C (0.0), $CH_{4(g)}$ (−50.6), H_2CO_3 (−623.3), HCO_3^- (−587.1), CO_3^{-2} (−528.3), HOH (−237.2), H^+ (0.0), and OH^- (−157.3)

Equations for the lines:

C/CH_4	$E = 0.13 - 0.059\,pH - 0.015 \log P_{CH4}$
CO_3^{-2}/CH_4	$E = 0.30 - 0.074\,pH + 0.007 \log [CO_3^{-2}] - 0.007 \log P_{CH4}$
H_2CO_3/C	$E = 0.23 - 0.059\,pH + 0.015 \log [H_2CO_3]$
HCO_3^-/C	$E = 0.32 - 0.074\,pH + 0.015 \log [HCO_3^-]$
CO_3^{-2}/C	$E = 0.48 - 0.089\,pH + 0.015 \log [CO_3^{-2}]$
CO_3^{-2}/HCO_3^-	$pH = 10.3$
HCO_3^-/H_2CO_3	$pH = 6.4$

eighteenth century as numerous chemists realized that it was a basic material which could not be split into simpler materials. In 1789, Antoine Lavoisier published the first classification of elements including C.

d. The element. The most common form of C is graphite, a crystalline form which is shiny gray-black and a number of amorphous forms which are black. C is not attacked by dilute acids or bases, and is inert toward air at room temperature. The element in its graphitic forms is prepared by pyrolysis of coal, coke, petroleum, or pitch at temperatures up to 3000°C. The latter temperature is used to produce crystalline graphite. When burned in an excess of air or O_2, C yields CO_2, when burned in limited air or O_2 the product is CO.

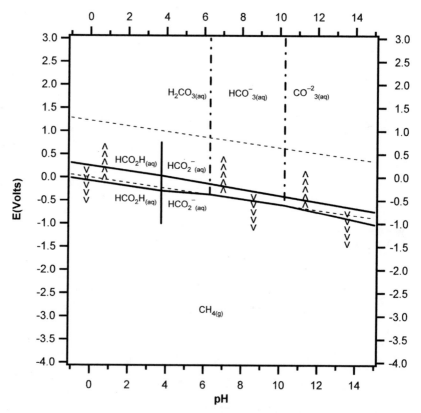

Figure 8.2 E–pH diagram for formate species oxidation and reduction. Soluble species concentrations (except H^+) $= 10^{-1.0}$ M. Gaseous species pressure $= 1.00$ atm. Soluble species and most solids are hydrated. No agents producing complexes or insoluble compounds are present other than HOH and OH^-.

Species ($\Delta G°$ in kJ/mol): $CH_{4(g)}$ (-50.6), H_2CO_3 (-623.3), HCO_3^- (-587.1), CO_3^{-2} (-528.3), HCO_2H (-372.0), HCO_2^- (-350.6), HOH (-237.2), H^+ (0.0), and OH^- (-157.3).

Figure 8.2 (Continued)

Equations for formate species reductions to methane:

HCO_2H/CH_4 $E = 0.26 - 0.059\,pH - 0.010\,\log P_{CH4} + 0.010\,\log [HCO_2H]$

HCO_2^-/CH_4 $E = 0.30 - 0.069\,pH - 0.010\,\log P_{CH4} + 0.010\,\log [HCO_2^-]$

Equations for formate species oxidations to carbonate species:

H_2CO_3/HCO_2H $E = -0.07 - 0.059\,pH$

H_2CO_3/HCO_2^- $E = -0.18 - 0.030\,pH$

HCO_3^-/HCO_2^- $E = 0.00 - 0.059\,pH$

CO_3^{-2}/HCO_2^- $E = 0.31 - 0.089\,pH$

Equations for protonation reactions:

CO_3^{-2}/HCO_3^- $pH = 10.3$

HCO_3^-/H_2CO_3 $pH = 6.4$

HCO_2^-/HCO_2H $pH = 3.8$

e. Oxides and acids.   When CO_2 is dissolved in HOH, carbonic acid H_2CO_3 is produced. And when CO is dissolved in an NaOH solution, sodium formate results. Acidification with a strong acid gives HOOCH. The two other acids are not derived from oxides. However, the less-familiar oxide C_3O_2 yields malonic acid when dissolved in water. Table 8.1 presents a number of species important as inorganic C aqueous species.

f. Carbonic acid and carbonates. H_2CO_3 is obtained as a by-product of H_2 production. In this process CH_4 is treated at high temperature with HOH to yield CO_2 and H_2. The CO_2 is then absorbed in K_2CO_3 solution to give $KHCO_3$ which is then heated to release the purified CO_2. CO_2 may also be made by treating carbonate minerals, such as $CaCO_3$ with acid. Absorption of CO_2 in HOH gives the hypothetical H_2CO_3 which cannot be isolated in a pure form. The amount of H_2CO_3 which exists in solution following the dissolution of CO_2 is actually very small, less than 1%. The most prevalent species is dissolved CO_2. However, for simplification, it will be treated here as the hydrated form H_2CO_3 ($CO_2 \cdot HOH$). The mixture of the two species, represented solely as H_2CO_3, acts as a weak diprotic acid which is colorless, and gives rise to two anions: hydrogen carbonate HCO_3^- and carbonate CO_3^{-2}. The logarithms of the protonation constants ($\log K_p$) relating the three species are 10.3 for CO_3^{-2} and 6.4 for HCO_3^-. In general, alkali metal carbonates are soluble while almost all others are insoluble. The hydrogen carbonates of the alkali metals are usually less soluble than the carbonates, whereas those of other metals tend to be more soluble. Carbonates and hydrogen carbonates dissolve in acids which are stronger than H_2CO_3 to give carbonic acid, which breaks up into CO_2 and HOH. When an alkali metal carbonate is added to HOH, the CO_3^{-2} ion takes up H^+ from the HOH which increases the pH in the solution. This means that some cations, such as Sn^{+2}, Al^{+3}, Cr^{+3}, and Fe^{+3} will precipitate a basic carbonate rather than the carbonate. Many insoluble carbonates, for example those of Ag^+, Cu^{+2}, Cd^{+2}, Ni^{+2}, and Zn^{+2}, will dissolve in excess carbonate due to the formation of complexes. An insoluble carbonate will dissolve in a complexing agent, such as edta^{-4} if the agent attaches more tightly to the cation than the cation attaches to the carbonate in the solid. Predictions of such solubilizations can be made using values of the

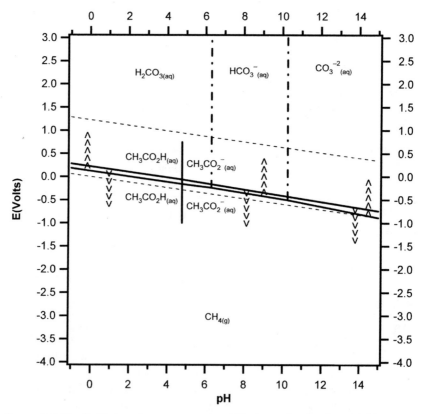

Figure 8.3 E–pH diagram for acetate species oxidation and reduction. Soluble species concentrations (except H^+) $= 10^{-1.0}$ M. Gaseous species pressure $= 1.00$ atm. Soluble species and most solids are hydrated. No agents producing complexes or insoluble compounds are present other than HOH and OH$^-$.

Species ($\Delta G°$ in kJ/mol): $CH_{4(g)}$ (-50.6), H_2CO_3 (-623.3), HCO_3^- (-587.1), CO_3^{-2} (-528.3), CH_3CO_2H (-396.6), $CH_3CO_2^-$ (-369.5), HOH (-237.2), H^+ (0.0), and OH$^-$ (-157.3).

Figure 8.3   (Continued)

Equations for acetate species reductions to methane:

CH_3CO_2H/CH_4 $E = 0.23 - 0.059 \, pH - 0.015 \log P_{CH4} + 0.007 \log [CH_3CO_2H]$

$CH_3CO_2^-/CH_4$ $E = 0.27 - 0.066 \, pH - 0.015 \log P_{CH4} + 0.007 \log [CH_3CO_2^-]$

Equations for acetate species oxidations to carbonate species:

H_2CO_3/CH_3CO_2H $E = 0.13 - 0.059 \, pH + 0.015 \log[H_2CO_3]$
$$- 0.007 \log [CH_3CO_2H]$$

$H_2CO_3/CH_3CO_2^-$ $E = 0.09 - 0.052 \, pH + 0.015 \log [H_2CO_3]$
$$- 0.007 \log [CH_3CO_2^-]$$

$HCO_3^-/CH_3CO_2^-$ $E = 0.19 - 0.066 \, pH + 0.015 \log [HCO_3^-]$
$$- 0.007 \log [CH_3CO_2^-]$$

$CO_3^{-2}/CH_3CO_2^-$ $E = 0.34 - 0.081 \, pH + 0.015 \log [CO_3^{-2}]$
$$- 0.007 \log [CH_3CO_2^-]$$

Equations for protonation reactions:

CO_3^{-2}/HCO_3^- $pH = 10.3$

HCO_3^-/H_2CO_3 $pH = 6.4$

$CH_3CO_2^-/CH_3CO_2H$ $pH = 4.8$

solubility product K_{sp} for the carbonate and the complexation constant β_n of the complex.

g. Formic acid and formates.   Formic acid, HOOCH, is a colorless weak monoprotic acid with a formate ion $OOCH^-$ log K_p value of 3.8. To produce HOOCH, carbonylation of $HOCH_3$ with CO is first carried out to give methyl formate

$$HOCH_3 + CO \rightarrow H_3COOCH,$$

then the H_3COOCH_3 is hydrolyzed to give HOOCH. Careful control of concentrations, temperature, and catalyst must be employed since this reaction can lead to acetic acid under different conditions. Neutralization of the acid with bases leads to the formation of salts. Formate compounds are usually soluble, with those of Pb^{+2}, Ag^+, and Hg_2^{+2} being exceptions. And the salts are white unless the cation is colored. Formic acid is a fairly good reducing agent, the carbonic-acid/formic-acid couple having an $E°$ value of -0.07 v, but kinetic retardation must be borne in mind. Formate is not a very strong complexing agent, showing only moderate strength in a few cations like Th^{+4}, Fe^{+3}, and In^{+3}.

h. Acetic acid and acetates.

Acetic acid $HC_2H_3O_2$ (HOOCCH$_3$) is a colorless weak monoprotic acid, the anion of which has a log K_p of 4.8. It is produced by the carbonylation of HOCH$_3$ with CO.

$$HOCH_3 + CO \rightarrow HOOCCH_3$$

Careful control of concentrations, temperature, and catalyst must be employed since this reaction can lead to methyl formate under different conditions. Neutralization of the acid with bases produces the salts. These compounds are usually soluble with the Ag$^+$ and the Hg$_2^{+2}$ salts being only slightly soluble. Some basic acetates, for example those of Al^{+3} and Fe^{+3}, are insoluble. Most of the acetates are white except those with a colored cation. The acetate ion is

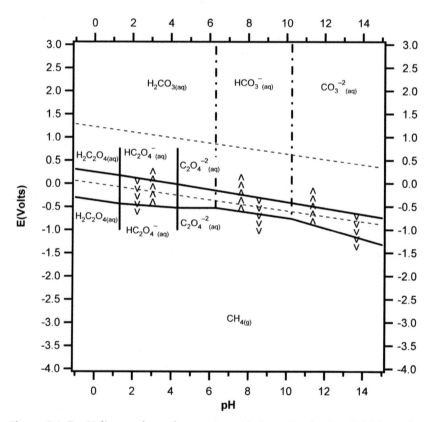

Figure 8.4 E–pH diagram for oxalate species oxidation and reduction. Soluble species concentrations (except H$^+$) $= 10^{-1.0}$ M. Gaseous species pressure $= 1.00$ atm. Soluble species and most solids are hydrated. No agents producing complexes or insoluble compounds are present other than HOH and OH$^-$.

Species ($\Delta G°$ in kJ/mol): CH$_{4(g)}$ (-50.6), H$_2$CO$_3$ (-623.3), HCO$_3^-$ (-587.1), CO$_3^{-2}$ (-528.3), H$_2$C$_2$O$_4$ (-705.4), HC$_2$O$_4^-$ (-698.3), C$_2$O$_4^{-2}$ (-674.0), HOH (-237.2), H$^+$ (0.0), and OH$^-$ (-157.3)

Figure 8.4 (Continued)

Equations for oxalate species reductions to methane:

$H_2C_2O_4/CH_4$ $\quad E = 0.26 - 0.059\ pH - 0.008 \log P_{CH4} + 0.004 \log [H_2C_2O_4]$

$HC_2O_4^-/CH_4$ $\quad E = 0.26 - 0.063\ pH - 0.008 \log P_{CH4} + 0.004 \log [HC_2O_4^-]$

$C_2O_4^{-2}/CH_4$ $\quad E = 0.28 - 0.067\ pH - 0.008 \log P_{CH4} + 0.004 \log [C_2O_4^{-2}]$

Equations for oxalate species oxidations to carbonate species:

$H_2CO_3/H_2C_2O_4$ $\quad E = -0.35 - 0.059\ pH + 0.059 \log [H_2CO_3]$
$$- 0.030 \log [H_2C_2O_4]$$

$H_2CO_3/HC_2O_4^-$ $\quad E = -0.38 - 0.030\ pH + 0.059 \log [H_2CO_3]$
$$- 0.030 \log [HC_2O_4^-]$$

$H_2CO_3/C_2O_4^{-2}$ $\quad E = -0.51 + 0.059 \log [H_2CO_3]$
$$- 0.030 \log [C_2O_4^{-2}]$$

$HCO_3^-/C_2O_4^{-2}$ $\quad E = -0.13 - 0.059\ pH + 0.059 \log [HCO_3^-]$
$$- 0.030 \log [C_2O_4^{-2}]$$

$CO_3^{-2}/C_2O_4^{-2}$ $\quad E = 0.48 - 0.118\ pH + 0.059 \log [CO_3^{-2}]$
$$- 0.030 \log [C_2O_4^{-2}]$$

Equations for protonation reactions:

CO_3^{-2}/HCO_3^- $\quad pH = 10.3$

HCO_3^-/H_2CO_3 $\quad pH = 6.4$

$C_2O_4^{-2}/HC_2O_4^-$ $\quad pH = 4.3$

$HC_2O_4^-/H_2C_2O_4$ $\quad pH = 1.2$

only a moderate complexing agent for some cations, Th^{+4}, Cr^{+3}, Fe^{+3}, In^{+3}, Tl^{+3}, and Sn^{+2} being among the strongest.

i. Oxalic acid and oxalates. Oxalic acid $H_2C_2O_4$ or $(HCOO)_2$ is manufactured by the action of HNO_3 on carbohydrates; for example

$$C_6H_{12}O_6 + 6HNO_3 \rightarrow 3H_2C_2O_4 + 6NO + 6HOH.$$

The acid is a colorless moderately strong diprotic acid which gives anions $C_2O_4^{-2}$ and $HC_2O_4^-$ with log K_p values of 4.3 and 1.2. Neutralization of the acid with bases produces oxalate and hydrogen oxalate salts which are white unless a colored cation is involved. Most of these salts are insoluble except those of the alkali metal cations. The acid and oxalates can be readily oxidized to CO_2 and HOH by MnO_4^- and CrO_4^{-2}, but lesser oxidants may be

Table 8.1
Carbon Species

Name	Formula	State	Color	Solubility	$\Delta G°$ (kJ/mole)
Carbon	C	s	Black	Insol	0.0
Acetate	$C_2H_3O_2^-$	aq	Colorless		−369.5
Acetic acid	$HC_2H_3O_2$	aq	Colorless		−396.6
Carbon dioxide	CO_2	g	Colorless	Sol	−394.5
Carbon monoxide	CO	g	Colorless	Decomp	−137.2
Carbonate	CO_3^{-2}	aq	Colorless		−528.3
Carbonic acid	H_2CO_3	aq	Colorless		−623.3
Hydrogen carbonate	HCO_3^-	aq	Colorless		−587.1
Hydrogen oxalate	$HC_2O_4^-$	aq	Colorless		−698.3
Formate	$OOCH^-$	aq	Colorless		−350.6
Formic acid	HOOCH	aq	Colorless		−372.0
Methane	CH_4	g	Colorless	Sl sol	−50.6
Oxalate	$C_2O_4^{-2}$	aq	Colorless		−674.0
Oxalic acid	$H_2C_2O_4$	aq	Colorless		−705.4

kinetically hindered. The oxalate ion is a bidentate complexing group which attaches fairly strong to many cations. Many insoluble oxalates dissolve in excess $C_2O_4^{-2}$ by virtue of complex formation. Notable among the strongest complexes are those of Sc^{+3}, Th^{+4}, VO^{+2}, UO_2^{+2}, PuO_2^{+2}, Mn^{+3}, Fe^{+3}, Hg^{+2}, Al^{+3}, and Ga^{+3}.

j. Other organic acids.   Among the other organic acids which are important to solution chemistry are tartaric acid and citric acid. Both are good tetradentate complexing agents. Tartaric acid is a weak diprotic acid HOOC–CH(OH)–CH(OH)–COOH with log K_p for the tartrate^{-2} anion equal to 4.3 and the log K_p for the hydrogen tartrate ion Htartrate$^-$ being 3.0. Citric acid is a weak triprotic acid, the anions exhibiting log K_p values of 6.4, 4.1, and 3.1. Neutralization of both acids with bases, and often with oxides and carbonates, usually produces white salts, unless the cation is colored. Tartrates and citrates are generally insoluble, except for those of the alkali metals, and the insoluble compounds often dissolve in excess tartrate or citrate due to complex formation. The formation of tartrate and citrate complexes with many metal ions moves the pH of hydroxide precipitation to higher values.

k. Redox reactions.   As developed above, the major redox consideration for C is to recognize that many thermodynamically predicted reactions (negative ΔG values) go very slowly, that is, they are kinetically hindered. This circumstance accounts for the metastabilities of many of the above entities as Figures 8.1 through 8.4 illustrate.

l. Caution about concentrations.   Take a look at the equation in the legend of Figure 8.3 which describes the H_2CO_3/CH_3CO_2H line. Notice that there are two C atoms in the CH_3CO_2H and only one C atom in H_2CO_3. E–pH diagrams are based on equal numbers of C atoms on both sides. Therefore, to describe the H_2CO_3/CH_3CO_2H line at 0.10 M, one needs to enter the CH_3CO_2H concentration at one-half of the H_2CO_3 concentration. For example, if $[H_2CO_3]$ is entered as 0.10 M, $[CH_3CO_2H]$ should be entered as 0.05 M. The reason is that 0.05 M CH_3CO_2H is 0.10 M in C atoms.

m. Analysis.   Analysis for C involves the conversion of the C-containing materials to CO_2 by various types of oxidizers, then measurement of the CO_2 by a conductometric or infrared detector. The methodolgy has been developed such that 1 ppb can be measured.

n. Health aspects.   The LD50 (oral rat) for $NaC_2H_3O_2$ is 3.5 g/kg, that for Na_2CO_3 is 4.1 g/kg, that for sodium citrate is 7.2 g/kg, that for sodium formate is 11 g/kg, that for sodium oxalate is 375 mg/kg, and that for sodium tartrate is 4.3 g/kg.

3. Silicon (Si) $3s^2 3p^2$

a. E–pH diagram.   The E–pH diagram for $10^{-1.0}$ M Si is presented in Figure 8.5. Its outstanding feature is the prevalence of silicon dioxide SiO_2, indicating a strong tendency of Si and its compounds to go to this substance. According to the diagram, Si is attacked by acids, HOH, and bases to yield SiO_2 over most of the pH range and to give SiO_3^{-2} in the strongly basic region. Experimentally Si is resistant to most acids and HOH, but is transformed by bases with the liberation of H_2. The inactivity toward acids and HOH is due to a resilient oxide coating. The species SiO_3^{-2} is a gross simplification of a very complex situation. In actuality the species represents a large collection of polysilicates plus forms of colloidal hydrated SiO_2. The main species present in dilute solution is probably $SiO_2(OH)_2^{-2}$, but the line between SiO_2 and SiO_3^{-2} is crossed only slowly by simple treatment of SiO_2 with aqueous strong base. The synthesis of soluble SiO_3^{-2} is best carried out by fusion of SiO_2 with excess Na_2CO_3 followed by dissolution in HOH.

An important consideration regarding the Si E–pH diagram is the $\Delta G°$ value which is used for SiO_2. When SiO_2 is prepared in aqueous solution, such as acidification of SiO_3^{-2}, it precipitates as a hydrated amorphous solid (or gel or colloid). As the solid ages, it becomes more crystalline and hence more stable (unless special means are employed to prevent it). That is, the $\Delta G°$ value becomes more negative. The area on the E–pH diagram which SiO_2 occupies increases as its $\Delta G°$ value becomes more negative. In other words, the SiO_2/SiO_3^{-2} line moves to higher pH values. This means that the

SiO_2 requires more base to transform it to SiO_3^{-2}. The $\Delta G°$ used for SiO_2 in the diagram is -785 kJ/mole which corresponds to freshly prepared SiO_2, but crystalline SiO_2 has a $\Delta G°$ of around -825 kJ/mole. Hence freshly prepared SiO_2 dissolves in base more readily (at a lower pH) than aged SiO_2. Similar considerations apply to some other insoluble oxides when they are freshly prepared as compared to their crystalline counterparts.

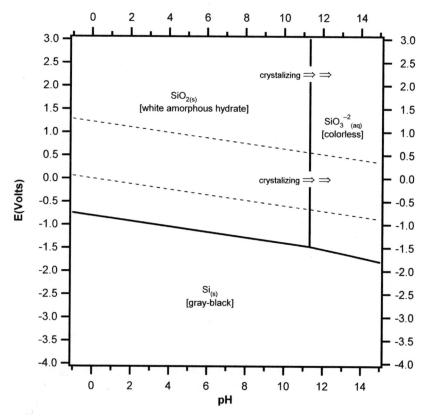

Figure 8.5 E–pH diagram for Si species. Soluble species concentrations (except H^+) $= 10^{-1.0}$ M. Soluble species and most solids are hydrated. No agents producing complexes or insoluble compounds are present other than HOH and OH^-.

Species ($\Delta G°$ in kJ/mol): Si (0.0), SiO_2 (-785.0), SiO_3^{-2} (-887.0), HOH (-237.2), H^+ (0.0), and OH^- (-157.3).

Equations for the lines:

$$SiO_2/Si \qquad E = -0.80 - 0.059\,pH$$

$$SiO_3^{-2}/Si \qquad E = -0.45 - 0.089\,pH + 0.015 \log [SiO_3^{-2}]$$

$$SiO_3^{-2}/SiO_2 \qquad 2\,pH = 23.7 + \log [SiO_3^{-2}]$$

b. Discovery.   The element Si was first isolated by Joseph Louis Gay-Lussac and Louis Jacques Thenard, who reacted SiF_4 with K to obtain a red-brown amorphous solid. In 1824, Jöns Jacob Berzelius repeated the experiment, purified the solid, and definitely identified it as an element. Thomas Thomson in 1831 named the element silicon after the Latin word silex, which means flint.

c. Extraction.   The major sources of Si are the many mineral forms of SiO_2 which range all the way from highly pure quartz to quartz with numerous impurities to silicates. The insolubility and non-reactivity of SiO_2 lead to its ready separation from many of its mineral forms. Purification of SiO_2 may be carried out in numerous ways such as the fusion of the mineral with Na_2CO_3, then dissolution in water, then acidification with concentrated acid to precipitate the SiO_2.

d. The element.   The element Si is prepared by the reduction of SiO_2 with C at high temperature. Very pure Si can be prepared by treating crude Si with Cl_2 to obtain $SiCl_4$, reduction with Mg or Zn, melting the resulting amorphous Si, then purification by zone refining. Si is a gray-black solid which is resistant to most acids and HOH, due to its non-reactive oxide coat, but can be dissolved in a mixture of HNO_3 and HF. The element is attacked by concentrated base to liberate H_2 and SiO_2 and/or SiO_3^{-2}. Si is attacked by air or O_2, but not readily, because of the formation of a protective layer of resistant oxide. Elevated temperature and powdering of the Si enhance the reaction.

e. Oxides and acids.   The main oxide of Si is SiO_2. Silica SiO_2 is attacked by strong alkalis to yield SiO_3^{-2}. Upon treatment with acid, this anion yields SiO_2 in various hydrated forms. Neither of the acids H_2SiO_3 nor H_4SiO_4 has been isolated.

f. Compounds and solubilities.   Table 8.2 lists important aqueous Si species along with pertinent properties. SiO_2 is a highly insoluble compound which undergoes acid attack only with HF. The products are HOH and gaseous SiF_4. The silicates of the alkali metals are soluble, but when other cations are added to the solution, insoluble salts usually result. These precipitates are readily decomposed by acid to yield SiO_2. The tetrahalides of Si, other than SiF_4, undergo hydrolysis producing SiO_2 and the hydrohalic acid. SiF_4 reacts with HOH to give SiF_6^{-2} and SiO_2.

g. Redox reactions.   The E–pH diagram shows that in an aqueous medium the only important states of Si are 0 and IV. Thermodynamically Si should function as a good reducing agent, but it is inert partly due to a silica coat.

Table 8.2
Silicon Species

Name	Formula	State	Color	Solubility	$\Delta G°$ (kJ/mole)
Silicon	Si	s	Gray-black	Insol	0.0
Silicate	SiO_3^{-2}	aq	Colorless		−887.0
hydrogen silicate	$HSiO_3^-$	aq	Colorless		−955.6
Silcon (IV)					
dioxide (amorph)	SiO_2	s	White	Insol	−785.0
dioxide (cryst)	SiO_2	s	White	Insol	−825.0
monoxide	SiO	s	White	Decomp	−386.6
tetraacetate	$Si(C_2H_3O_2)_4$	s	White	Decomp	
tetrabromide	$SiBr_4$	l	Colorless	Decomp	
tetrachloride	$SiCl_4$	l	Colorless	Decomp	−572.8
tetrafluoride	SiF_4	g	Colorless	Decomp	−1574.8
tetraiodide	SiI_4	s	White	Decomp	
tetrathiocyanate	$Si(SCN)_4$	s	White	Decomp	

h. Complexes. The major complex of importance to HOH solutions is SiF_6^{-2} which is made by introducing SiF_4 into HOH or a solution of NaOH, or by dissolving SiO_2 in HF. The equations are

$$3SiF_4 + 2HOH \rightarrow 4H^+ + 2SiF_6^{-2} + SiO_2$$

$$3SiF_4 + 4Na^+ + 4OH^- \rightarrow 4Na^+ + 2SiF_6^{-2} + SiO_2$$

$$SiO_2 + 6HF \rightarrow 2H^+ + SiF_6^{-2} + 2HOH$$

The hexafluorosilicate salts of Na^+, K^+, Rb^+, Cs^+, Ba^{+2}, and Hg_2^{+2} are slightly soluble, and those of Sc^{+3} Y^{+3}, and the lanthanoid ions are insoluble, but those of most other cations are soluble.

i. Analysis. Atomic absorption, emission, and mass spectrographic separation are the most sensitive methods for the analysis of Si. Electrothermal atomization-atomic absorption spectroscopy ETAAS has a sensitivity of 10 ppb, ICPAES 1 ppb, and ICPMS 10 ppb. Colorimetric agents permit spectrometric analysis down to about 10 ppb.

j. Health aspects. Laboratory manipulation of Si and its solid inorganic compounds poses little health problem. Of course, caution must be exercised with halide hydrolysis reactions, especially of SiF_4, and with the caustic property of Na_2SiO_3 solutions. The LD50 (oral rat) of Na_2SiO_3 is 1.2 g/kg.

4. Germanium (Ge) 4s²4p²

a. E–pH diagrams. The E–pH diagram as shown in Figure 8.6 for $10^{-1.0}$ M Ge resembles that for Si in most ways. The white solid dioxide GeO_2 dominates the diagram, oxidation states of 0 and IV are seen, the GeO_2/Ge

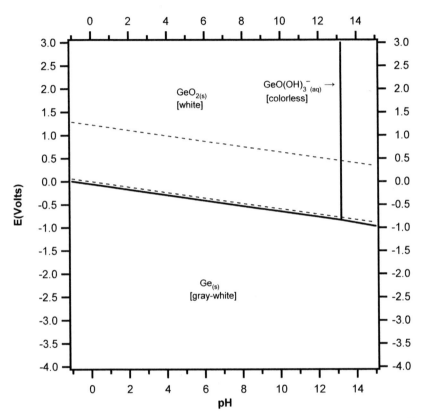

Figure 8.6 E–pH diagram for Ge species. Soluble species concentrations (except H^+) = $10^{-1.0}$ M. Soluble species and most solids are hydrated. No agents producing complexes or insoluble compounds are present other than HOH and OH^-.

Species ($\Delta G°$ in kJ/mol): Ge (0.0), GeO_2 (−495.8), H_2GeO_3 (−724.7), $GeO(OH)_3^-$ (−889.1), HOH (−237.2), H^+ (0.0), and OH^- (−157.3).

Equations for the lines:

GeO_2/Ge $\quad\quad\quad$ $E = -0.06 - 0.059\,pH$

$GeO(OH)_3^-/Ge$ $\quad$ $E = 0.15 - 0.074\,pH + 0.015 \log [GeO(OH)_3^-]$

$GeO(OH)_3^-/GeO_2$ $\quad$ $pH = 14.2 + \log [GeO(OH)_3^-]$

and $GeO(OH)_3^-$/Ge lines are higher than in Si, and the $GeO(OH)_3^-$/GeO_2 line is at a higher pH than in Si. The closeness of the lines involving Ge to the $HOH \equiv H^+/H_2$ line indicates that Ge, if there is a resistant oxide coat, does not dissolve in non-oxidizing acids or HOH or dilute alkali. Such is the case. $GeO(OH)_3^-$ is a simplified formula representing a combination of species such as GeO_3^{-2} and $GeO_2(OH)_2^{-2}$. GeO_2 as made in aqueous medium is the hexagonal crystal form and is soluble to the extent of $10^{-1.4}$ M. The soluble species will be represented as H_2GeO_3. Hence, the predominant species will be solid GeO_2 on the $10^{-1.0}$ M diagram (Figure 8.6) and soluble H_2GeO_3 on the $10^{-2.0}$ M diagram which is shown in Figure 8.7.

In common with SiO_2, freshly prepared GeO_2 dissolves more readily in base than aged GeO_2. This occurs because the $\Delta G°$ of freshly prepared GeO_2 is about -477.8 kJ/mole, whereas that of crystalline GeO_2 is -495.8 kJ/mole. Aged GeO_2 has a value somewhere in between.

b. Discovery. Clemens Alexander Winkler in 1886 isolated Ge from the mineral argyrodite $(Ag_2S)_4GeS_2$. He fused the mineral with Na_2CO_3 and S, dissolved it in HOH, and filtered off the residue of Ag_2S. A large excess of HCl was added to the filtrate and GeS_2 precipitated. Winkler had discovered that GeS_2 was soluble in basic solution, but insoluble in acid. Ge was prepared by heating the GeS_2 in a stream of H_2.

c. Extraction. The major source which is processed for Ge is flue dust obtained from the smelting of Zn ores. This flue dust is treated with H_2SO_4, then the solution is made basic with NaOH, which precipitates impure GeO_2. Concentrated HCl is added to the precipitate, the mixture is heated, and gaseous $GeCl_4$ comes off. The $GeCl_4$ is hydrolyzed to give pure GeO_2, which can be reduced to Ge with H_2 at elevated temperature.

d. The element. Elemental Ge is inert in air, insoluble in HOH, and in dilute acids and bases. The element burns in air or O_2 at high temperatures to give GeO_2. Ge is attacked by concentrated oxidizing acids or mixtures of acids and H_2O_2 to give GeO_2 and H_2GeO_3, and is attacked by hot concentrated strong bases to yield GeO_3^{-2}. Ge generally shows some resistance to reactions because of its oxide coat. The tetrahalides and dihalides may be made by proper control of combinations of the elements.

e. Oxides, acids, and hydroxides. Ge forms two oxides, GeO_2 and the much less stable GeO. GeO_2 shows two crystalline forms: soluble H_2GeO_3 (hexagonal) and insoluble GeO_2 (tetragonal). The insoluble form of GeO_2 is prepared by heating the soluble form to very high temperatures. It is far less reactive than the soluble form. The yellow hydroxide $Ge(OH)_2$ can be prepared by the reduction of GeO_2 dissolved in HCl with hypophosphorus acid $H(H_2PO_2)$, then making the solution basic with NH_4OH. The hydroxide can then be dehydrated in an inert atmosphere to give brown GeO.

f. Compounds and solubilities. Ge species of importance to aqueous chemistry are given in Table 8.3. The GeO$_2$ prepared in aqueous solution is made up of the soluble form, which is assumed to dissolve as follows: GeO$_2$(solid) + HOH $\rightarrow$ H$_2$GeO$_3$(aqueous). GeO$_2$ and H$_2$GeO$_3$ react with strong bases to give GeO$_3^{-2}$ and strong acids increase the solubility of GeO$_2$. White germanates are prepared by dissolution of GeO$_2$ in the several

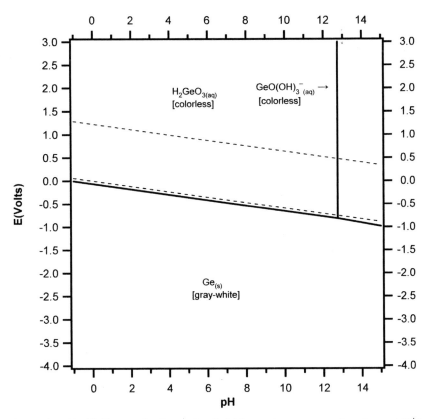

Figure 8.7 E–pH diagram for Ge species. Soluble species concentrations (except H$^+$) = 10$^{-2.0}$ M. Soluble species and most solids are hydrated. No agents producing complexes or insoluble compounds are present other than HOH and OH$^-$.

Species ($\Delta G°$ in kJ/mol): Ge (0.0), GeO$_2$ (−495.8), H$_2$GeO$_3$ (−724.7), GeO(OH)$_3^-$ (−889.1), HOH (−237.2), H$^+$ (0.0), and OH$^-$ (−157.3)

Equations for the lines:

H$_2$GeO$_3$/Ge	E = −0.03 − 0.059 pH + 0.015 log [H$_2$GeO$_3$]
GeO(OH)$_3^-$/Ge	E = 0.15 − 0.074 pH + 0.015 log [GeO(OH)$_3^-$]
GeO(OH)$_3^-$/H$_2$GeO$_3$	pH = 12.8

Table 8.3

Name	Formula	State	Color	Solubility	$\Delta G°$ (kJ/mole)
Germanium	Ge	s	Gray-white	Insol	0.0
Germanate	$GeO(OH)_3^-$	aq	Colorless		−889.1
Germanium(II) ion	Ge^{+2}	aq	Colorless		47.7
Hexafluorogermate(IV)	GeF_6^{-2}	aq	Colorless		
Germanic(IV) acid	H_2GeO_3	aq	Colorless		−724.7
Germanium(II)					
dibromide	$GeBr_2$	s	Yellow	Decomp	
dichloride	$GeCl_2$	s	White	Decomp	
difluoride	GeF_2	s	White	Decomp	
diiodide	GeI_2	s	Orange	Decomp	
hydroxide	$Ge(OH)_2$	s	Yellow	Insol-decomp	−425.1
monoxide	GeO	s	Brown	Insol-decomp	−218.9
sulfide	GeS	s	Red-brown	Decomp	−71.0
Germanium(IV)					
acetate	$Ge(C_2H_3O_2)_4$	s	White	Decomp	
dioxide hydrated	GeO_2	s	White	Sl sol	−480.0
dioxide	GeO_2	s	White	Insol	−495.8
sulfide	GeS_2	s	White	Sol	−152.5
tetrabromide	$GeBr_4$	s	White	Decomp	−323.0
tetrachloride	$GeCl_4$	l	Colorless	Decomp	−498.0
tetrafluoride	GeF_4	g	Colorless	Decomp	−1149.9
tetraiodide	GeI_4	s	Orange	Decomp	−171.0

strong bases. When GeO_2 is heated with concentrated HCl, the volatile $GeCl_4$ is given off. The treatment of GeO_2 with oxalic acid results in dissolution due to the complex $Ge(C_2O_4)_3^{-2}$. When GeO_2 is dissolved in concentrated HF, the complex GeF_6^{-2} results. The monoxide GeO dissolves in dilute acids to give Ge^{+2} and in dilute bases to produce GeO_2^{-2}, all three entities being unstable in water. All the di- and tetrahalides are subject to hydrolysis.

g. Redox reactions. Ge would be a fairly good reducing agent except for its passivity. The theoretical E° value for GeO/Ge is 0.10 v, and that for Ge^{+2}/Ge is 0.25 v. The II oxidation state compounds of Ge are unstable with respect to the IV oxidation state, and hence do not appear on the E–pH diagram. The E° values for several couples illustrate this: GeO_2/GeO (−0.21 v), GeO_2/Ge^{+2} (−0.36 v), GeO_2/GeO_2^{-2} (−0.57 v), GeO_2/$Ge(OH)_2$ (−0.37 v). Kinetic retardation, however, permits them to exist for sometime in solution.

h. Complexes.   Complexes of Ge(II) bring about some stabilization of the state. These complexes include ones with Cl^- ($GeCl_3^-$) and tartrate^{-2}. Ge(IV) complexes include GeF_6^{-2}, $GeCl_6^{-2}$, $Ge(C_2O_4)_3^{-2}$, and ones with polyhydric alcohol compounds.

i. Analysis.   The most sensitive techniques for the analysis of Ge include ICPAES (1 ppb) and ICPMS (0.1 ppb). Colorimetric agents (phenylfluoronone) and molecular absorption spectroscopy can detect 100 ppb. One of the better spot tests involves treatment of a slightly basic germanate solution with phenolphthalein to produce a red color, then the addition of mannitol which turns the solution colorless. Sensitivity is 100 ppm. A major interference is B.

j. Health aspects.   Laboratory operations with Ge and solid Ge compounds pose few health problems. However, caution must be exercised with hydrolysis and volatilities of Ge halides. LD50 (oral rat) values for Ge compounds range from 500 to 5000 mg/kg.

5. Tin (Sn) $5s^2 5p^2$

a. E–pH diagram.   Figure 8.8 is the E–pH diagram for $10^{-1.0}$ M Sn. Oxidation states of 0, II, and IV are shown. The designations Sn^{+2} and Sn^{+4} are simplifications, a number of hydrolytic and polymeric species being more accurate. It is to be further noted that the $\Delta G°$ value for the freshly prepared, amorphous, hydrated SnO_2 has been employed in constructing the diagram. As the SnO_2 ages and crystallization sets in, the $\Delta G°$ value becomes more negative and all three lines making up the borders of SnO_2 expand. The expansion is such for the crystalline mineral SnO_2 that the lines almost eliminate the Sn^{+2} and the SnO species. Note that the diagram indicates that both the oxides are amphoteric.

b. Discovery.   Sn is one of the oldest metals known, probably due to the reaction between the mineral cassiterite (SnO_2) and hot C (charcoal) to liberate the metal. Its symbol Sn comes from the Latin word stannum. The name tin is found in Middle English as an Anglo-Saxon term which is probably akin to the German word zinn.

c. Extraction.   The major source of Sn is the mineral cassiterite which is almost pure SnO_2 (major impurity often Fe). Its extraction is essentially the same as in ancient times, namely, the reduction with C. Refining of the Sn so obtained is done with several processes, including melting, then bubbling air through the melt to oxidize the iron which often accompanies the initial reduction product.

d. The element. Sn is stable in air and HOH. The metal has some tendency to be passive, but it dissolves in dilute acids to give Sn^{+2} and H_2 and in concentrated base to yield $HSnO_2^-$ and H_2. The products of these dissolutions are subject to oxidation by the air. Sn burns in air to SnO_2.

e. Oxides. Sn shows two oxides, SnO and SnO_2. The Sn(II) compound is prepared by the dissolution of Sn in acid to provide Sn^{+2}, then the precipitation of SnO by adding base. The hydrated amorphous oxide which first comes down is white, but it changes over to black. SnO_2 can be made by treating Sn with HNO_3 or by burning the metal in air. SnO is soluble in both acids and bases giving Sn^{+2} salts and tinate(II) compounds. SnO_2 is resistant to both acids and bases when it is in the crystalline form, but

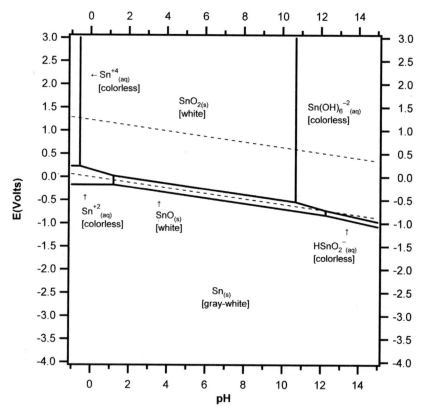

Figure 8.8 E–pH diagram for Sn species. Soluble species concentrations (except H^+) $= 10^{-1.0}$ M. Soluble species and most solids are hydrated. No agents producing complexes or insoluble compounds are present other than HOH and OH^-.

Species ($\Delta G°$ in kJ/mol): Sn (0.0), Sn^{+2} (−27.6), Sn^{+4} (16.7), SnO (−254.0), SnO_2 (−476.1), $Sn(OH)_6^{-2}$ (−1297.0), $HSnO_2^-$ (−418.1), HOH (−237.2), H^+ (0.0), and OH^- (−157.3)

Figure 8.8 (Continued)

Equations for the lines:

Sn^{+2}/Sn	$E = -0.14 + 0.030 \log [Sn^{+2}]$
SnO/Sn	$E = -0.09 - 0.059 \, pH$
$HSnO_2^-/Sn$	$E = 0.29 - 0.089 \, pH + 0.030 \log [HSnO_2^-]$
Sn^{+4}/Sn^{+2}	$E = 0.23$
SnO_2/Sn^{+2}	$E = 0.13 - 0.118 \, pH - 0.030 \log [Sn^{+2}]$
SnO_2/SnO	$E = 0.08 - 0.059 \, pH$
$Sn(OH)_6^{-2}/SnO$	$E = 0.74 - 0.118 \, pH + 0.030 \log [Sn(OH)_6^{-2}]$
$Sn(OH)_6^{-2}/HSnO_2^-$	$E = 0.36 - 0.089 \, pH$
$Sn(OH)_6^{-2}/SnO_2$	$2 \, pH = 22.4 + \log [Sn(OH)_6^{-2}]$
SnO_2/Sn^{+4}	$4 \, pH = -3.2 - \log [Sn^{+4}]$
$HSnO_2^-/SnO$	$pH = 12.8 + \log [HSnO_2^-]$
SnO/Sn^{+2}	$2 \, pH = 1.9 - \log [Sn^{+2}]$

the amorphous form (shown in the E-pH diagram) is subject to attack by both, especially by base. Neither of the hydroxides is known, all forms being hydrated oxides.

f. Compounds and solubilities. Table 8.4 is a presentation of Sn compounds and ions which are pertinent to the aqueous chemistry. Sn forms two sulfides, brown SnS and yellow SnS_2, both of which dissolve in concentrated acid. Excess S^{-2} oxidizes SnS to SnS_2 and reacts with SnS_2 to give SnS_3^{-2}. When Sn^{+2} salts in solution are reacted with OH^- there is a tendency to precipitate basic salts which contain both OH^- and the anion of the salt, for example $Sn(OH)Cl$. All Sn(IV) compounds hydrolyze readily, except for those which are insoluble. Alkali metal tinate(II) and tinate(IV) compounds are soluble, but those with most other cations are not.

g. Redox reactions. Sn would be a good reducing agent, but its capability is somewhat moderated by its passivity. Sn does reduce ions of Ag^+, Hg^{+2}, Bi^{+3}, and Cu^{+2} to the metal. Sn^{+2} can be produced by reacting Sn with Sn^{+4}. And Sn^{+2} can be stabilized against air oxidation to Sn^{+4} by keeping metallic Sn in the solution. Sn^{+2} solutions are effective as mild reductants in a number of reactions, for example, Hg^{+2} in Cl^- is first reduced to white Hg_2Cl_2, then to black Hg.

Table 8.4
Tin Species

Name	Formula	State	Color	Solubility	$\Delta G°$ (kJ/mole)
Tin	Sn	s	Gray-white	Insol	0.0
Tin(II) ion	Sn^{+2}	aq	Colorless		−27.6
Tin(IV) ion	Sn^{+4}	aq	Colorless		16.7
Hydrogen tinate(II)	$HSnO_2^-$	aq	Colorless		−418.1
Hexahydroxotinate(IV)	$Sn(OH)_6^{-2}$	aq	Colorless		−1297.0
Tin(II)					
acetate	$Sn(C_2H_3O_2)_2$	s	White	Decomp	
bromide	$SnBr_2$	s	White	Sol	−248.9
chloride	$SnCl_2$	s	White	Sol	−302.1
chromate	$SnCrO_4$	s	Yellow	Insol	
cyanide	$Sn(CN)_2$	s	White	Insol	
fluoride	SnF_2	s	White	Sol	
formate	$Sn(OOCH)_2$	s	White	Decomp	
iodide	SnI_2	s	Red	Sol	−143.9
oxalate	SnC_2O_4	s	White	Insol	
oxide	SnO	s	Brown	Insol	−256.9
oxide (hydrated)	SnO	s	White	Insol	−254.0
perchlorate	$Sn(ClO_4)_2$	s	White	Sol	
phosphate	$Sn_3(PO_4)_2$	s	White	Insol	−2540.6
sulfate	$SnSO_4$	s	White	Sol	−1056.0
sulfide	SnS	s	Brown-black	Insol	−108.4
Tin(IV)					
acetate	$Sn(C_2H_3O_2)_4$	s	White	Decomp	
bromide	$SnBr_4$	s	White	Decomp	−456.1
chloride	$SnCl_4$	l	Colorless	Decomp	−474.0
chromate	$Sn(CrO_4)_2$	s	Yellow	Decomp	
fluoride	SnF_4	s	White	Decomp	
formate	$Sn(OOCH)_4$	s	White	Decomp	
iodide	SnI_4	s	Red	Decomp	
nitrate	$Sn(NO_3)_4$	s	White	Decomp	
oxide (amorph)	SnO_2	s	White	Insol	−476.1
oxide (cryst)	SnO_2	s	White	Insol	−520.1
sulfate	$Sn(SO_4)_2$	s	White	Decomp	−1451.0
sulfide	SnS_2	s	Yellow	Insol	−179.5

h. Complexes. Complexes of Sn(II) include $SnCl_4^{-2}$, SnF_3^-, $Sn(C_2H_3O_2)_3^-$, $Sn(tartrate)_2^{-2}$, and $Sn(edta)^{-2}$. Those of Sn(IV) include SnX_6^{-2} with $X = F$, Cl, Br, I, $Sn(C_2O_4)_3^{-2}$, and SnS_3^{-2}. Complexation brings about stabilization of the involved Sn oxidation state. Investigations on numerous complexes of Sn have provided these values (log β_1, log β_2, log β_3, log β_4) for Sn^{+2}: $P_2O_7^{-4}$ (14.0, 16.4), OH^- (10.6, 20.9), F^- (4.1, 6.7, 9.5), Cl^- (1.5, 2.3, 2.0, 1.5), Br^- (1.2, 1.7, 1.2, 0.4), I^- (0.7, 1.1, 2.1, 2.3),

$C_2H_3O_2^-$ (3.3, 6.0, 7.3), tartrate^{-2} (5.2, 9.9), edta^{-4} (18.3); for Sn^{+4}: OH$^-$ (14.0, 28.0, 42.0, 54.0), F$^-$ (log $\beta_6 = 25.0$).

i. Analysis. Colorimetry with proper reagents (such as nitrophenylfluo-ronone) permits analysis down to about 100 ppb. ETAAS detects Sn down to 1 ppb, and ICPMS is effective down to 0.1 ppb, as is anodic stripping voltammetry. Spot testing involves the use of cacotheline or diazine green (a dye made by reacting diazotized safranine with dimethylaniline). Sensitivity of these is about 50 ppm.

j. Health aspects. Organic compounds of Sn are highly toxic, but inor-ganic compounds tend to be less so. Part of the reason for this is that they are poorly resorbed in animals and humans. The LD50 (oral rat) for $SnCl_2$ is 700 mg/kg.

6. Lead (Pb) $6s^2 6p^2$

a. E–pH diagram. The E–pH diagram for $10^{-1.0}$ M Pb is presented in Figure 8.9. In the acidic region, there is Pb^{+2} ion, but as the pH is raised, the Pb(II) species become hydrolytic variations of Pb(OH)$^+$, one of the chief ones probably being Pb$_4$(OH)$_4^{+4}$. In the extreme basic region for Pb(IV) there is evidence for the Pb(OH)$_6^{-2}$ anion, but it has not been included in the diagram because the anion is best produced by fusion of PbO$_2$ with alkali hydroxides. Bases greater than 10.0 M would be needed to make it.

b. Discovery. Pb, as Sn, dates back into antiquity. This is due to its occurrence in relatively pure form as galena, PbS, and to its simple preparation by roasting the galena in air (2PbO + PbS → 3Pb + SO$_2$).

c. Extraction. Galena is the major source of Pb, although there are other ores which are not as rich in the element. The roasting process is still employed for the production of Pb. If a limited supply of air is employed in the process, the PbS is oxidized to PbO. Limestone CaCO$_3$ (acting as a flux) and coke C are added, the mixture is smelted, and the reduction product Pb results. An alternate method is to roast the ore to the oxide, then to add unroasted PbS, then to heat again PbS + 2PbO → 3Pb + SO$_2$. Impurities are removed by controlled oxidation, with final refining by electrolysis.

d. The element. Pb is a gray metallic solid with a slight bluish tinge. It has some tendency to be passive due to an oxide coat which has been produced in moist air and its overvoltage for H$_2$ liberation. The result is that the metal is not attacked further by air or HOH. However air-saturated HOH does slowly react. Pb is not soluble in dilute non-oxidizing acids, but is attacked by dilute HNO$_3$ or hot, concentrated H$_2$SO$_4$ to give the Pb^{+2} ion.

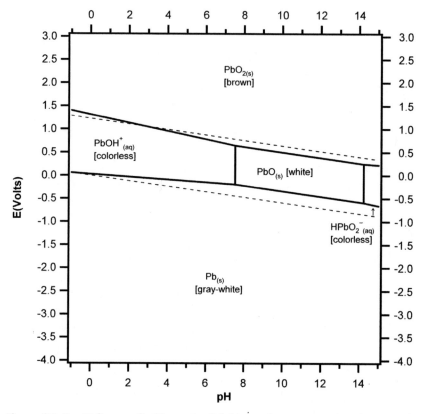

Figure 8.9 E–pH diagram for Pb species. Soluble species concentrations (except H$^+$) = $10^{-1.0}$ M. Soluble species and most solids are hydrated. No agents producing complexes or insoluble compounds are present other than HOH and OH$^-$.

Species ($\Delta G°$ in kJ/mol): Pb (0.0), Pb^{+2} (−24.7), Pb(OH)$_2$ (−421.3), PbO$_2$ (−215.5), HPbO$_2^-$ (−338.9), HOH (−237.2), H$^+$ (0.0), and OH$^-$ (−157.3).

Equations for the lines:

Pb^{+2}/Pb	E = −0.13 + 0.030 log [Pb^{+2}]
Pb(OH)$_2$/Pb	E = 0.28 − 0.059 pH
HPbO$_2^-$/Pb	E = 0.70 − 0.089 pH + 0.030 log [HPbO$_2^-$]
PbO$_2$/Pb^{+2}	E = 1.47 − 0.118 pH − 0.030 log [Pb^{+2}]
PbO$_2$/Pb(OH)$_2$	E = 1.07 − 0.059 pH
PbO$_2$/HPbO$_2^-$	E = 0.64 − 0.030 pH − 0.030 log [HPbO$_2^-$]
HPbO$_2^-$/Pb(OH)$_2$	pH = 14.5 + log [HPbO$_2^-$]
Pb(OH)$_2$/Pb^{+2}	2 pH = 13.7 − log [Pb^{+2}]

Strong concentrated bases slowly react with Pb giving H_2 and $HPbO_2^-$. Upon heating in air, Pb yields PbO. The reactivity of Pb to certain reagents is enhanced by powdering it. The metal is subject to developing thin non-reactive layers of carbonate, basic carbonate, sulfide, and sulfate, as well as oxide, all of which increase its inertness.

e. Oxides. Four oxides of Pb are important: PbO, Pb_3O_4, Pb_2O_3, and PbO_2. The two intermediate oxides may be viewed as being made up of PbO and PbO_2, such as $2PbO + PbO_2 = Pb_3O_4$ and $PbO + PbO_2 = Pb_2O_3$. That is, they do not represent other oxidation states of Pb, but they contain Pb(II) and Pb(IV) in different combinations. PbO is obtained by heating Pb in air. This compound is amphoteric as shown in the E–pH diagram. Pb_3O_4 is produced by heating PbO in air. The dissolution of this Pb_3O_4 in HNO_3 gives PbO_2, or it may be prepared by treating a Pb^{+2} solution with ClO^-. This dioxide is resilient to attack by either acids or bases, but is subject to attack by reducing agents. The oxide Pb_2O_3 can be prepared by heating PbO under O_2 pressure.

f. Compounds and solubilities. Listed in Table 8.5 are the Pb entities which are of importance in aqueous chemistry. Notice that there are three forms of PbO and two of PbO_2. The $\Delta G°$ values show a larger range for PbO, which means that some forms of PbO differ chemically from others to a greater

Table 8.5
Lead Species

Name	Formula	State	Color	Solubility	$\Delta G°$ (kJ/mole)
Lead	Pb	s	Gray-white	Insol	0.0
Lead(II) ion	Pb^{+2}	aq	Colorless		-24.7
Hydrogen leadate(II)	$HPbO_2^-$	aq	Colorless		-338.9
Lead(II)					
acetate	$Pb(C_2H_3O_2)_2$	s	White	Sol	
arsenate	$Pb_3(AsO_4)_2$	s	White	Insol	
basic carbonate	$Pb_3(OH)_2(CO_3)_2$	s	White	Insol	
borate	$Pb(BO_2)_2$	s	White	Insol	
bromate	$Pb(BrO_3)_2$	s	White	Insol	
bromide	$PbBr_2$	s	White	Insol	-260.4
carbonate	$PbCO_3$	s	White	Insol	-625.9
chlorate	$Pb(ClO_3)_2$	s	White	Sol	
chloride	$PbCl_2$	s	White	Insol	-314.0
chlorite	$Pb(ClO_2)_2$	s	White	Insol	
chromate	$PbCrO_4$	s	Yellow	Insol	-851.9
cyanate	$Pb(CNO)_2$	s	White	Sl sol	
cyanide	$Pb(CN)_2$	s	White	Insol	

Table 8.5
(Continued)

Name	Formula	State	Color	Solubility	$\Delta G°$ (kJ/mole)
Lead(II)—*cont'd*					
fluoride	PbF_2	s	White	Insol	−619.6
formate	$Pb(OOCH)_2$	s	White	Sol	
hydrogen phosphate	$PbHPO_4$	s	White	Insol	−1182.7
hydroxide	$Pb(OH)_2$	s	White	Insol	−421.3
hydroxoacetate	$PbOH(C_2H_3O_2)_2$	s	White	Sol	
hydroxochloride	$Pb(OH)Cl$	s	White	Insol	
hydroxonitrate	$PbOH(NO_3)$	s	White	Sol	−303.7
iodate	$Pb(IO_3)_2$	s	White	Insol	
iodide	PbI_2	s	Yellow	Insol	
metavanadate	$Pb(VO_3)_2$	s	White	Sl sol	
molybdate	$PbMoO_4$	s	Yellow	Insol	−969.4
nitrate	$Pb(NO_3)_2$	s	White	Sol	−252.3
oxalate	PbC_2O_4	s	White	Insol	−754.4
oxide	PbO	s	Yellow	Insol	−188.5
oxide	PbO	s	Red	Insol	−189.3
oxide (amorphous)	PbO	s	White	Insol	−183.7
perchlorate	$Pb(ClO_4)_2$	s	White	Sol	
phosphate	$Pb_3(PO_4)_2$	s	White	Insol	−2371.6
pyrophosphate	$Pb_2P_2O_7$	s	White	Insol	
sulfate	$PbSO_4$	s	White	Insol	−813.8
sulfide	PbS	s	Black	Insol	−98.8
thiocyanate	$Pb(CNS)_2$	s	White	Insol	
tungstate	$PbWO_4$	s	White	Insol	−1050.0
Lead(II,II,IV)					
oxide	Pb_3O_4	s	Red	Insol	−616.2
Lead(II,IV)					
oxide	Pb_2O_3	s	Orange	Insol	−411.8
Lead(IV)					
acetate	$Pb(C_2H_3O_2)_4$	s	White	Decomp	
chloride	$PbCl_4$	l	Yellow	Decomp	
fluoride	PbF_4	s	White	Decomp	−745.2
oxide (amorph)	PbO_2	s	Brown	Insol	−215.5
oxide (cryst)	PbO_2	s	Brown	Insol	−219.0
sulfate	$Pb(SO_4)_2$	s	White	Decomp	

extent than do the PbO_2 forms. PbO dissolves in both acids and bases to give Pb^{+2} and $HPbO_2^{-}$, respectively. The leadate(II) salts of the alkali metals are soluble, but those of most other cations are not. As the diagram shows, PbO_2 is non-reactive toward acids and bases, but it is a strong oxidizing agent. Most compounds of Pb(IV) hydrolyze readily to PbO_2, the insolubility of this oxide accounting for its stability.

g. Redox reactions. Pb should be quite active toward acids, but it yields only to oxidizing acids or non-oxidizing acids plus an oxidant, such as O_2 or H_2O_2. PbO_2 is a strong oxidizing agent.

h. Complexes. Pb(II) forms numerous complexes, many of them being indicated here by their complexation constants (log β_1, log β_2, log β_3, log β_4): $B(OH)_4^-$ (5.2, 8.7, 11.2), CO_3^{-2} (log $\beta_2 = 9.1$), CN^- (log $\beta_4 = 10.3$), NO_3^- (1.2, 1.4), $P_2O_7^{-4}$ (7.3, 10.2), OH^- (6.3, 10.9, 13.9), SO_4^{-2} (2.8, 2.0), $S_2O_3^{-2}$ (2.4, 4.9, 6.2, 6.1), F^- (1.4, 2.5), Cl^- (1.6, 1.8, 1.7, 1.4), Br^- (1.8, 2.6, 3.0, 2.3), I^- (1.3, 2.8, 3.4, 3.9), ethylenediamine (7.0, 8.5), oxinate$^-$ (9.0), $HCOO^-$ (1.2, 2.0, 1.8), $C_2O_4^{-2}$ (3.3, 5.5), $C_2H_3O_2^-$ (2.7, 4.1, 3.6, 2.9), malonate^{-2} (2.6, 3.6, 4.3), lactate$^-$ (2.0, 2.8, 4.3), maleate^{-2} (2.8, 4.0, 4.4), succinate^{-2} (2.4, 3.7, 4.1), tartrate^{-2} (2.6, 4.0), citrate^{-3} (4.3, 6.1, 7.0), benzoate$^-$ (2.0, 3.3), aminoacetate$^-$ (5.5, 8.9), iminodiacetate^{-2} (7.4), nta^{-3} (11.4), edta^{-4} (18.0), pyridine-2,6-dicarboxylate^{-2} (8.7, 11.6).

i. Analysis. Colorimetric analysis of Pb with dithizone can be performed down to 200 ppb, with ETAAS and ICPAES going down to 1 ppb, and ICPMS to 0.1 ppb. Spot testing to a detection limit of almost 1 ppm can be carried out using dithizone and KCN.

j. Health aspects. Pb is exceptionally toxic as the element and in all of its compounds, particularly those that are soluble or gaseous. The values of LD50 given for lead compounds can be misleading, since far less Pb than that for LD50 will produce marked symptoms of poisoning. It is generally agreed that greater than 0.4 mg/L in the blood stream of an adult will give indications of poisoning. The ingestion of about 100 mg of a Pb compound could provide this level for a short time. The level will subsequently drop as Pb distributes to other tissues in the body and is excreted.

9

The Nitrogen Group

1. Introduction

The Nitrogen Group of the Periodic Table contains the elements nitrogen N, phosphorus P, arsenic As, antimony Sb, and bismuth Bi. The outer electron structure ns^2np^3 characterizes all five of the elements, with n representing principal quantum numbers 2, 3, 4, 5, and 6, respectively. The ns^2np^3 indicates the possibility of oxidation states V, III, and -III. As one goes down the group, the metallic character increases, with N and P being distinctly non-metals, As a metalloid, and Sb and Bi metals. However, the major bonding in most of the compounds of the group is covalent, aqueous cationic species being formed only by Sb and Bi. A covalency of 5 is exhibited by all the elements except N, this being assignable to the considerable energy required to place 10 electrons around the atom. The pentavalent state is the most stable for P, with its stability falling off down the group, as the trivalent state stability increases. Covalent radii in pm are as follows: N (75), P (110), As(122), and Sb(143). Ionic radii (most hypothetical) in pm are these: Sb^{+3} (90), Sb^{+5} (74), Bi^{+3} (117), and Bi^{+5} (90).

2. Nitrogen (N) $2s^22p^3$

a. E–pH diagram. Figure 9.1 depicts the E–pH diagram for N with the soluble species (except H^+) at $10^{-1.0}$ M. Equations for the lines that separate

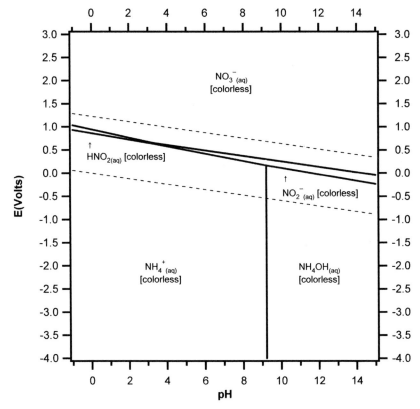

Figure 9.1 E–pH diagram for N species. Soluble species concentrations (except H^+) = $10^{-1.0}$ M. Soluble species and most solids are hydrated. No agents producing complexes or insoluble compounds are present other than HOH and OH^-.

Species ($\Delta G°$ in kJ/mol): N_2 (0.0), HNO_3 (−103.4), NO_3^- (−110.9), HNO_2 (−55.7), NO_2^- (−36.8), NH_4^+ (−79.5), NH_4OH (−264.0), HOH (−237.2), H^+ (0.0), and OH^- (−157.3)

Equations for the lines:

$$HNO_2/NH_4^+ \qquad E = 0.86 - 0.069\,pH$$

$$NO_2^-/NH_4^+ \qquad E = 0.89 - 0.079\,pH$$

$$NO_2^-/NH_4OH \qquad E = 0.80 - 0.069\,pH$$

$$NO_3^-/HNO_2 \qquad E = 0.94 - 0.089\,pH$$

$$NO_3^-/NO_2^- \qquad E = 0.85 - 0.059\,pH$$

$$NO_2^-/HNO_2 \qquad pH = 3.2$$

$$NH_4OH/NH_4^+ \qquad pH = 9.2$$

the species are displayed in the legend. The colorless strong acid nitric acid HNO_3, its colorless anion nitrate NO_3^-, the colorless weak acid nitrous acid HNO_2, its colorless anion NO_2^-, the colorless ammonium ion NH_4^+, and the colorless hypothetical compound ammonium hydroxide NH_4OH are involved. It is known that NH_4OH is actually hydrated NH_3, but for convenience, the formula NH_4OH will be used to reflect its basic character. Many reactions of N_2 and compounds of N are controlled primarily by kinetics, and hence conclusions drawn from thermodynamically based E–pH diagrams are likely not to be in accord with experiment. Taking this into account, N_2 has not been introduced into Figure 9.1, and even though N_2 may be liberated in certain aqueous reactions, the oxidation or reduction of N_2 in HOH is kinetically hindered. In spite of these reservations, the diagram presented here can be used to rationalize and predict many reactions of the represented species.

b. Discovery and occurrence. Nitrogen was discovered by Daniel Rutherford in 1772, when he burned a candle in confined air, bubbled the resulting gas through KOH solution, and identified the residual gas as a pure substance which will not support combustion or life. He termed it noxious air. Two other investigators, at about the same time, Carl Wilhelm Scheele and Henry Cavendish, isolated the same substance using similar methods. The name nitrogen was proposed by Jean Antoine Claude Chaptal in 1790. This naming was based upon the recognition that N was a constituent of nitrates (Greek nitron and Greek genes: nitrate former). Antoine Lavoisier called it azote (Greek azotikos: no life), and the Germans Stickstoff (German sticken: to suffocate). The major sources of N are the atmosphere (78% by volume) and the minerals KNO_3 (nitre or saltpeter) and $NaNO_3$ (sodanitre or Chile saltpeter).

c. The element. Nitrogen N_2 is a colorless, odorless gas which is quite inert at room temperature, which does not react with HOH, and whose compounds are largely covalent. N_2 is generally prepared by liquefaction of air, and then fractional distillation. At higher temperatures, N_2 has an increased reactivity. This permits the production of NH_3 which is the basis for the preparation of most other N compounds. The reaction is $N_2 + 3H_2 \rightarrow 2NH_3$ carried out at moderately high temperature, high pressure, and with an Fe or Ru catalyst. The NH_3 may then be converted to NH_4^+ salts by neutralization, and to nitrites, nitrates, and HNO_3 by oxidation.

d. Oxides. Numerous oxides are formed by N including N_2O, NO, N_2O_3, NO_2 and its dimer N_2O_4, and N_2O_5. Of chief pertinence to aqueous chemistry are the colorless gas nitrogen oxide NO and the brown gas NO_2. NO is produced commercially by the oxidation of NH_3 at moderately high temperature with a Pt catalyst. NO is very reactive combining rapidly with O_2 to form NO_2. When NO_2 is dissolved in HOH, the main

Table 9.1
Nitrogen Species

Name	Formula	State	Color	Solubility	$\Delta G°$ (kJ/mole)
Nitrogen	N_2	g	Colorless	Insol	0.0
Ammonium ion	NH_4^+	aq	Colorless		−79.5
Cyanate ion	OCN^-	aq	Colorless		−98.7
Cyanide ion	CN^-	aq	Colorless		166.0
Nitrate ion	NO_3^-	aq	Colorless		−110.9
Nitrite ion	NO_2^-	aq	Colorless		−36.8
Thiocyanate ion	SCN^-	aq	Colorless		88.7
Ammonia	NH_3	g	Colorless	Sol	−16.5
Ammonium hydroxide	NH_4OH	aq	Colorless	Sol	−264.0
Cyanic acid	$HOCN$	aq	Colorless	Sol	−121.0
Dintrogen monoxide	N_2O	g	Colorless	Decomp	104.2
Dinitrogen trioxide	N_2O_3	g	Brown	Decomp	139.4
Dintrogen pentoxide	N_2O_5	s	White	Decomp	114.0
Hydrocyanic acid	HCN	aq	Colorless	Sol	112.0
Nitric acid	HNO_3	aq	Colorless	Sol	−103.4
Nitrogen dioxide	NO_2	g	Red-brown	Decomp	51.3
Nitrogen monoxide	NO	g	Colorless	Insol	86.6
Nitrous acid	HNO_2	aq	Colorless	Sol	−55.7
Thiocyanic acid	$HSCN$	aq	Colorless	Sol	83.6

reaction is: $3NO_2 + HOH \rightarrow 2H^+ + 2NO_3^- + NO$. Table 9.1 presents the more important N species for aqueous chemistry.

e. Nitric acid and nitrates. Colorless HNO_3 is prepared by the oxidation of NH_3 to NO, then the oxidation of NO to NO_2, then the dissolution of the NO_2 in HOH in accordance with the equation given in the previous paragraph. The acid is very strong, the colorless NO_3^- anion having an estimated log K_p of approximately −1.3. As the concentration of the acid increases, so does its oxidizing power. Neutralizations of HNO_3 with metal hydroxides, oxides, and carbonates give the corresponding nitrate salts, most of which are soluble. The attack of HNO_3 on metals is usually a redox phenomenon which results in the reduction of the acid to NO_2, NO, N_2, and/or NH_3 (NH_4^+) and the metal nitrate or the oxide (with As, Sb, Sn, W, Mo). Metals which are not attacked include those with high $E°$ values (Rh, Ir, Pt, Au) and a few which are sometimes rendered passive (Al, Cr, Fe, Cu).

f. Nitrites. Salts of nitrous acid HNO_2 are usually prepared by heating nitrates with a reducing agent such as C or Fe, for example, $2NaNO_3 + C \rightarrow 2NaNO_2 + CO_2$. Nitrites of alkali metals can also be made by bubbling a mixture of NO and NO_2 into a solution of the hydroxide. Acidification of a solution of a soluble nitrite gives colorless nitrous acid HNO_2 in solution, but the acid itself has not been isolated. HNO_2 is a weak unstable acid subject to decomposition into NO and NO_2. The log K_p of the colorless nitrite ion NO_2^- is 3.2. The acid is a good reducing agent with $E°(NO_3^-/HNO_2) = 0.94$ v, but it can also act as a good oxidizing agent with $E°(HNO_2/NO) = 0.98$ v and $E°(HNO_2/NH_4^+) = 0.86$ v. Metal nitrites are prepared by dissolving oxides or hydroxides in HNO_2. Most nitrites are soluble, except $AgNO_2$, but many basic nitrites are insoluble. The colorless NO_2^- anion is a good coordinating agent for many transition metal cations.

g. Hydrocyanic acid and cyanides. Passage of CH_4, NH_3, and O_2 through a heated Pt gauze yields colorless hydrogen cyanide gas HCN and HOH. When the gas is dissolved in HOH, it gives hydrocyanic acid, whose anion cyanide CN^- has a log K_p value of 9.2, indicating a very weak acid. The gas and its solution and salts are all deadly poisons, the fatal dose often being as low as 60 mg. The acid dissolves metal oxides and hydroxides to yield cyanide salts. Most of these salts are insoluble except those of the alkali metals, the alkaline earths, and Hg(II). Solutions of the soluble salts are alkaline due to hydrolysis. The CN^- anion forms many very stable complexes. A cyanide solution along with the O_2 of the air will dissolve Au, Pb, Hg, As, Sb, Sn, Bi, and Cd. A typical equation for the dissolution is $4Au + 8CN^- + O_2 + 2HOH \rightarrow 4Au(CN)_2^- + 4OH^-$. Other metals are dissolved with the evolution of H_2 (Cu, Al, Co, Ni, Zn, Mg), an exemplary equation being $2Cu + 2CN^- + 2HOH \rightarrow 2CuCN + 2OH^- + H_2$.

h. Cyanates and thiocyanates. Gentle oxidation of an alkali cyanide gives a cyanate MOCN, and heating an alkali cyanide with S gives a thiocyanide MSCN. Many of the salts of the colorless anions are stable, but the acids are not, decomposition occurring readily. The log K_p for OCN^- is 3.5 and that for SCN^- is 0.9. Cyanates of the alkali metals and Ca are soluble; most others are insoluble. Thiocyanates of the alkali metals, the alkaline earths, Fe(II), Mn(II), Cu(II), Zn(II), and Co(II) are soluble with most others being insoluble. The OCN^- anion is a moderately strong complexing agent, and the SCN^- is usually somewhat stronger.

i. Redox reactions. The redox chemistry of N is complicated by kinetic factors. These influences give quasi-stability to numerous species which are thermodynamically unstable. This is well illustrated by the action of HNO_3 on metals, which is described above in Section 2e. Changes in the metal, the concentration of the acid, the temperature of the reaction, and the presence of various impurities can alter the products, often resulting in mixtures of various species.

j. Analysis. Total nitrogen analysis may be carried out by combusting the sample to NO_2, then reacting the NO_2 with ozone to produce an excited state of NO_2. The light emitted when this excited molecule returns to the ground state is measured by a chemiluminescence detector. The method is sensitive down to about 20 ppb.

k. Health aspects.   Nitrates are relatively safe with the LD50 (oral rat) for $NaNO_3$ being about 1.3 g/kg. The corresponding LD50 for $NaNO_2$ is about 85 mg/kg. For potassium cyanide, the LD50 (oral rat) is 5 mg/kg, with the adult human fatal dose being quoted as 250 mg. Hence extreme care must be exercised with soluble cyanides, since adsorption through the skin can occur. In working with cyanides, one should avoid allowing the solution to go acid, since this liberates HCN, for which inhalation must be avoided. HCN has a mild almond smell which warns of its presence. 200 ppm of HCN in air inhaled for a few minutes results in death. The LD50 for NaOCN is 260 mg/kg and the minimum lethal dose in humans for NH_4SCN is 80 mg/kg.

3. Phosphorus (P) $3s^2 3p^3$

a. E–pH diagram.   Figure 9.2 presents the $10^{-1.0}$ M E–pH diagram for P. The diagram shows the most important simple aqueous species displayed by the element. Equations for the lines that separate the species are displayed in the legend. The species include phosphoric acid $H_3(PO_4)$ and its successively deprotonated anions, the same for phosphorous acid $H_2(HPO_3)$, and the same for hypophosphorous acid $H(H_2PO_2)$. The formulas have been written with the ionizable hydrogen atoms preceding the parentheses. In the molecules, the non-ionizable hydrogens are attached to P and the ionizable hydrogens are attached to O.

b. Discovery and occurrence.   Elemental phosphorus was discovered by Hennig Brand in 1669. He obtained it from the distillation of the residue resulting from the boiling down of aged urine. The gaseous phosphorus was collected under water as a solid white substance. When exposed to air, it glowed with a greenish-white light. Somewhat later, the material was given the name phosphorus from the Greek words phos for light and phoros for bringing. Even though occurring in animal and plant tissues, the major sources of phosphorus compounds are minerals. The most employed minerals for industrial extraction are the apatites and phosphorites. Among the more common apatites are $Ca_5(PO_4)_3F$, $Ca_5(PO_4)_3Cl$, and $Ca_5(PO_4)_3OH$. Phosphorites are amorphous phosphate rocks which have the approximate formula $Ca_5(PO_4)_3F$ and contain varied amounts of impurities.

c. The element.   Elemental phosphorus is presently produced by heating phosphate rock with sand and coke, as follows: $2Ca_3(PO_4)_2 + 6SiO_2 + 10C \rightarrow 6CaSiO_3 + 10CO + P_4$. The P_4 so obtained is a waxy, white solid, which is highly poisonous, 50 mg taken internally leading to death. The P_4 is insoluble in HOH and oxidizes rapidly in air. By heating this white phosphorus for several days in the absence of air, red phosphorus is produced. It is much less reactive and is relatively non-toxic, this being assignable to its polymeric character. Other allotropic forms of elemental phosphorus have been prepared.

d. Oxides.
The two most important oxides of P are P_4O_6 and P_4O_{10}. The former is made by controlled oxidation of P_4 with a N_2/O_2 mixture at low pressure with gentle heating. The colorless liquid (freezing point 23.9°C) hydrolyzes in HOH to produce phosphorous acid $H_2(HPO_3)$. The white solid P_4O_{10} forms when P_4 is burned in air. Upon introduction into HOH,

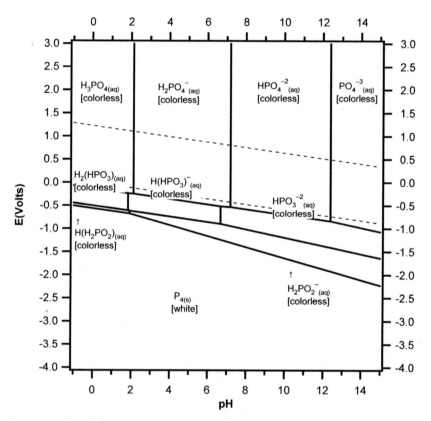

Figure 9.2 E–pH diagram for P species. Soluble species concentrations (except H^+) $= 10^{-1.0}$ M. Soluble species and most solids are hydrated. No agents producing complexes or insoluble compounds are present other than HOH and OH^-.

Species ($\Delta G°$ in kJ/mol): P_4 (0.0), H_3PO_4 (−1118.8), $H_2PO_4^-$ (−1106.7), HPO_4^{-2} (−1065.7), PO_4^{-3} (−995.0), $H_2(HPO_3)$ (−856.9), $H(HPO_3^-)$ (−846.4), HPO_3^{-2} (−808.3), $H(H_2PO_2)$ (−523.0), $H_2PO_2^-$ (−512.1), HOH (−237.2), H^+ (0.0), and OH^- (−157.3).

Figure 9.2   (Continued)

Equations for the lines:

$H(H_2PO_2)/P_4$	$E = -0.50 - 0.059 \text{ pH} + 0.059 \log[H(H_2PO_2)]$
$H_2PO_2^-/P_4$	$E = -0.39 - 0.118 \text{ pH} + 0.059 \log[H_2PO_2^-]$
$H_3PO_4/H_2(HPO_3)$	$E = -0.13 - 0.059 \text{ pH}$
$H_3PO_4/H(HPO_3^-)$	$E = -0.18 - 0.030 \text{ pH}$
$H_2PO_4^-/HPO_3^{-2}$	$E = -0.32 - 0.030 \text{ pH}$
$H_2PO_4^-/H(HPO_3^-)$	$E = -0.12 - 0.059 \text{ pH}$
$H_2(HPO_3)/H(H_2PO_2)$	$E = -0.50 - 0.059 \text{ pH}$
$H(HPO_3^-)/H(H_2PO_2)$	$E = -0.45 - 0.089 \text{ pH}$
$H(HPO_3^-)/H_2PO_2^-$	$E = -0.51 - 0.059 \text{ pH}$
$HPO_3^{-2}/H_2PO_2^-$	$E = -0.31 - 0.089 \text{ pH}$
HPO_4^{-2}/HPO_3^{-2}	$E = -0.11 - 0.059 \text{ pH}$
PO_4^{-3}/HPO_3^{-2}	$E = 0.26 - 0.089 \text{ pH}$
PO_4^{-3}/HPO_4^{-2}	$\text{pH} = 12.4$
$HPO_4^{-2}/H_2PO_4^-$	$\text{pH} = 7.2$
$H_2PO_4^-/H_3PO_4$	$\text{pH} = 2.1$
$HPO_3^{-2}/H(HPO_3^-)$	$\text{pH} = 6.7$
$H(HPO_3^-)/H_2(HPO_3)$	$\text{pH} = 1.8$
$H_2PO_2^-/H(H_2PO_2)$	$\text{pH} = 1.9$

it hydrolyzes to phosphoric acid $H_3(PO_4)$. The two oxoacids just mentioned and a third hypophosphorous acid $H(H_2PO_2)$ are the most important in aqueous chemistry, although many others have been characterized, there being more for P than for any other element. See Table 9.2 for P species.

e. Hypophosphorous acid and hypophosphites.   The heating of P_4 in water solutions of alkali metal or alkaline earth hydroxides produces the hypophosphite anion: $P_4 + 4Na^+ + 4OH^- + 4HOH \rightarrow 4H_2PO_2^- + 2H_2 + 4Na^+$ (very toxic PH_3 also produced). Acidification of the solution gives hypophosphorous acid, but the acid cannot be isolated by evaporation, since this decomposes it. Solvent extraction, however, permits the white solid to be obtained. The acid is monobasic with an anionic log K_p value of 1.9. The acid and its salts are strong reducing agents in both acid and basic solution.

Table 9.2
Phosphorus Species

Name	Formula	State	Color	Solubility	$\Delta G°$ (kJ/mole)
Phosphorus	P_4	s	White	Insol	0.0
Dihydrogenphosphate	$H_2(PO_4)^-$	aq	Colorless		−1106.7
Dihydrogenphosphite	$H(HPO_3)^-$	aq	Colorless		−846.4
Hypophosphosphite	$H_2PO_2^-$	aq	Colorless		−512.1
Monohydrogenphosphate	$H(PO_4)^{-2}$	aq	Colorless		−1065.7
Monohydrogenphosphite	HPO_3^{-2}	aq	Colorless		−808.3
Phosphate	PO_4^{-3}	aq	Colorless		−995.0
Hypophosphorous acid	$H(H_2PO_2)$	aq	Colorless		−523.0
Phosphine	PH_3	g	Colorless	Decomp	13.4
Phosphoric acid	$H_3(PO_4)$	s	Colorless	Sol	−1118.8
Phosphorous acid	$H_2(HPO_3)$	aq	Colorless		−856.9
Phosphorus(III)					
bromide	PBr_3	l	Colorless	Decomp	−175.5
chloride	PCl_3	l	Colorless	Decomp	−272.1
fluoride	PF_3	g	Colorless	Decomp	−896.6
oxide	P_4O_6	s	White	Decomp	
Phosphorus(V)					
fluoride	PF_5	g	Colorless	Decomp	
oxide	P_4O_{10}	s	White	Decomp	−634.2

The alkali metal and alkaline earth salts are soluble, as are many of the salts of other metals.

f. Phosphorous acid and phosphites.   Direct combination of P_4 and Cl_2 gives PCl_3 which hydrolyzes to phosphorous acid $H_2(HPO_3)$. The white solid acid may also be prepared by the hydrolysis of P_4O_6. $H_2(HPO_3)$ is dibasic, one of the hydrogen atoms being non-ionizable, with log K_p for HPO_3^{-2} being 6.7 and that for $H(HPO_3)^-$ being 1.8. Two series of salts can be made, one series of the type $NaH(HPO_3)$ and the other $Na_2(HPO_3)$. The alkali phosphites are soluble, but most others are insoluble. The acid is a moderately strong reducing agent with the E° for the $H_3(PO_4)/H_2(HPO_3)$ couple being −0.13 v.

g. Phosphoric acid and phosphates.   When P_4O_{10} is dissolved in water the colorless tribasic acid H_3PO_4 is obtained; evaporation yields the white solid. Its three anions are also colorless, their log K_p values being $H_2(PO_4)^-$ (2.1), $H(PO_4)^{-2}$ (7.2), and PO_4^{-3} (12.4). As the anions indicate, three series of salts are known. The phosphates and dihydrogen phosphates are insoluble except for the alkali metal ones, but more of the monohydrogenphosphates are soluble. The phosphate anions are strong complexing agents for many transition and inner-transition cations.

h. Redox reactions. The dominant species in the E–pH diagram are P and the phosphates. The others occupy a narrow band just below the lower HOH line. This position indicates that thermodynamically these other species are unstable with regard to the liberation of H_2 from HOH. However, there is sufficient kinetic stabilization to permit these species to be worked with. Both the hypophosphites and the phosphites are good reducing agents. Although not shown, the gaseous species phosphine PH_3 resides along the bottom of the E–pH diagram. This can be produced in cases of extreme reduction. The ready oxidation of the P(I) and P(III) species to phosphate species should be noted, but the possibility of kinetic retardation must be kept in mind.

i. Analysis. Phosphorus is generally measured as phosphate. HNO_3 oxidizing all other species to this form. The colorimetric method involving malachite green and phosphomolybdate permits the determination of P down to 10 ppb. ICPMS gives about the same sensitivity.

j. Health aspects. The most poisonous of the inorganic P compounds which might occur as a by-product in investigations involving the above materials is the gas phosphine PH_3, especially in the preparation of hypophosphites from P. Concentrations above 100 ppm in air can be fatal if breathed for less than an hour. Other P-containing compounds mentioned above are not considered to be very toxic. LD50 (oral rat) for NaH_2PO_2 is 1.6 g/kg, that for Na_2HPO_3 is 5.0 g/kg, and that for Na_3PO_4 is 7.4 g/kg.

4. Arsenic (As) $4s^24p^3$

a. E–pH diagram. Figure 9.3, the E–pH diagram for As at $10^{-1.0}$ M, resembles that of P in several respects: the position of the element, the prevalence and transformations of the arsenate species, and the placement of the lower oxidation state species in between. The compound As_2O_3 is soluble in HOH to the extent of about $10^{-1.0}$ M to form H_3AsO_3, which is what appears in the diagram. Hence, all species in the diagram, except As, are soluble. Were the diagram to be for $10^{0.0}$ M As, the solid species As_2O_3 would need to be included, and it would replace H_3AsO_3 as the predominant species. Equations for the lines that separate the species are displayed in the legend.

b. Discovery and occurrence. Indications are that Albertus Magnus isolated As in 1250 by heating the mineral orpiment As_2S_3 with soap and distilling off the element. Others may have obtained As before this since its ores are readily reduced and it volatilizes easily. Numerous minerals of As occur in nature, but the main ones used for industrial production are arsenopyrite

FeAsS, realgar As_4S_4, and orpiment As_2S_3. The name arsenic derives from the Greek arsenicon, which in turn comes from the Persian azzarnikh, both terms referring to orpiment.

c. The element. Since As occurs in Cu and Pb minerals, flue dusts from smelters treating these minerals contain As_2O_3. These dusts are the principal industrial source of As. Separation is relatively easy since As_2O_3 sublimes at 193°C. The As_2O_3 is then reduced to As using C. As is extracted from FeAsS by heating and collection of the sublimed As, leaving behind FeS. The sulfide ores, realgar and orpiment are roasted in air to give the oxide As_2O_3 which is then reduced with C at elevated temperatures to yield the element and CO_2. As is a gray, brittle, semi-metallic solid, which also takes several other

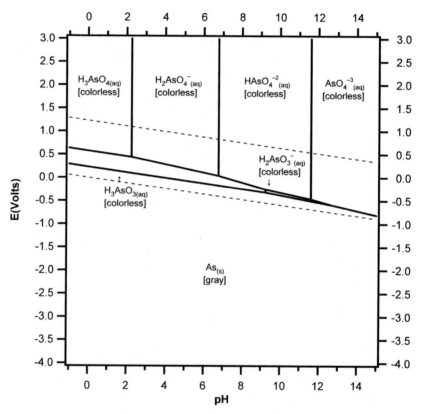

Figure 9.3 E–pH diagram for As species. Soluble species concentrations (except H^+) = $10^{-1.0}$ M. Soluble species and most solids are hydrated. No agents producing complexes or insoluble compounds are present other than HOH and OH^-.

Species ($\Delta G°$ in kJ/mol): As (0.0), H_3AsO_4 (−766.0), $H_2AsO_4^-$ (−753.1), $HAsO_4^{-2}$ (−714.4), AsO_4^{-3} (−648.2), As_2O_3 (−576.6), H_3AsO_3 (−639.6), $H_2AsO_3^-$ (−586.9), HOH (−237.2), H^+ (0.0), and OH^- (−157.3).

Figure 9.3 (Continued)

Equations for the lines:

H_3AsO_3/As	$E = 0.25 - 0.059\ pII + 0.020\ log[H_3AsO_3]$
$H_2AsO_3^-/As$	$E = 0.43 - 0.079\ pH + 0.020\ log[H_2AsO_3^-]$
AsO_4^{-3}/As	$E = 0.62 - 0.094\ pH + 0.012\ log[AsO_4^{-3}]$
H_3AsO_4/H_3AsO_3	$E = 0.57 - 0.059\ pH$
$H_2AsO_4^-/H_3AsO_3$	$E = 0.64 - 0.089\ pH$
$HAsO_4^{-2}/H_3AsO_3$	$E = 0.84 - 0.118\ pH$
$HAsO_4^{-2}/H_2AsO_3^-$	$E = 0.57 - 0.089\ pH$
$AsO_4^{-3}/H_2AsO_3^-$	$E = 0.91 - 0.118\ pH$
$AsO_4^{-3}/HAsO_4^{-2}$	$pH = 11.6$
$HAsO_4^{-2}/H_2AsO_4^-$	$pH = 6.8$
$H_2AsO_4^-/H_3AsO_4$	$pH = 2.3$
$H_2AsO_3^-/H_3AsO_3$	$pH = 9.2$

allotropic modifications. The element is stable in air, but upon heating burns to As_2O_3. Non-oxidizing acids do not attack As, but treatment with HNO_3 gives H_3AsO_4. Hot concentrated alkali dissolves As to give arsenite $H_2AsO_3^-$ and some gaseous arsine AsH_3.

d. Oxides. As forms two oxides arsenic(III) oxide As_2O_3 and arsenic(V) oxide As_2O_5. The former is prepared by heating As in air, it is soluble in water to some extent, namely about 22 g/L, and it dissolves in base. As_2O_3 exists in several other modifications, the more crystalline forms showing diminished solubilities. As_2O_5 can be prepared by treating As or As_2O_3 with O_2 under pressure and heating or by dehydration of solid H_3AsO_4. It is soluble in water, acids, and bases, and functions as a strong oxidizing agent. These and other compounds are listed in Table 9.3.

e. Arsenious acid and arsenites. The dissolution of As_2O_3 in HOH gives arsenious acid H_3AsO_3. Above a concentration of about $10^{-1.0}$ M, no more As_2O_3 will dissolve. H_3AsO_3 is a very weak monoprotic acid, with the anion $H_2AsO_3^-$ having a log K_p value of 9.2. Arsenites of the alkali metals are soluble, those of the alkaline earths slightly soluble, and most others are insoluble. Most of the salts have the anion AsO_2^- ($H_2AsO_3^- - H_2O$), while others have AsO_3^{-3}. The salts are generally highly hydrolyzed due to the weakness of the acid.

Table 9.3
Arsenic Species

Name	Formula	State	Color	Solubility	$\Delta G°$ (kJ/mole)
Arsenic	As_4	s	Gray	Insol	0.0
Arsenate	AsO_4^{-3}	aq	Colorless		−648.2
Dihydrogenarsenate	$H_2AsO_4^-$	aq	Colorless		−753.1
Dihydrogenarsenite	$H_2AsO_3^-$	aq	Colorless		−586.9
Monohydrogenarsenate	$HAsO_4^{-2}$	aq	Colorless		−714.4
Arsenic acid	H_3AsO_4	aq	Colorless		−766.0
Arsenious acid	H_3AsO_3	aq	Colorless		−639.6
Arsenic(III)					
bromide	$AsBr_3$	s	Colorless	Decomp	−160.2
chloride	$AsCl_3$	l	Colorless	Decomp	−256.7
fluoride	AsF_3	l	Colorless	Decomp	−908.9
iodide	AsI_3	s	Red	Decomp	−59.4
oxide	As_4O_6	s	White	Decomp	−1151.1
oxychloride	$AsOCl$	s	Tan	Decomp	
sulfide	As_2S_3	s	Yellow/red	Insol	−168.4
Arsenic(V)					
fluoride	AsF_5	g	Colorless	Decomp	
oxide	As_4O_{10}	s	White	Decomp	−1562.8
sulfide	As_2S_5	s	Yellow	Insol	
Arsine	AsH_3	g	Colorless		68.9

f. Arsenic acid and arsenates. Arsenic acid H_3AsO_4 is formed by the oxidation of As or As_2O_3 with HNO_3 or by adding As_2O_5 to water. The tribasic acid is completely soluble in water, and its anions show the following log K_p values: $H_2AsO_4^-$ (2.3), $HAsO_4^{-2}$ (6.8), and AsO_4^{-3} (11.6). These possible anions give rise to three series of salts. The arsenates and monohydrogenarsenates are insoluble except for the alkali metal salts, with the dihydrogenarsenates being somewhat more soluble.

g. Redox reactions. One of the major differences between the aqueous chemistry of P and that of As is the stronger oxidizing capability of the arsenates. The $E°$ for the H_3AsO_4/H_3AsO_3 couple is 0.57 v as compared to the $E°$ of the corresponding P couple of −0.13 v. The ready oxidation of As(III) species to As(V) species should be noted, especially the possibility of O_2 doing this. The very poisonous gaseous species arsine AsH_3 resides along the bottom of the E–pH diagram, and may be produced under strong reducing conditions.

h. Analysis. Colorimetric analysis for As using reduction by Zn to AsH_3, then color development with $HgBr_2$ to give AsH_2HgBr permits determination down to 10 ppb. ETAAS and ICPAES have a detection limit of 10 ppb, and ICPMS goes down to 1 ppb.

i. Health aspects. As and its compounds are highly toxic. The minimum lethal dose for adult humans for various simple compounds has been reported to be from 100 to 200 mg. For most compounds, the danger is ingestion, but for AsH_3, it is inhalation. The LD50 (oral rat) for As_2O_3 is 15 mg/kg, that for As_2O_5 is 8 mg/kg, and that for Na_3AsO_3 is 10 mg/kg.

5. Antimony (Sb) $5s^2 5p^3$

a. E–pH diagram. Oxidation states of 0, III, and V are seen in Figure 9.4 which is the E–pH diagram for Sb with all soluble species, except H^+, at $10^{-1.0}$ M. For the first time in the group, cation chemistry appears with $Sb(OH)_2{}^+$, it being partially hydrolyzed and strongly subject to further hydrolysis. Unlike As_2O_3, Sb_2O_3 is not very soluble, so no soluble species H_3SbO_3 is ever predominant. Equations for the lines that separate the species are displayed in the legend.

b. Discovery and occurrence. Antimony has been known since ancient times, extraction occurring from its black mineral stibnite Sb_2S_3 by heating with charcoal. Early on the mineral and the element were called by the Latin stibium and also by antimonium. Many ores of Sb are known, including complex sulfides with Ni, Hg, Cu, Fe, Pb, and Ag, and some rarer oxides.

c. The element. Poorer ores of Sb are smelted and the oxide is recovered from flue dust, followed by reduction with C; richer ores such as Sb_2S_3 are reduced with Fe. Further purification is often attained by electrolysis. The Sb obtained is a brittle, silvery-white metal. It is not attacked by dry air at room temperature, but when heated, it burns to white Sb_2O_3. The metal is stable in HOH, but slowly oxidizes in HOH containing dissolved O_2. Non-oxidizing acids will not react with Sb unless there is O_2 present. The action of HNO_3 on Sb produces Sb_2O_3 or Sb_2O_5 depending upon the acid concentration. Hot concentrated H_2SO_4 reacts with Sb to yield $Sb_2(SO_4)_3$ and SO_2. Aqua regia ($3HCl + 1HNO_3$) dissolves Sb to yield $SbCl_6{}^-$.

d. Oxides. Three oxides are formed by Sb, Sb_2O_3, Sb_2O_4, and Sb_2O_5, the first and the last being of import to solution chemistry. Antimony(III) oxide is a white solid, practically insoluble in HOH, soluble in strong concentrated acids, and soluble in strong bases. This oxide is soluble in tartaric acid to form complexes with the acid anion: $Sb_2O_3 + 2H_2C_4H_4O_6 \rightarrow 2H^+ + HOH + 2SbO(C_4H_4O_6)^-$. Similar action is seen with citric acid. Antimony(V) oxide is a white solid, insoluble in HOH and acids, but soluble in strong bases. See Table 9.4 for these and additional Sb species.

e. Antimonites and antimonates. Antimonous acid and antimonic acid have not been isolated. When alkaline solutions of $Sb(OH)_4^-$ are evaporated antimonite salts precipitate. In a similar fashion, when alkaline solutions of $Sb(OH)_6^-$ are evaporated, antimonate salts precipitate. Some antimonates of alkali metals and alkaline earths tend to contain the $Sb(OH)_6^-$ ion, but most others are probably structures of mixed oxides ($ZnSb_2O_6 = ZnO \cdot Sb_2O_5$).

f. Cationic antimony. Antimony is just sufficiently enough a metal that it forms compounds which may be considered as salts. The halides SbX_3 are readily hydrolyzed in HOH, so they only exist in solutions with high concentrations of the corresponding hydrohalic acids. The preparation of

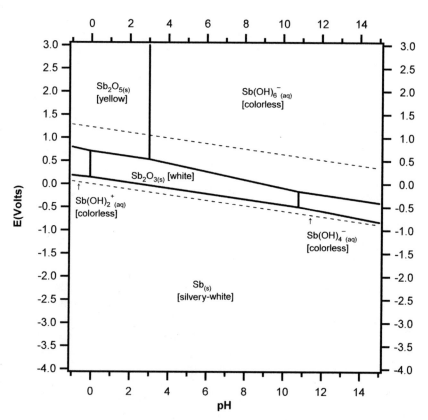

Figure 9.4 E–pH diagram for Sb species. Soluble species concentrations (except H^+) $= 10^{-1.0}$ M. Soluble species and most solids are hydrated. No agents producing complexes or insoluble compounds are present other than HOH and OH^-.

Species ($\Delta G°$ in kJ/mol): Sb (0.0), $Sb(OH)_6^-$ (−1221.7), Sb_2O_5 (−829.1), $Sb(OH)_4^-$ (−839.7), $Sb(OH)_2^+$ (−426.8), Sb_2O_3 (−628.4), HOH (−237.2), H^+ (0.0), and OH^- (−157.3)

Figure 9.4 (Continued)

Equations for the lines:

$Sb_2O_5/Sb(OH)_2{}^+$	$E = 0.68 - 0.089 \text{ pH} - 0.030 \log[Sb(OH)_2{}^+]$
Sb_2O_5/Sb_2O_3	$E = 0.71 - 0.059 \text{ pH}$
$Sb(OH)_6{}^-/Sb_2O_3$	$E = 0.83 - 0.089 \text{ pH} + 0.030 \log[Sb(OH)_6{}^-]$
$Sb(OH)_6{}^-/Sb(OH)_4{}^-$	$E = 0.48 - 0.059 \text{ pH}$
$Sb(OH)_2{}^+/Sb$	$E = 0.16 - 0.039 \text{ pH} + 0.020 \log[Sb(OH)_2{}^+]$
Sb_2O_3/Sb	$E = 0.14 - 0.059 \text{ pH}$
$Sb(OH)_4{}^-/Sb$	$E = 0.38 - 0.079 \text{ pH} + 0.020 \log[Sb(OH)_4{}^-]$
$Sb(OH)_6{}^-/Sb_2O_5$	$2 \text{ pH} = 8.3 + 2 \log[Sb(OH)_6{}^-]$
$Sb(OH)_4{}^-/Sb_2O_3$	$2 \text{ pH} = 23.7 + 2 \log[Sb(OH)_4{}^-]$
$Sb_2O_3/Sb(OH)_2{}^+$	$2 \text{ pH} = -2.1 - 2 \log[Sb(OH)_2{}^+]$

soluble $Sb_2(SO_4)_3$ has been mentioned, and it is subject to hydrolysis to the insoluble basic salt and the oxide. $Sb(NO_3)_3$ must be prepared in concentrated HNO_3, but dilution leads to hydrolysis. Insoluble $SbPO_4$ is obtained by the reaction of Sb_2O_3 and concentrated H_3PO_4.

g. Complexes. Complex fluorides of Sb may be prepared by treating a mixture of an alkali carbonate and Sb_2O_3 with HF to give $SbF_4{}^-$. Similar preparations lead to $SbCl_n{}^{3-n}$ species (n = 4, 5, 6). Some complexation constants for Sb(III) are as follows (log β_1, log β_2, log β_3, log β_4, log β_5, log β_6): OH^- (−, −, 21.5, 32.9, 35.1), F^- (3.0, 5.7, 8.3, 10.9), Cl^- (2.3, 3.5, 4.2, 4.7, 4.7, 4.1). Sb(III) also forms soluble stable complexes $SbBr_4{}^-$, $SbO(tart)^-$, $Sb(C_2O_4)^-$, $SbO(cit)^{-2}$, $SbS_2{}^-$, $Sb(CN)_4{}^-$, $SbO(Hnta)^-$, and $Sbedta^-$. Especially noteworthy are the complexes which Sb(III) forms with polyhydroxy compounds with two adjacent hydroxy groups: catechol, α-hydroxy acids, 1,2-dihydroxybenzene, tartaric acid, lactic acid, malic acid, mandelic acid, and oxalic acid. Sb(V) forms stable complexes $SbF_6{}^{-3}$ and $SbCl_6{}^{-3}$, and also is complexed by some of the polyhydroxy compounds indicated above.

h. Redox reactions. The Sb(V) species are moderately good oxidizing agents. Under extreme reducing conditions the poisonous, colorless gas stibine SbH_3 can be generated from Sb and Sb compounds. This substance rests at the bottom of the E–pH diagram.

Table 9.4
Antimony Species

Name	Formula	State	Color	Solubility	$\Delta G°$ (kJ/mole)
Antimony	Sb	s	Silvery-white	Insol	0.0
Dihydroxoantimony(III)	$Sb(OH)_2{}^+$	aq	Colorless		−426.8
Tetrahydroxoantimonite(III)	$Sb(OH)_4{}^-$	aq	Colorless		−839.7
Hexahydroxoantimonate(V)	$Sb(OH)_6{}^-$	aq	Colorless		−1221.7
Antimony(III)					
bromide	$SbBr_3$	s	White	Decomp	−239.3
chloride	$SbCl_3$	s	White	Sol	−323.7
fluoride	SbF_3	s	Colorless	Sol	−836.0
iodide	SbI_3	s	Red	Decomp	−94.0
oxide	Sb_2O_3	s	White	Insol	−628.4
oxychloride	$SbOCl$	s	White	Insol	
oxysulfate	$Sb_2O_2SO_4$	s	White	Decomp	
sulfate	$Sb_2(SO_4)_3$	s	White	Insol	
sulfide	Sb_2S_3	s	Black	Insol	−173.7
tartrate	$Sb_2(C_4H_4O_6)_3$	s	White	Sol	
Antimony(V)					
chloride	$SbCl_5$	l	Colorless	Decomp	−350.2
fluoride	SbF_5	l	Colorless	Sol	
oxide	Sb_2O_5	s	Yellow	Insol	−829.1
sulfide	Sb_2S_5	s	Yellow	Insol	
Stibine	SbH_3	g	Colorless	Decomp	147.7

i. Analysis. Colorimetric analysis using rhodamine B for color development permits the determination of Sb down to 50 ppb. ETAAS and ICPAES are capable of a sensitivity of 5 ppb, and ICPMS extends this to 0.1 ppb.

j. Health aspects. Sb and its compounds are toxic, especially SbH_3. LD50 (oral rat) for potassium antimonyl tartrate is 115 mg/kg.

6. Bismuth (Bi) $6s^26p^3$

a. E–pH diagram. Figure 9.5 is the E–pH diagram for $10^{-1.0}$ M Bi species. Special considerations apply to several of the species. Bi^{+3} may be an oversimplification since the anion influences the hydrolysis of the simple ion. As the pH is increased, the ion $BiOH^+$ appears, but it too may be a simplification. As the pH is raised further the dominant species is believed to be $Bi_6(OH)_{12}{}^{+6}$ which has been written as $Bi(OH)_2{}^+$. Other polyhydroxy species are probably present. $Bi(OH)_3$ is the compound precipitated at increasing pH, but as the precipitate is aged it forms Bi_2O_3, the $\Delta G°$ value of the equation

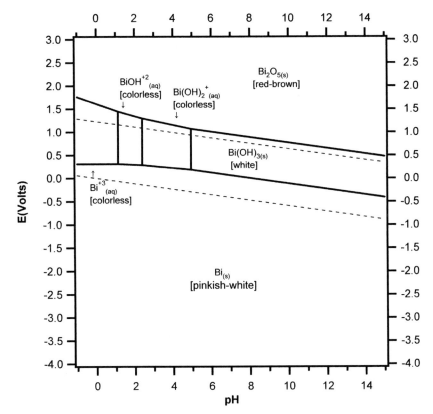

Figure 9.5 E–pH diagram for Bi species. Soluble species concentrations (except H^+) $= 10^{-1.0}$ M. Soluble species and most solids are hydrated. No agents producing complexes or insoluble compounds are present other than HOH and OH^-.

Species ($\Delta G°$ in kJ/mol): Bi (0.0), Bi_2O_5 (-383.3), Bi^{+3} (95.6), $Bi(OH)_3$ (-573.2), $BiOH^{+2}$ (-135.1), $Bi(OH)_2^+$ (-358.6), HOH (-237.2), H^+ (0.0), and OH^- (-157.3).

Equations for the lines:

Bi^{+3}/Bi	$E = 0.33 + 0.020 \log[Bi^{+3}]$
$BiOH^{+2}/Bi$	$E = 0.35 - 0.020 \, pH + 0.020 \log[BiOH^{+2}]$
$Bi(OH)_2^+/Bi$	$E = 0.40 - 0.039 \, pH + 0.020 \log[Bi(OH)_2^+]$
$Bi(OH)_3/Bi$	$E = 0.48 - 0.059 \, pH$
Bi_2O_5/Bi^{+3}	$E = 1.60 - 0.148 \, pH - 0.030 \log[Bi^{+3}]$
$Bi_2O_5/BiOH^{+2}$	$E = 1.55 - 0.118 \, pH - 0.030 \log[BiOH^{+2}]$
$Bi_2O_5/Bi(OH)_2^+$	$E = 1.48 - 0.089 \, pH - 0.030 \log[Bi(OH)_2^+]$
$Bi_2O_5/Bi(OH)_3$	$E = 1.36 - 0.059 \, pH$
$Bi(OH)_3/Bi(OH)_2^+$	$pH = 4.0 - \log[Bi(OH)_2^+]$
$Bi(OH)_2^+/BiOH^{+2}$	$pH = 2.4$
$Bi(OH)^{+2}/Bi^{+3}$	$pH = 1.1$

$2Bi(OH)_3 \rightarrow Bi_2O_3 + 3HOH$ being -62.3 kJ. The species Bi_2O_5 is an idealized formula of a poorly characterized red-brown substance which is produced by the oxidation of Bi_2O_3. When a mixture of Na_2O and Bi_2O_3 is heated in O_2, a compound $NaBiO_3$ is produced. This can be considered as $Na_2O \cdot Bi_2O_5$ and is used as a strong oxidizing agent. Equations for the lines that separate the species are displayed in the legend.

b. Discovery and occurrence.
Bi was prepared quite early since numerous objects made of it can be dated long prior to 1400. Reports give evidence that it was treated by many as a kind of Pb, but Georgius Agricola in 1529 clearly describes its preparation and identifies it as a separate substance. A fire was built on top of charcoal and bismuth ore was added; liquid Bi flowed out. Bi occurs as bismite Bi_2O_3, bismutite $(BiO)_2CO_3$, bismuthinite Bi_2S_3, and in a number of Pb, Cu, Sn, Ag, and Co ores. The name derives from the German wis mat (white mass), later Wismuth; Agricola's designation was the Latin bisemutum.

c. The element.
The main industrial source of the element is as a by-product of refining of Pb, Cu, Sn, Ag, and Co ores. In some cases, Bi_2O_3 collects in flue dust, in others alloys are formed and then separated by various chemical processes, some involving the volatilization of $BiCl_3$, some using electrolysis. Bi is a brittle pinkish-white metal. It is stable in air and in air-free HOH, does not dissolve in non-oxidizing acids unless O_2 is present, dissolves in oxidizing acids like HNO_3 and hot H_2SO_4, and does not dissolve in base. When strongly heated in air, it goes over to Bi_2O_3.

d. Oxides and hydroxides.
Bi shows two oxides, yellow white Bi_2O_3 and an ill-characterized red-brown substance of the approximate formula Bi_2O_5. The former is produced by burning Bi, and the latter by heating Bi_2O_3 in the presence of O_2. Bi_2O_3 is insoluble in water and bases, but dissolves in acids to give the Bi^{+3} ion and/or soluble hydroxy complexes or insoluble hydroxo compounds with the acid anion. Bi_2O_5 is insoluble in acid, HOH, and base, and functions as a strong oxidizing agent. The substance is generally used as $NaBiO_3$ ($= Na_2O + Bi_2O_5$). When a solution of Bi^{+3} is treated with strong base, white $Bi(OH)_3$ precipitates. This compound is insoluble in water and bases, soluble in many acids, and upon aging is converted into $BiO(OH)$, then Bi_2O_3.

e. Compounds and solubilities.
Table 9.5 presents the most important Bi compounds for aqueous chemical considerations. The prevalence of decomposing and insoluble salts is due to the strong propensity of Bi(III) compounds to hydrolyze to insoluble oxy or hydroxy species. See for examples, $BiCl_3$ and $BiOCl$, also $Bi_2(SO_4)_3$ and $Bi_2O_2SO_4$. This indicates that the non-hydrolyzed compounds are stable only in the presence of their respective acids, that is, in the low pH range.

Table 9.5
Bismuth Species

Name	Formula	State	Color	Solubility	$\Delta G°$ (kJ/mole)
Bismuth	Bi	s	Pinkish-white	Insol	0.0
Bismuth(III) ion	Bi^{+3}	aq	Colorless		95.6
Dihydroxobismuth(III)	$Bi(OH)_2^+$	aq	Colorless		−358.6
Hydroxobismuth(III)	$BiOH^{+2}$	aq	Colorless		−135.1
Bismuth(III)					
acetate	$Bi(C_2H_3O_2)_3$	s	White	Insol	
arsenate	$BiAsO_4$	s	White	Insol	
bromide	$BiBr_3$	s	Yellow	Decomp	−255.5
chloride	$BiCl_3$	s	White	Decomp	−314.6
citrate	$BiC_6H_5O_7$	s	White	Sl sol	
fluoride	BiF_3	s	Gray	Insol	
hydroxide	$Bi(OH)_3$	s	White	Insol	−573.2
iodate	$Bi(IO_3)_3$	s	White	Insol	
iodide	BiI_3	s	Black	Insol	−148.7
nitrate	$Bi(NO_3)_3$	s	White	Decomp	
oxalate	$Bi_2(C_2O_4)_3$	s	White	Decomp	
oxide	Bi_2O_3	s	Yellow white	Insol	−497.1
oxybromide	$BiOBr$	s	White	Insol	−297.1
oxycarbonate	$Bi_2O_2CO_3$	s	White	Insol	
oxychloride	$BiOCl$	s	White	Insol	−319.3
oxyfluoride	$BiOF$	s	White	Insol	
oxyiodide	$BiOI$	s	Red	Insol	
oxynitrate	$BiONO_3$	s	White	Insol	
oxysulfate	$Bi_2O_2SO_4$	s	White	Insol	
phosphate	$BiPO_4$	s	White	Insol	
sulfate	$Bi_2(SO_4)_3$	s	White	Decomp	
sulfide	Bi_2S_3	s	Brown-black	Insol	−140.6
tartrate	$Bi_2(C_4H_4O_6)_3$	s	White	Insol	
Bismuth(V)					
oxide	Bi_2O_5	s	Red-brown	Insol	−383.3
Sodium bismuthate	$NaBiO_3$	s	Red-brown	Insol	

f. Redox reactions. The E–pH diagram shows the strong oxidizing power of Bi_2O_5 and the ease with which Bi(III) can be reduced to the metal. Among the metals that will accomplish this are Pb, Sn, Cu, Cd, Fe, Zn, and Mg.

g. Complexes. Notable complexes of Bi(III) include $Bi(SCN)_5^{-2}$, $Bi(NO_3)_2^+$, $Bi(SO_4)_2^-$, BiF^{+2}, $BiCl_5^{-2}$, $BiBr_5^{-2}$, BiI_6^{-3}, $BiO(cit)^{-2}$, $Bi(tart)_2^-$, $Bi(nta)_2^{-3}$, and $Biedta^-$. Complexation constants for several species include (log β_1, log β_2, log β_3, log β_4, log β_5, log β_6): SCN^- (2.2, 2.7, 4.4, 5.2, 5.8, 5.4),

NO_3^- (1.7, 2.5, 0.7), OH^- (12.9, 24.0, 34.2), SO_4^{-2} (2.0, 3.4, 4.1, 4.3, 4.6), F^- (1.4), Cl^- (2.4, 3.5, 5.4, 6.1, 6.7, 6.6), Br^- (3.1, 5.6, 7.4, 8.6, 9.2, 8.7), I^- (3.6, 7.7, 11.4, 15.0, 16.8, 18.8), nta^{-3} (17.5, 26.0), $edta^{-4}$ (27.8). Note that the stabilities of the complexes of the halide ions increase as the halide gets heavier. This is characteristic of heavy cations in the region of Hg, whereas the opposite trend is seen with most other cations.

h. Analysis. Color development with iodide and ascorbic acid permits the analysis of Bi down to about 0.5 ppm. ETAAS and ICPAES go down to 10 ppb, and ICPMS extends the low limit to 0.1 ppb.

i. Health aspects. The LD50 (oral rat) for $BiCl_3$ is 3 g/kg, that for $Bi(NO_3)_3$ is 3.3 g/kg, and that for $Bi_2(SO_4)_3$ is 2.5 g/kg. This indicates that Bi compounds are relatively non-toxic as compared to some other members of this group.

10

The Oxygen Group

1. Introduction

The Oxygen Group of the Periodic Table consists of the elements oxygen O, sulfur S, selenium Se, tellurium Te, and polonium Po. The outer electron structure ns^2np^4 characterizes all five of the elements, with n representing principal quantum numbers 2, 3, 4, 5, and 6, respectively. The ns^2np^4 indicates the possibility of oxidation states of $-II$, II, IV, and VI in all of the elements, with O almost always showing a value of $-II$. As one descends the group, non-metallic character gradually diminishes with only Po being distinctly metallic, and the crossover occurring with the metalloid or semi-metal Te. The II and the VI oxidation states decrease in stability down the group, and the IV oxidation state increases. Oxygen stands out as considerably different from the other elements, there being numerous discontinuities in properties between it and S. Covalent radii in pm are as follows: O(73), S(102), Se(117), Te(135), and Po (149). Ionic radii in pm are as: $O^{-2}(126)$, $S^{-2}(170)$, $Se^{-2}(184)$, and $Te^{-2}(207)$.

2. Oxygen (O) $2s^2 2p^4$

a. E–pH diagrams. Figure 10.1 depicts the E–pH diagram for O with the soluble species (except H^+) at $10^{-1.0}$ M. The upper region is occupied by O_2, the lower region by H_2, and the intermediate area by HOH and its equilibrium species H^+ and OH^-. This diagram functions as the background for all chemical reactions in HOH solution and in an air or O_2 atmosphere. The compound hydrogen peroxide H_2O_2 is another compound of O and H which is of importance. Because of complicated kinetic behavior, H_2O_2 can act as either an oxidant or a reductant. Figure 10.2 displays the E–pH diagram for H_2O_2 at $10^{-1.0}$ M when it is functioning as an oxidizing agent. Figure 10.3 is the E–pH diagram for H_2O_2 at $10^{-1.0}$ M when it is acting as a reducing agent. Equations for the lines that separate the species are displayed in the legends of the diagrams.

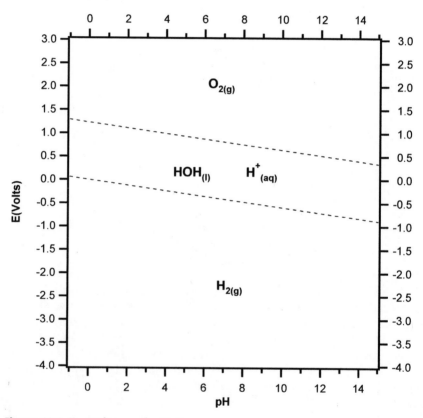

Figure 10.1 E–pH diagram for H_2O.

Species ($\Delta G°$ in kJ/mol): H_2 (0.0), O_2 (0.0), HOH (−237.2), H^+ (0.0), and OH^- (−157.3).

Equations for the lines:

$$H_2O/H_2 \text{ or } H^+/H_2 \qquad E = -0.059\,pH$$

$$O_2/OH^- \text{ or } O_2/H_2O \quad E = 1.23 - 0.059\,pH$$

b. Discovery and occurrence. Oxygen was isolated in the 1630s by a Dutch pharmacist Cornelius Drebble who demonstrated explosions of mixed H_2 and O_2. However, it was Joseph Priestley and Carl Wilhelm Scheele who, independently and about simultaneously, studied it in detail and are generally named as its discoverers. Lavoisier shortly thereafter recognized oxygen as an element and named it from the Greek oxys which means sharp or acid and the Greek genes which means forming. This name of acid-former was assigned because it was then thought that all acids contain oxygen. The occurrence

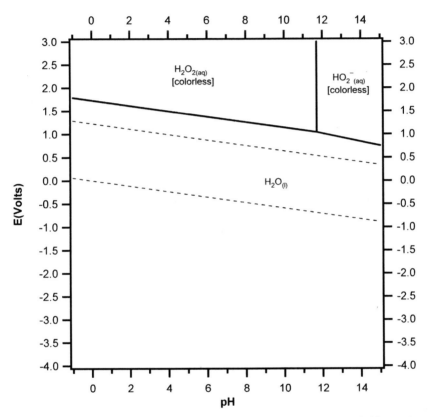

Figure 10.2 E–pH diagram for H_2O_2 acting as an oxidizing agent. Soluble species concentrations (except H^+) = $10^{-1.0}$ M.

Species ($\Delta G°$ in kJ/mol): H_2O_2 (−134.1), HO_2^- (−67.4), HOH (−237.2), H^+ (0.0), and OH^- (−157.3)

Equations for the lines:

$$H_2O_2/H_2O \quad E = 1.76 - 0.059\, pH + 0.030 \log [H_2O_2]$$

$$HO_2^-/H_2O \quad E = 2.11 - 0.089\, pH + 0.030 \log [HO_2^-]$$

$$HO_2^-/H_2O_2 \quad pH = 11.7$$

of the element is in the free state as O_2 which makes up 21% by volume of the atmosphere and in combined forms as HOH, living tissue, rocks, and minerals. It is the most common element in the crust of the earth.

c. The element.   Oxygen O_2 is a colorless, odorless, tasteless gas which is highly reactive. This reactivity is reflected in the fact that O_2 reacts directly or indirectly with all elements except He, Ne, and Ar. The molecule is sparingly soluble in HOH. It is obtained commercially by the fractional distillation of

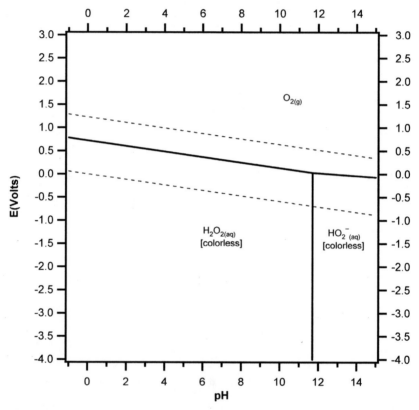

Figure 10.3 E–pH diagram for H_2O_2 acting as a reducing agent. Soluble species concentrations (except H^+) $= 10^{-1.0}$ M.

Species ($\Delta G°$ in kJ/mol): H_2O_2 (-134.1), HO_2^- (-67.4), HOH (-237.2), H^+ (0.0), and OH^- (-157.3)

Equations for the lines:

$$O_2/H_2O_2 \qquad E = 0.69 - 0.059 \, pH - 0.030 \log [H_2O_2]$$

$$O_2/HO_2^- \qquad E = 0.35 - 0.030 \, pH - 0.030 \log [HO_2^-]$$

$$HO_2^-/H_2O_2 \quad pH = 11.7$$

Table 10.1
Oxygen Species

Name	Formula	State	Color	Solubility	$\Delta G°$(kJ/mole)
Oxygen	O_2	g	Colorless	Sl sol	0.0
Ozone	O_3	g	Blue	Sol-decomp	163.2
Hydrogen ion	H^+	aq	Colorless		0.0
Hydrogen peroxide	H_2O_2	aq	Colorless		−134.1
Hydrogen peroxide	H_2O_2	l	Colorless	Sol	−120.4
Hydroperoxide ion	HO_2^-	aq	Colorless		−67.4
Hydroxide	OH^-	aq	Colorless		−157.3
Water	HOH	l	Colorless		−237.2

liquid air. Metals form oxides which are basic; some borderline metals form oxides which are amphoteric; non-metals form oxides which are acidic; and a few oxides are neutral. An allotropic form of oxygen is ozone O_3, a toxic, pungent-smelling blue gas, which can be produced by an electrical discharge in O_2, and is a very powerful oxidizing agent. Most oxygen-containing compounds are presented in tables under the other elements, but Table 10.1 presents those which are not.

d. Water. HOH is a colorless, odorless, tasteless, volatile liquid which has a remarkable ability to dissolve a large number of substances. It occurs in the atmosphere and widely on the surface of the earth. HOH is slightly ionized as described by HOH $\rightarrow$ H^+ + OH^-, an equation which has a log K value of −14.0 at room temperature. When salts crystallize out of HOH solution, they often carry HOH with them as HOH coordinated to a cation and/or an anion, or as HOH trapped in the crystal lattice. Water can interact with many cations, particularly those of high charge, to undergo hydrolysis, for example Al^{+3} + HOH $\rightarrow$ $AlOH^{+2}$ + H^+. HOH can also interact with anions, particularly those of weak acids to hydrolyze, as this example illustrates: CN^- + HOH $\rightarrow$ HCN + OH^-. E–pH diagrams of many elements indicate the important role that HOH plays in acid–base and redox reactions. HOH further functions as a coordinating group toward both cations and anions in aqueous solutions.

e. Hydrogen peroxide. H_2O_2 was first prepared by Louis Jacques Thenard in 1818 who treated BaO_2 with acid and then pumped off the excess HOH under reduced pressure. The compound is a pale blue liquid which is soluble in HOH. The pure substance is strongly prone to explosion by disproportionation into HOH and O_2. H_2O_2 in HOH solution is a weak acid, its anion HO_2^- having a log K_p value of 11.7. Hydroperoxides (salts of HO_2^-) are known for a number of metals, peroxides (salts of O_2^{-2}) for some, and a number of peroxoanions (containing –O–O–groups) can be prepared (for examples, peroxoborate, peroxonitrate, peroxophosphate, and peroxosulfate). As the E–pH diagrams illustrate, H_2O_2 can act as a strong oxidizing agent and as a somewhat poor reducing agent. It usually functions as a reducing agent in the presence of very strong oxidizing species in acid solution. Oxidation by H_2O_2 is sometimes slow in acid, but generally more rapid in base.

f. Acids and bases. Intimately connected with water and water solutions are the ions H^+ and OH^-. Those substances which produce H^+ in water solution are acids and those which produce OH^- in water solution are bases. Acids can be strong or weak, strong acids dissociating almost completely in water and weak acids dissociating only partially. The degree of dissociation of an acid is generally indicated by the anionic protonation constant K_p. For an acid HA, the K_p is associated with the equation $H^+ + A^- \rightarrow HA$ where $K_p = [HA] / [H^+] [A^-]$. Anions of strong acids, such as HNO_3, H_3PO_4, H_2SO_4, H_2SeO_4, HXO_3, HXO_4, and HX (X = Cl, Br, I) have log K_p values less than 1, and anions of weak acids have log K_p values greater than 1. Most acids other than the listed strong acids are weak. Table 10.2 gives protonation constants for numerous anions. Most soluble metal hydroxides are strong bases, dissociating almost completely, most other hydroxides are weak bases. Strong bases include LiOH, NaOH, KOH, RbOH, CsOH, Ca(OH)_2, Sr(OH)_2, and Ba(OH)_2.

Table 10.2
Log Protonation Constants (Log K_p)

$As(OH)_4^-$	9.3		dimethglyoximate$^-$	10.6
AsO_4^{-3}	11.6		edta^{-4}	10.3
BF_4^-	0.5		F^-	3.2
$B(OH)_4^-$	9.3		glycinate$^-$	9.6
Br^-	−9.0		glycine	2.3
BrO^-	8.6		$HAsO_4^{-2}$	6.8
BrO_3^-	−0.7		Hcitrate^{-2}	4.8
citrate^{-3}	6.4		$HCOO^-$	3.8
Cl^-	−7.0		HCO_3^-	6.4
ClO^-	7.5		HCS_3^-	2.7
ClO_2^-	2.0		$HC_2O_4^-$	1.2
ClO_3^-	−2.7		$HCrO_4^-$	−0.9
ClO_4^-	−7.5		Hedta^{-3}	6.2
CN^-	9.2		$HMoO_4^-$	3.5
CNO^-	3.5		Hnitrilotriacetate^{-2}	2.9
CNS^-	0.9		HPO_3^{-2}	6.7
CO_3^{-2}	10.3		HPO_4^{-2}	7.2
CS_3^{-2}	8.2		HS^-	6.9
$C_2H_3O_2^-$	4.8		Hsalicylate$^-$	3.0
$C_2O_4^{-2}$	4.3		HSO_3^-	1.9
CrO_4^-	6.4		HSO_4^-	−3.0

Table 10.2
(*Continued*)

$HS_2O_3^-$	0.6	MnO_4^-	−0.3
HSe^-	3.8	MoO_4^{-2}	3.7
$HSeO_3^-$	2.6	NH_3	9.2
$HSeO_4^-$	−2.0	NH_4OH	9.2
Htartrate$^-$	3.0	nitrilotriacetate^{-3}	10.3
HTe^-	2.6	NO_2^-	3.2
HWO_4^-	3.5	NO_3^-	−1.3
8-hydoxquinolinate$^-$	9.8	N_3^-	4.7
8-hydoxyquinoline	4.9	OH^-	14.0
$H_2AsO_4^-$	2.3	OOH^-	11.7
H_2citrate$^-$	3.1	2,4-pentanedionate$^-$	8.2
H_2edta^{-2}	2.7	PO_4^{-3}	12.4
$H_2GeO_4^{-2}$	12.6	ReO_4^-	−1.3
H_2nitrilotriacetate$^-$	1.7	S^{-2}	12.9
$H_2PO_3^-$	1.8	salicylate^{-2}	12.4
$H_2PO_4^-$	2.1	$Sb(OH)_6^-$	4.2
$H_2SiO_4^{-2}$	13.1	SO_3^{-2}	7.3
H_3edta$^-$	2.0	SO_4^{-2}	2.1
$H_3TeO_4^-$	2.7	$S_2O_3^{-2}$	1.6
$H_3TeO_6^{-3}$	14.4	Se^{-2}	15.0
$H_4TeO_6^{-2}$	11.1	SeO_3^{-2}	7.3
$H_5TeO_6^-$	7.7	SeO_4^{-2}	1.9
I^-	−9.5	tartrate^{-2}	4.4
IO^-	10.6	TcO_4^-	−0.4
IO_3^-	0.9	Te^{-2}	12.1
$H_3IO_6^{-2}$	6.8	TeO_3^{-2}	9.2
$H_4IO_6^-$	3.0	WO_4^{-2}	4.6

g. Analyses. O_2 can be analyzed in gases and/or in solution by several different sensors including coulometric, paramagnetic, polarographic, and ZrO_2-based electrochemical devices. Some of these techniques are sensitive down to 10 ppb. H_2O_2 and O_3 may be analyzed by their oxidizing characters as they produce colored species as the result of redox reactions.

h. Health aspects. The two major dangers of substances presented in this section are the high toxicity of inhaled gaseous O_3 and the attack of concentrated solutions of H_2O_2 on skin. The LD50(mouse) for O_3 is breathing an air concentration of 22 ppm for 3 h. Strong acids and strong bases are very dangerous substances because of the intense corrosive action of H^+ and OH^- on human tissue. Weaker acids and bases are often still corrosive, so caution is advised. In addition to action of the H^+ and the OH^-, the anions of acids and the cations of bases may also be deleterious.

3. Sulfur (S) $3s^2 3p^4$

a. E–pH diagram. The E–pH diagram for S at $10^{-1.0}$ M is presented as Figure 10.4. Table 10.3 shows the properties of the major sulfur species which are pertinent to aqueous chemistry. It will be seen that the species included in the E–pH diagram are the peroxodisulfate ion $S_2O_8^{-2}$, the hydrogen sulfate ion HSO_4^-, the sulfate ion SO_4^{-2}, elemental S_8, hydrosulfuric acid H_2S, the hydrogen sulfide ion HS^-, and the sulfide ion S^{-2}. The other species in Table 10.3 are thermodynamically unstable with regard to these, but they often evidence kinetic stability by their slowness to act. Equations for the lines that separate the species are displayed in the legend.

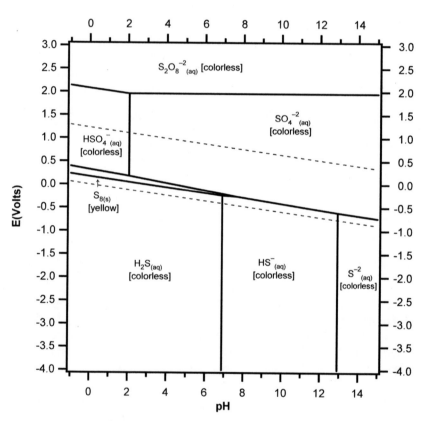

Figure 10.4 E–pH diagram for S species. Soluble species concentrations (except H^+) $= 10^{-1.0}$ M. Soluble species and most solids are hydrated. No agents producing complexes or insoluble compounds are present other than HOH and OH^-.

Species ($\Delta G°$ in kJ/mol): H_2S (−27.2), HS^- (12.1), S^{-2} (85.8), S (0.0), HSO_4^- (−756.0), SO_4^{-2} (−744.3), $S_2O_8^{-2}$ (−1115.0), HOH (−237.2), H^+ (0.0), and OH^- (−157.3).

Figure 10.4 (Continued)

Equations for the lines:

S/H_2S	$E = 0.14 - 0.059\ pH$
SO_4^{-2}/HS^-	$E = 0.25 - 0.066\ pH$
SO_4^{-2}/S^{-2}	$E = 0.15 - 0.059\ pH$
HSO_4^-/S	$E = 0.33 - 0.069\ pH + 0.010 \log [HSO_4^-]$
SO_4^{-2}/S	$E = 0.35 - 0.079\ pH + 0.010 \log [SO_4^{-2}]$
$S_2O_8^{-2}/HSO_4^{-2}$	$E = 2.06 - 0.059\ pH - 0.059 \log [HSO_4^-] + 0.030 \log [S_2O_8^{-2}]$
$S_2O_8^{-2}/SO_4^{-2}$	$E = 1.93 - 0.059 \log [SO_4^{-2}] + 0.030 \log [S_2O_8^{-2}]$
SO_4^{-2}/HSO_4^-	$pH = 2.1$
S^{-2}/HS^-	$pH = 12.9$
HS^-/H_2S	$pH = 6.9$

Table 10.3
Sulfur Species

Name	Formula	State	Color	Solubility	$\Delta\ G° kJ/mole)$
Sulfur	S_8	s	Yellow	Insol	0.0
Hydrogen sulfide	H_2S	g	Colorless	Sol	−33.6
Hydrogen sulfate ion	HSO_4^-	aq	Colorless		−756.0
Hydrogen sulfide ion	HS^-	aq	Colorless		12.1
Hydrogen sulfite ion	HSO_3^-	aq	Colorless		−527.8
Hydrogen thiosulfate ion	$HS_2O_3^-$	aq	Colorless		−532.1
Hydrosulfuric acid	H_2S	aq	Colorless		−27.2
Peroxodisulfate ion	$S_2O_8^{-2}$	aq	Colorless		−1115.0
Peroxodisulfuric acid	$H_2S_2O_8$	s	Colorless	Sol	
Sulfate ion	SO_4^{-2}	aq	Colorless		−744.3
Sulfide ion	S^{-2}	aq	Colorless		85.8
Sulfite ion	SO_3^{-2}	aq	Colorless		−486.6
Sulfur					
dioxide	SO_2	g	Colorless	Decomp	−300.2
trioxide	SO_3	l	Colorless	Decomp	−371.1
Sulfuric acid	H_2SO_4	l	Colorless	Sol	−690.1
Sulfuric acid	H_2SO_4	aq	Colorless		−744.6
Sulfurous acid	H_2SO_3	aq	Colorless		−537.9
Thiosulfate ion	$S_2O_3^{-2}$	aq	Colorless		−518.8
Thiosulfuric acid	$H_2S_2O_3$	aq	Colorless		−535.6

b. Discovery and occurrence. The recognition of S as a particular substance dates to pre-historic times, with references to it in Egypt about 1500 BC. The element occurs in the free state, as H_2S in natural gas, and in the form of sulfide minerals such as pyrite FeS_2 and copper pyrite $CuFeS_2$, and of sulfate minerals such as gypsum $CaSO_4 \cdot 2HOH$ and barite $BaSO_4$.

c. The element. The three major processes for extracting S include treating H_2S with O_2 under suitable temperature and catalyst conditions according to $24H_2S + 12O_2 \rightarrow 3S_8 + 24HOH$, forcing S_8 out of the earth with superheated HOH which melts it, and roasting FeS_2 to SO_2 which is used as is. Solid yellow elemental sulfur is the ring structure S_8. It is insoluble in HOH and non-oxidizing acids, but dissolves in hot NaOH to produce sulfide S^{-2}, polysulfides (S_n^{-2}) and thiosulfate $S_2O_3^{-2}$. Sulfur reacts with CN^- to give SCN^-, with HNO_3 to give H_2SO_4 and NO, with hot concentrated H_2SO_4 to give SO_2, and with $HClO_3$ to yield HCl and H_2SO_4. S burns in air to give gaseous SO_2, which combines with O_2 at elevated temperatures and under the influence of a catalyst to form SO_3.

d. Hydrosulfuric acid and sulfides. S combines with many metals to give sulfides. Upon treatment of these sulfides with acid, the colorless, odoriferous, poisonous gas H_2S is liberated. H_2S dissolves in HOH to the extent of about $10^{-1.0}$ M, where it acts as a weak diprotic acid, with the log K_p for hydrogen sulfide HS^- being 6.9 and that for sulfide S^{-2} being 12.9. As the anions indicate, the acid gives rise to two types of salts. Most metal sulfides are insoluble in HOH, with the heavy metal sulfides being extremely insoluble. The sulfides of the Li and Be Groups and those of Al(III), Ga(III), and Cr(III) tend to dissolve and/or decompose in water. The sulfides of the Li and Be Groups tend to be white or yellow, but most others are highly colored, often brown or black. The precipitation of sulfides can be regulated according to their K_{sp} values by varying the pH which controls the concentration of S^{-2}. Log K_{sp} values for a number of metal sulfides are: Ag_2S (−49.2), CdS (−26.1), CoS (−20.1), CuS (−47.6), FeS (−17.2), HgS (−47.0), MnS (−9.6), NiS (−18.5), PbS (−27.1), SnS (−25.0), Tl_2S (−20.3), ZnS (−23.8).

e. Sulfurous acid and sulfites. SO_2 is a colorless, toxic gas with a sharp acidic odor. It has a high solubility in water, and a small fraction of it forms sulfurous acid H_2SO_3. For simplicity, both the hydrated SO_2 and the acid will be designated H_2SO_3 (= $SO_2 \cdot HOH$), even though the species H_2SO_3 cannot be isolated. The acid acts as dibasic with log K_p for HSO_3^- being 1.9 and that for SO_3^{-2} being 7.3. This dibasic character indicates that two series of salts can be formed, hydrogen sulfites and sulfites. Most hydrogen sulfites are largely known only in solution, but sulfites of many metals have been isolated. The sulfites of the alkali metals and the ammonium ion are soluble, but most others are not. The sulfite species are good reducing agents with the SO_4^{-2}/H_2SO_3

couple having an $E°$ value of 0.16 v. Hence, sulfites of Ag(I), Hg(II), Cu(II), and Fe(III) are unstable because the cation is reduced.

f. Sulfuric acid and sulfates. The dissolution of SO_3 in HOH gives colorless sulfuric acid H_2SO_4, a diprotic acid with a log K_p value for HSO_4^- of -3.0 and for SO_4^{-2} of 2.1. Both hydrogen sulfates and sulfates are known, these being produced by dissolution of metals, oxides, hydroxides, and carbonates in the acid. Most sulfates are soluble with the exception of those of Ca, Sr, Ba, Pb, and Hg(I). Almost all the hydrogen sulfates that have been characterized are those of the alkali metals and the alkaline earths. The sulfate species occupy considerable area on the E–pH diagram, including a large overlap with the HOH region, indicating that they are quite stable toward oxidation or reduction.

g. Thiosulfuric acid and thiosulfates. When an aqueous solution of a sulfite has H_2S added to it, the thiosulfate ion is produced: $2HS^- + 4HSO_3^- \rightarrow 3S_2O_3^{-2} + 3HOH$. Thiosulfate ion may also be prepared by heating a sulfite solution with S_8: $8HSO_3^- + S_8 \rightarrow 8S_2O_3^{-2} + 8H^+$. If these solutions are acidified in an attempt to produce thiosulfuric acid $H_2S_2O_3$, the free acid decomposes. Thiosulfate salts of Pb, Ag, Tl(I), and Ba are insoluble; most others are soluble. The thiosulfate ion is a moderately strong reducing agent, it going over to tetrathionate $S_4O_6^{-2}$, with this relation applying to the reaction: $S_4O_6^{-2}/S_2O_3^{-2}$ with $E° = 0.08$ v. As the chapters on the metals indicate, $S_2O_3^{-2}$ is a good complexing agent for heavy metal cations.

h. Peroxodisulfates. When a solution of K_2SO_4 is anodically oxidized in acid with Pt electrodes and at high current density, peroxodisulfate ion is produced: $2HSO_4^- \rightarrow S_2O_8^{-2} + 2H^+ + 2e^-$. Upon cooling the solution, $K_2S_2O_8$ precipitates. From it, the acid can be produced as a white solid, which is subject to explosion. The acid itself and the peroxodisulfates of the alkali metals and the alkaline earths are soluble. The peroxodisulfate ion is a powerful oxidizing agent, the couple $S_2O_8^{-2}/HSO_4^-$ having an $E°$ value of 2.06 v, and the couple $S_2O_8^{-2}/SO_4^{-2}$ having an $E°$ value of 1.93 v, these values being more positive than the vast majority of inorganic aqueous couples. Oxidation by $S_2O_8^{-2}$ is often very slow, but it can often be made rapid by the presence of a small amount of Ag^+.

i. Analysis. Both ICPAES and ICPMS are capable of detecting S down to about 10 ppb. Ion chromatography can detect SO_4^{-2} down to about 1 ppm.

j. Health aspects. Sulfates pose little health hazard, the oral LD50(oral mouse) for Na_2SO_4 being 5 g/kg. The value for Na_2SO_3 is 800 mg/kg. That for Na_2S is 200 mg/kg. But the major danger is the inhalation of the very poisonous H_2S. The lethal concentration for 50% human deaths (LC50) is 600 ppm in air breathed for 30 min. Fortunately, about 0.1 ppm in air can be smelled.

4. Selenium (Se) $4s^2 4p^4$

a. E–pH diagram. Figure 10.5 depicts the E–pH diagram for $10^{-1.0}$ M Se. Its resemblance to that of S is to be noted. The major difference is in the presence of the Se(IV) species in between the Se(-II) and Se(VI) species. Also elemental Se occupies a larger area than does elemental S, and no peroxodiacid appears on the Se diagram. The position of Se within the HOH lines indicates that upper species are readily reduced to it and lower species are readily oxidized to it. Equations for the lines that separate the species are displayed in the legend.

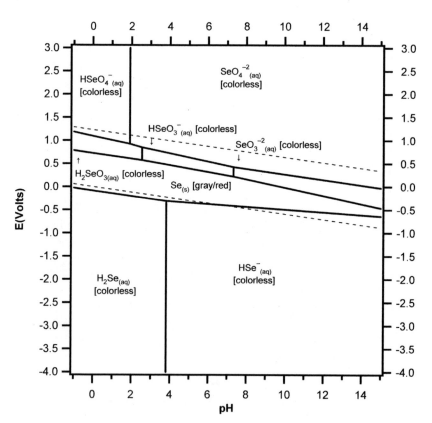

Figure 10.5 E–pH diagram for Se species. Soluble species concentrations (except H^+) = $10^{-1.0}$ M. Soluble species and most solids are hydrated. No agents producing complexes or insoluble compounds are present other than HOH and OH^-.

Species ($\Delta G°$ in kJ/mol): H_2Se (22.2), HSe^- (43.9), Se (0.0), $HSeO_4^-$ (−452.3), SeO_4^{-2} (−441.4), H_2SeO_3 (−426.3), $HSeO_3^-$ (−411.7), SeO_3^{-2} (−369.9), HOH (−237.2), H^+ (0.0), and OH^- (−157.3)

Figure 10.5 (Continued)

Equations for the lines:

Se/H_2Se	$E = -0.12 - 0.059\ pH - 0.030\ \log [H_2Se]$
Se/HSe^-	$E = -0.23 - 0.030\ pH - 0.030\ \log [HSe^-]$
H_2SeO_3/Se	$E = 0.74 - 0.059\ pH + 0.015\ \log [H_2SeO_3]$
$HSeO_3^-/Se$	$E = 0.78 - 0.074\ pH + 0.015\ \log [HSeO_3^-]$
SeO_3^{-2}/Se	$E = 0.89 - 0.089\ pH + 0.015\ \log [SeO_3^{-2}]$
$HSeO_4^-/H_2SeO_3$	$E = 1.09 - 0.089\ pH$
SeO_4^{-2}/H_2SeO_3	$E = 1.15 - 0.118\ pH$
$SeO_4^{-2}/HSeO_3^-$	$E = 1.08 - 0.089\ pH$
SeO_4^{-2}/SeO_3^{-2}	$E = 0.86 - 0.059\ pH$
HSe^-/H_2Se	$pH = 3.8$
$SeO_3^{-2}/HSeO_3^-$	$pH = 7.3$
$HSeO_3^-/H_2SeO_3$	$pH = 2.6$
$SeO_4^{-2}/HSeO_4^-$	$pH = 1.9$

b. Discovery and occurrence. In 1817 Jöns Jacob Berzelius and Johan Gottlieb Gahn discovered Se in the lead chamber residual mud of the process for obtaining SO_2 from copper pyrite. The element resembled tellurium which had been named from the Latin tellus, which means earth, so it was named after the Greek selene which means moon. The mud was heated and SeO_2 sublimed at about 315°C. Se occurs in a number of selenide minerals, but they are very rare. Its most important occurrence is in Cu sulfide ores.

c. The element. The major source of Se is in the anodic slime which remains after the electrolytic refining of Cu from sulfidic ores. The slimes contain 5–25% Se and 2–10% Te mostly in the forms of Cu_2Se, Ag_2Se, Cu_2Te, and Ag_2Te. The slime is fused with Na_2CO_3 to produce Na_2SeO_3 and Na_2TeO_3: $Cu_2Se + Na_2CO_3 + 2O_2 \rightarrow 2CuO + Na_2SeO_3 + CO_2$. The sodium salts are leached out with HOH, and addition of acid precipitates TeO_2. The remaining Na_2SeO_3 solution is reduced with SO_2 to give Se. Se is a gray red solid, stable in air, HOH, dilute acid, and dilute base, but which dissolves in strong alkali. The element burns in air to give white SeO_2. Se species are listed in Table 10.4 along with pertinent properties.

Table 10.4
Selenium Species

Name	Formula	State	Color	Solubility	$\Delta G°$(kJ/mole)
Selenium	Se	s	Gray/red	Insol	0.0
Hydrogen selenide	H_2Se	g	Colorless	Sol	15.9
Hydrogen selenate ion	$HSeO_4^-$	aq	Colorless		−452.3
Hydrogen selenide ion	HSe^-	aq	Colorless		43.9
Hydrogen selenite ion	$HSeO_3^-$	aq	Colorless		−411.7
Hydroselenic acid	H_2Se	aq	Colorless		22.2
Selenate ion	SeO_4^{-2}	aq	Colorless		−441.4
Selenide ion	Se^{-2}	aq	Colorless		129.3
Selenite ion	SeO_3^{-2}	aq	Colorless		−369.9
Selenium dioxide	SeO_2	s	White	Decomp	−173.6
trioxide	SeO_3	s	White	Decomp	−94.6
Selenic acid	H_2SeO_4	s	White	Sol	−441.4
Selenous acid	H_2SeO_3	s	White	Sol	−406.3
Selenous acid	H_2SeO_3	aq	Colorless		−426.3

d. Hydroselenic acid and selenides. Al combines with Se to form the light brown Al_2Se_3. This compound decomposes in HOH or dilute acid to give hydrogen selenide H_2Se, a colorless, odoriferous, highly poisonous gas which is soluble in HOH. H_2Se as a gas and in HOH solution is thermodynamically unstable and undergoes decomposition into the elements slowly. Aqueous H_2Se is a dibasic acid which is slightly more acidic than H_2S, the log K_p value for protonation of HSe^- being 3.8, but that for Se^{-2} is estimated to be about 15.0. Two series of salts corresponding to the two anions are known, those of the alkali metals and the alkaline earths being soluble, and others insoluble. These salts are hydrolyzed in solution. All selenides, both soluble and insoluble, are readily oxidized to Se by the O_2 in the air.

e. Selenous acid and selenites. When SeO_2 is dissolved in HOH it produces selenous acid H_2SeO_3. This diprotic acid is weak, the log K_p value for $HSeO_3^-$ being 2.6 and that for SeO_3^{-2} being 7.3. The acid forms two series of salts of the types $NaHSeO_3$ and Na_2SeO_3. Selenites are usually insoluble, except for those of the alkali metals.

f. Selenic acid and selenates. H_2SeO_4 is generally produced by the oxidation of H_2SeO_3 with strong oxidizing agents such as H_2O_2, Cl_2, MnO_4^-, or ClO_3^-. The acid is strong, the log K_p value for $HSeO_4^-$ being −2.0, and that for SeO_4^{-2} being 1.9. These anions give rise to two series of salts, their solubilities resembling those of their sulfate counterparts.

g. Analysis.   Colorimetic analysis for H_2SeO_3 using 3,3′-diaminobenzidine as the color-developing reagent goes down to a limit of about 50 ppb. This can be extended to 5 ppb by the use of ETAAS, ICPAES, or ICPMS.

h. Health aspects.   Se is an essential element in very small quantities, namely 70 μg/day. Larger quantities can be dangerous, the LD50(oral rat) for SeO_2 being 68 mg/kg and that for Na_2SeO_3 is 7 mg/kg. The gas H_2Se is very toxic, its poisonous character resembling that of H_2S.

5. Tellurium (Te) $5s^25p^4$

a. E–pH diagrams.   When the species listed in the caption to Figure 10.6 are considered the diagram depicted therein results. The diagram is for all soluble species except H^+ at $10^{-1.0}$ M. Take note of the presence of the slightly

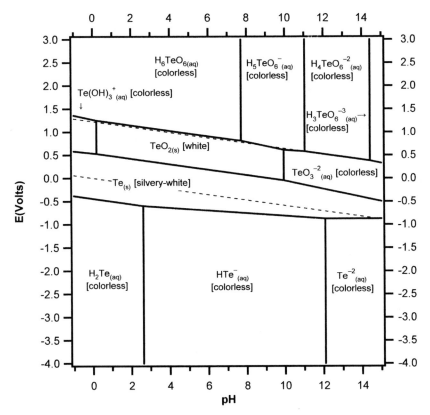

Figure 10.6 E–pH diagram for Te species. Soluble species concentrations (except H^+) $= 10^{-1.0}$ M. Soluble species and most solids are hydrated. No agents producing complexes or insoluble compounds are present other than HOH and OH^-.

Figure 10.6 (Continued)

Species ($\Delta G°$ in kJ/mol): H_2Te (91.0), HTe^- (106.0), Te^{-2} (175.0), Te (0.0), TeO_2 (−266.1), H_2TeO_3 (−477.0), $HTeO_3^-$ (−436.4), TeO_3^{-2} (−384.0), $Te(OH)_3^+$ (−498.7), H_6TeO_6 (−965.0), $H_5TeO_6^-$ (−921.0), $H_4TeO_6^{-2}$ (−858.0), $H_3TeO_6^{-3}$ (−775.7), HOH (−237.2), H^+ (0.0), and OH^- (−157.3).

Equations for the lines:

$H_6TeO_6/Te(OH)_3^+$ $E = 1.27 - 0.089 \, pH$

H_6TeO_6/TeO_2 $E = 1.29 - 0.059 \, pH + 0.030 \log [H_6TeO_6]$

$H_5TeO_6^-/TeO_2$ $E = 1.52 - 0.089 \, pH + 0.030 \log [H_5TeO_6^-]$

$H_5TeO_6^-/TeO_3^{-2}$ $E = 0.90 - 0.030 \, pH$

$H_4TeO_6^{-2}/TeO_3^{-2}$ $E = 1.23 - 0.059 \, pH$

$H_3TeO_6^{-3}/TeO_3^{-2}$ $E = 1.66 - 0.089 \, pH$

$Te(OH)_3^+/Te$ $E = 0.55 - 0.044 \, pH + 0.015 \log [Te(OH)_3^+]$

TeO_2/Te $E = 0.54 - 0.059 \, pH$

TeO_3^{-2}/Te $E = 0.85 - 0.089 \, pH + 0.015 \log [TeO_3^{-2}]$

Te/H_2Te $E = -0.47 - 0.059 \, pH - 0.029 \log [H_2Te]$

Te/HTe^- $E = -0.55 - 0.029 \, pH - 0.029 \log [HTe^-]$

Te/Te^{-2} $E = -0.91 - 0.029 \log [Te^{-2}]$

TeO_3^{-2}/TeO_2 $2 \, pH = 20.9 + \log [TeO_3^{-2}]$

$TeO_2/Te(OH)_3^+$ $pH = -0.81 - \log [Te(OH)_3^+]$

$H_5TeO_6^-/H_6TeO_6$ $pH = 7.7$

$H_4TeO_6^{-2}/H_5TeO_6^-$ $pH = 11.1$

$H_3TeO_6^{-3}/H_4TeO_6^{-2}$ $pH = 14.4$

Te^{-2}/HTe^- $pH = 12.1$

HTe^-/H_2Te $pH = 2.6$

soluble compound TeO_2 and the absence of H_2TeO_3. TeO_2 is soluble to the extent of about $10^{-5.0}$ M, and upon dissolution it goes over to soluble H_2TeO_3. Figure 10.7 is the E–pH diagram for soluble species (except H^+) at $10^{-5.0}$ M. The insoluble TeO_2 has been replaced by the soluble H_2TeO_3 as the predominant species. Equations for the lines that separate the species are displayed in the legends.

b. Discovery and occurrence. Te was first isolated in 1782 by Franz Joseph Müller von Reichenstein who named it metallum problematicum. He examined a gold-containing ore from Transylvania which had previously been thought to be an antimony compound. His work indicated that the substance was actually gold and a new substance. A sample was sent to Martin Heinrich Klaproth who confirmed the discovery of the new element and named it tellurium after the Latin word tellus which means earth. Te is rare, sometimes found free in nature, and in ores such as sylvanite ($AgAuTe_4$) and calaverite ($AuTe_2$). It also occurs in low concentrations in sulfidic copper ores.

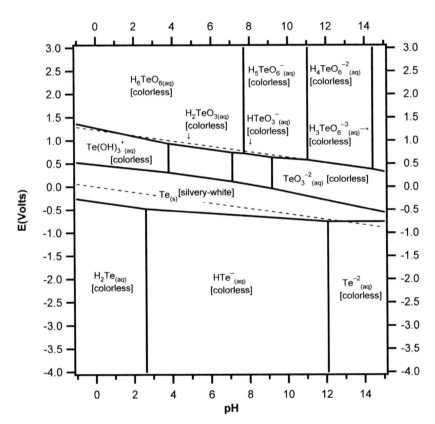

Figure 10.7 E–pH diagram for Te species. Soluble species concentrations (except H^+) = $10^{-5.0}$ M. Soluble species and most solids are hydrated. No agents producing complexes or insoluble compounds are present other than HOH and OH^-.

Species ($\Delta G°$ in kJ/mol): H_2Te (91.0), HTe^- (106.0), Te^{-2} (175.0), Te (0.0), TeO_2 (−266.1), H_2TeO_3 (−477.0), $HTeO_3^-$ (−436.4), TeO_3^{-2} (−384.0), $Te(OH)_3^+$ (−498.7), H_6TeO_6 (−965.0), $H_5TeO_6^-$ (−921.0), $H_4TeO_6^{-2}$ (−858.0), $H_3TeO_6^{-3}$ (−775.7), HOH (−237.2), H^+ (0.0), and OH^- (−157.3).

Figure 10.7 (Continued)

Equations for the lines:

$H_6TeO_6/Te(OH)_3{}^+$	$E = 1.27 - 0.089\ pH$
H_6TeO_6/H_2TeO_3	$E = 1.16 - 0.059\ pH$
$H_6TeO_6/HTeO_3{}^-$	$E = 0.95 - 0.030\ pH$
$H_5TeO_6{}^-/HTeO_3{}^-$	$E = 1.18 - 0.059\ pH$
$H_5TeO_6{}^-/TeO_3{}^{-2}$	$E = 0.90 - 0.030\ pH$
$H_4TeO_6{}^{-2}/TeO_3{}^{-2}$	$E = 1.23 - 0.059\ pH$
$H_3TeO_6{}^{-3}/TeO_3{}^{-2}$	$E = 1.66 - 0.089\ pH$
$Te(OH)_3{}^+/Te$	$E = 0.55 - 0.044\ pH + 0.015\ \log\ [Te(OH)_3{}^+]$
H_2TeO_3/Te	$E = 0.61 - 0.059\ pH + 0.015\ \log\ [H_2TeO_3]$
$HTeO_3{}^-/Te$	$E = 0.71 - 0.074\ pH + 0.015\ \log\ [HTeO_3{}^-]$
$TeO_3{}^{-2}/Te$	$E = 0.85 - 0.089\ pH + 0.015\ \log\ [TeO_3{}^{-2}]$
Te/H_2Te	$E = -0.47 - 0.059\ pH - 0.029\ \log\ [H_2Te]$
Te/HTe^-	$E = -0.55 - 0.029\ pH - 0.029\ \log\ [HTe^-]$
Te/Te^{-2}	$E = -0.91 - 0.029\ \log\ [Te^{-2}]$
$H_5TeO_6{}^-/H_6TeO_6$	$pH = 7.7$
$H_4TeO_6{}^{-2}/H_5TeO_6{}^-$	$pH = 11.1$
$H_3TeO_6{}^{-3}/H_4TeO_6{}^{-2}$	$pH = 14.4$
$H_2TeO_3/Te(OH)_3{}^+$	$pH = 3.8$
$HTeO_3{}^-/H_2TeO_3$	$pH = 7.1$
$TeO_3{}^{-2}/HTeO_3{}^-$	$pH = 9.2$
HTe^-/H_2Te	$pH = 2.6$
Te^{-2}/HTe^-	$pH = 12.1$

c. The element. The major source of Te is in the anodic slime which remains after the electrolytic refining of Cu from sulfidic ores. The slimes contain 5–25% Se and 2–10% Te mostly in the forms of Cu_2Se, Ag_2Se, Cu_2Te, and Ag_2Te. The slime is fused with Na_2CO_3 to produce Na_2SeO_3 and Na_2TeO_3: $Cu_2Te + Na_2CO_3 + 2O_2 \rightarrow 2CuO + Na_2TeO_3 + CO_2$. The sodium salts are leached out with HOH, and addition of acid precipitates TeO_2. This precipitate is dissolved in NaOH and is electrolytically reduced

Table 10.5
Tellerium Species

Name	Formula	State	Color	Solubility	$\Delta G°$ (kJ/mole)
Tellurium	Te	s	Silvery-white	Insol	0.0
Hydrogen telluride	H_2Te	g	Colorless	Sol	138.5
Hydrogen telluride ion	HTe^-	aq	Colorless		106.0
Hydrogen tellurite ion	$HTeO_3^-$	aq	Colorless		−436.4
Hydrotelluric acid	H_2Te	aq	Colorless		91.0
Pentahydrogen tellurate ion	$H_5TeO_6^-$	aq	Colorless		−921.0
Telluride ion	Te^{-2}	aq	Colorless		175.0
Telluric acid	H_6TeO_6	aq	Colorless		−965.0
Tellurite ion	TeO_3^{-2}	aq	Colorless		−384.0
Tellurium dioxide	TeO_2	s	White	Sl sol	−266.1
Tellurium trioxide	TeO_3	s	Yellow-orange		
Tellurous acid	H_2TeO_3	aq	Colorless		−477.0
Tetrahydrogen tellurate ion	$H_4TeO_6^{-2}$	aq	Colorless		−858.0
Trihydrogen tellurate ion	$H_3TeO_6^{-3}$	aq	Colorless		−775.7
Trihydroxotellurium(IV) ion	$Te(OH)_3^+$	aq	Colorless		−498.7

to Te. The element is a silvery-white non-metallic solid which is stable in air, HOH, air-free non-oxidizing acids, and bases. Te burns in air to form white TeO_2. Te species are listed in Table 10.5 along with pertinent properties.

d. Hydrotelluric acid and tellurides. Al combines with Te to give Al_2Te_3, a compound which reacts with HOH to give H_2Te. This colorless, bad-smelling gas is very poisonous. It dissolves in HOH where it acts as a weak acid, the log K_p value for HTe^- being 2.6, and that for Te^{-2} 12.1. In HOH solution, H_2Te and its anions decompose slowly giving Te. The salts of H_2Te resemble those of the sulfides in their solubilities, except they are more readily oxidized.

e. Tellurous acid and tellurites. When Te is burned in air, TeO_2 results. This oxide is soluble only to the extent of about $10^{-5.0}$ M. The species occurring upon dissolution is tellurous acid H_2TeO_3, but it can exist only in low concentrations. TeO_2 dissolves in acid to yield $Te(OH)_3^+$ and dissolves in base to produce $HTeO_3^-$ and TeO_3^{-2}. The log K_p values for these species are 7.1 and 9.2 respectively. Two series of salts are formed, most being insoluble, the alkali metal ones excluded.

f. Telluric acid and tellurates. Orthotelluric acid H_6TeO_6 can be made by strong oxidation of Te with agents such as $HClO_3$ or H_2O_2. Only three of the hydrogen atoms can be readily replaced: the log K_p value for $H_5TeO_6^-$ is 7.7, that for $H_4TeO_6^{-2}$ is 11.1, and that for $H_3TeO_6^{-3}$ is 14.4. Tellurates can be made by the neutralization of hydroxides with H_6TeO_6. Those of the alkali metals are soluble, most others are insoluble. When H_6TeO_6 is isolated, then heated, it evolves HOH to give yellow-orange TeO_3. However, this oxide will not dissolve in neutral HOH to revert to the acid. To obtain the acid from it, it is necessary to dissolve the TeO_3 in base, then to acidify the resulting solution.

g. Analysis. Colorimetic analysis for H_2TeO_3 using diethyldithiocarbamate as the color-developing reagent goes down to a limit of about 100 ppb. This can be extended to 10 ppb by the use of ETAAS or ICPAES, or to 0.1 ppb by ICPMS.

h. Health aspects. LD50 (oral rat) for Na_2TeO_4 is 385 mg/kg and that for Na_2TeO_3 is 83 mg/kg. This indicates a notable toxicity.

6. Polonium (Po) $6s^26p^4$

a. Po isotopes. No stable isotope of Po is known. The most employed radioisotope in investigations of Po chemistry has been Po-210 which decays by α-particle emission with a half life of 138.4 days. This nuclide occurs in U ores as a decay product of U-238. A ton of high quality U ore contains about 100 μg of Po-210. For a time, therefore, Po-210 was available only in trace amounts, and so experiments were carried out with solutions in the range of $10^{-7.0}$ to $10^{-9.0}$ M. Most of these experiments followed the Po-210 by its radioactivity as the unweighable amounts accompanied compounds of similar elements in various reactions. Many of the investigations were burdened by the peculiar behavior of very low concentration substances in solution, the properties not being able to be extrapolated from those of more concentrated solutions. Later, milligram quantities of Po-210 were produced by neutron irradiation of Bi-210, and similar quantities of Po-208 (half life 3.0 years) and Po-209 (half life 103 years) were produced by charged particle irradiation of Bi-210. This permitted experiments with solutions in the range of $10^{-4.0}$ to $10^{-6.0}$ M, but these were burdened by their high radioactivities which produced chemical reactions in the solutions and mandated special handling procedures to protect the chemists.

b. Discovery. Po was discovered by Marie Sklodowska Curie in 1898. She and her husband processed several tons of residue from U mines, following the separated fractions by radioactivity. A very active substance accompanied Bi when it was isolated as the sulfide. This was a new element which Curie named polonium after her native land Poland.

c. E–pH diagram. Figure 10.8 is the E–pH diagram for Po with all soluble species except H^+ at $10^{-5.0}$ M, a concentration with which a person using special equipment is likely to be working with the element. The equations in the figure legend may be used to arrive at the diagram for any other Po soluble species concentration. The diagram is to be treated with much caution because of several points: (1) most of the species identifications have been made on the basis of insufficient experimental data, (2) the Po species assume no complexation with anions except OH^-, (3) most of the $\Delta G°$ values are estimates based upon what is compatible with most of the experimental information, (4) the influence of the radioactivity on reactions of Po is sometimes not known, and (5) very dilute solutions are known to behave strangely. This strangeness consists of the formation of colloids by

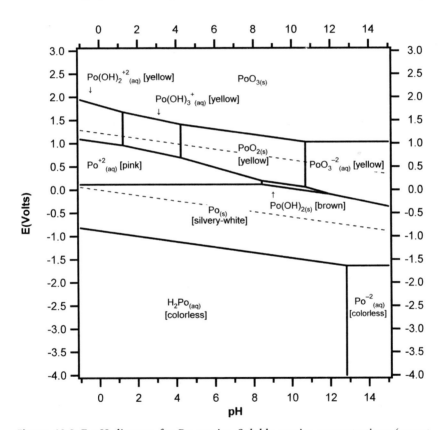

Figure 10.8 E–pH diagram for Po species. Soluble species concentrations (except H^+) = $10^{-4.0}$ M. Soluble species and most solids are hydrated. No agents producing complexes or insoluble compounds are present other than HOH and OH^-.

Species ($\Delta G°$ in kJ/mol): Po (0.0), H_2Po (192.5), Po^{-2} (338.9), Po^{+2} (45.2), $Po(OH)_2$ (−355.6), $Po(OH)_2^{+2}$ (−230.1), $Po(OH)_3^+$ (−460.2), PoO_2 (−221.8), PoO_3^{-2} (−313.8), PoO_3 (−138.1), HOH (−237.2), H^+ (0.0), and OH^- (−157.3).

Figure 10.8 (Continued)

Equations for the lines:

$PoO_3/Po(OH)_2^{+2}$	$E = 1.71 - 0.118\,pH - 0.030 \log [Po(OH)_2^{+2}]$
PoO_3/PoO_2	$E = 1.66 - 0.059\,pH$
$PoO_3/Po(OH)_3^{+}$	$E = 1.67 - 0.089\,pH - 0.030 \log [Po(OH)_3^{+}]$
PoO_3/PoO_3^{-2}	$E = 0.91 - 0.030 \log [PoO_3^{-2}]$
$PoO_3^{-2}/Po(OH)_2$	$E = 1.45 - 0.118\,pH + 0.030 \log [PoO_3^{-2}]$
PoO_3^{-2}/Po	$E = 1.03 - 0.089\,pH + 0.015 \log [PoO_3^{-2}]$
Po^{+2}/Po	$E = 0.23 + 0.030 \log [Po^{+2}]$
PoO_2/Po^{+2}	$E = 1.07 - 0.118\,pH - 0.030 \log [Po^{+2}]$
Po/H_2Po	$E = -1.00 - 0.059\,pH - 0.030 \log [H_2Po]$
Po/Po^{-2}	$E = -1.76 - 0.030 \log [Po^{-2}]$
$Po(OH)_2/Po$	$E = 0.62 - 0.059\,pH$
$Po(OH)_2^{+2}/Po^{+2}$	$E = 1.03 - 0.059\,pH$
$Po(OH)_3^{+}/Po^{+2}$	$E = 1.07 - 0.089\,pH$
$PoO_2/Po(OH)_2$	$E = 0.69 - 0.059\,pH$
PoO_3^{-2}/PoO_2	$2pH = 25.5 + \log [PoO_3^{-2}]$
$Po(OH)_2/Po^{+2}$	$2pH = 12.9 - \log [Po^{+2}]$
$PoO_2/Po(OH)_3^{+}$	$pH = 0.21 - \log [Po(OH)_3^{+}]$
Po^{-2}/H_2Po	$pH = 12.8$
$Po(OH)_3^{+}/Po(OH)_2^{+2}$	$pH = 1.2$

adsorption of the ions onto insoluble solution impurities and of failure of the Nernst equation to apply to extremely low concentrations. The species HPo^{-} probably appears in the lower portion of the diagram, but insufficient data are available to provide an estimate.

d. Po chemistry.   After Po has been produced by bombardment of Bi, the target is dissolved in acid and the silvery-white Po metal is plated out on Ag or Ni. The Po may then be isolated by volatilization. The metal dissolves in concentrated HNO_3 to give yellow $Po(OH)_2^{+2}$ and in dilute HCl to give pink Po^{+2}, and it burns in air to give yellow PoO_2. The metallic character of Po is indicated by the numerous salts it forms, some of which are listed in Table 10.6. The polonium(II) salts are readily oxidized, often by the effects

Table 10.6
Polonium Species

Name	Formula	State	Color	Solubility	$\Delta G°$(kJ/mole)
Polonium	Po	s	Silvery-white	Insol	0.0
Polonium(II) ion	Po^{+2}	aq	Pink		45.2
Polonate(IV) ion	PoO_3^{-2}	aq	Yellow		−313.8
Dihydroxopolonium(IV) ion	$Po(OH)_2^{+2}$	aq	Yellow		−230.1
Trihydroxopolonium(IV) ion	$Po(OH)_3^+$	aq	Yellow		−460.2
Polonide ion	Po^{-2}	aq	Colorless		338.9
Hydrogen polonide ion	HPo^-	aq	Colorless		
Hydropolonic acid	H_2Po	aq	Colorless		192.5
Polonium(II)					
bromide	$PoBr_2$	s	Purple	Sol	
chloride	$PoCl_2$	s	Red	Sol	
hydroxide	$Po(OH)_2$	s	Brown	Insol	−355.6
sulfate	$PoSO_4$	s	Pink	Sol	
sulfide	PoS	s	Black	Insol	218.0
Polonium(IV)					
bromide	$PoBr_4$	s	Red	Decomp	
chloride	$PoCl_4$	s	Yellow	Decomp	
chromate	$Po(CrO_4)_2$	s	Yellow	Decomp	
dioxide	PoO_2	s	Yellow	Insol	−221.8
iodate	$Po(IO_3)_4$	s	White	Decomp	
iodide	PoI_4	s	Black	Insol	
nitrate	$Po(NO_3)_4$	s	White	Decomp	
oxohydroxide	$PoO(OH)_2$	s	Yellow	Insol	
sulfate	$Po(SO_4)_2$	s	Purple	Decomp	
Polonium(VI)					
trioxide	PoO_3			Decomp	−138.1

of the radioactivity, and the polonium(IV) salts are subject to hydrolysis as indicated on the E–pH diagram. Strong complexes are formed between Po(IV) and many ligands. For example, Cl^- probably gives a series of complexes of the general formula $PoCl_a(OH)_b(HOH)_c^{4-a-b}$ with $a + b + c = 6$. In concentrated Cl^- at a very low pH, the value of a is high, the value of b is low, and c makes up the difference. In concentrated HCl the predominant species is $PoCl_6^{-2}$. As the Cl^- concentration decreases, and as the pH increases, species such as $PoCl_5(OH)^{-2}$, $PoCl_4(OH)_2^{-2}$, $PoCl_2(OH)_3(HOH)^-$, and $PoCl(OH)_3(HOH)_2$ occur, until finally $Po(OH)_6^{-2}$ (PoO_3^{-2}) is evidenced in solutions of low Cl^- concentration and high pH. The neutral hydroxy-containing species with $4 - a - b = 0$ are likely to be insoluble. Numerous other anions complex with Po(IV) in a similar fashion, with the strength of the Po-to-anion bonding decreasing in the series $OH^- > edta^{-4} > C_2O_4^{-2} >$

citrate^{-3} > tartrate^{-2} > nta^{-3} > $C_2H_3O_2^-$ > I^- > Br^- > Cl^- > SO_4^{-2} > NO_3^- >> ClO_4^-. This is an approximate series derived from somewhat inadequate data, so caution is advised. Complexation with Po(II) also occurs, but the complexes are considerably less stable than those with Po(IV). It is also to be noted that PoO_3 has been detected only for tracer amounts of the element.

e. *Analysis.* All three isotopic species which have been employed in investigations of Po chemistry are analyzed by their α-particle radioactivity. This is generally done by liquid scintillation counting.

f. *Health aspects.* Po is an exceptionally poisonous substance, both chemically and because of its radioactivity. The permissible body burden of Po-210 is $10^{-11.3}$ g, and the dose which gives a median survival time of 20 days in rats, mice, cats, and dogs is about 0.01 μg/kg. Hence, investigations must be carried out in tightly confined enclosures (glove boxes) to avoid ingestion or inhalation.

11

The Fluorine Group

1. Introduction

The Fluorine Group of the Periodic Table, whose elements are known as halogens (Greek halos and genes, meaning salt-forming), consists of fluorine F, chlorine Cl, bromine Br, iodine I, and astatine At. The outer electron structure ns^2np^5 characterizes all five of the elements, with n representing principal quantum numbers 2, 3, 4, 5, and 6, respectively. The ns^2np^5 indicates that oxidation number possibilities are $-I, I, III, V$, and VII with F showing only $-I$ (except for the unstable HOF). The bonding in the oxidation state of $-I$ is sometimes ionic and sometimes covalent, while that in the other states is covalent. Fluorine is the most electronegative element in the Periodic Table, and as one descends the group, the electronegativities decrease. Fluorine stands out as considerably different from the other elements, there being numerous discontinuities in properties between it and Cl. Astatine also differs from the other elements in that all its isotopes are radioactive, the long-lived At-210 having a half life of 8.1 days. Covalent radii in pm are as follows: F(71), Cl(99), Br(114), I(133), and At (147). Ionic radii in pm are as: F^-(119), Cl^-(167), Br^-(182), and I^-(206).

2. Fluorine (F) $2s^22p^5$

a. E–pH diagram. Figure 11.1 depicts the E–pH diagram for F with the soluble species (except H^+) at $10^{-1.0}$ M. The diagram is valid only in the absence of substances with which F forms soluble complexes or insoluble compounds. The species which have been considered are F_2, OF_2, F^-, HF, and HF_2^-. This last species is not very stable and will appear on the diagram only at higher F^- concentrations. It shows up in between HF and F^-. The E–pH diagram emphasizes the very strong oxidizing power of F_2 and indicates that it will easily attack HOH to produce OF_2. The species oxygen fluoride OF_2 is also unstable but persists in solution longer than F_2. The decomposition

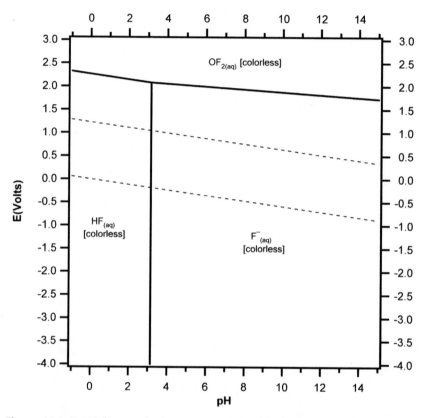

Figure 11.1 E–pH diagram for F species. Soluble species concentrations (except H^+) $= 10^{-1.0}$ M. Soluble species and most solids are hydrated. No agents producing complexes or insoluble compounds are present other than HOH and OH^-.

Species ($\Delta G°$ in kJ/mol): F_2 (0.0), OF_2 (39.3), HF (−296.6), F^- (−278.2), HF_2^- (−578.2) HOH (−237.2), H^+ (0.0), and OH^- (−157.3)

Equations for the lines:

OF_2/HF $\quad E = 2.25 - 0.059\ pH - 0.030 \log [HF] + 0.015 \log [OF_2]$

OF_2/F^- $\quad E = 2.16 - 0.030\ pH - 0.030 \log [F^-] + 0.015 \log [OF_2]$

F^-/HF $\quad pH = 3.2$

products of OF_2 are O_2 and F^-. The diagram also indicates that HF is a weak acid with the F^- exhibiting a log K_p value of 3.2. The potentials involving F_2 have been indirectly ascertained, no direct experimental data being available. Equations describing the lines between species are presented in the legend of Figure 11.1.

b. Discovery and occurrence. Because of its exceptional reactivity, F was discovered quite late. It was 1886 when Henri Moissan finally isolated it after many chemists had tried for numerous decades. He electrolyzed a solution of potassium hydrogen fluoride KHF_2 in HF. The name derives from the mineral fluorspar which was used as a flux (Latin fluor which means flowing). The major occurrences of F are the minerals fluorspar CaF_2, cryolite Na_3AlF_6, and apatite $Ca_5(PO_4)_3F$, but the first is the one that is generally employed for the production of F_2.

c. The element. The diatomic F_2 is commercially prepared by modifications of Moissan's method. The pale yellow, pungent, highly reactive gas reacts with almost everything, in many cases violently. Even the inert gases Kr and Xe form fluoride compounds. F_2 reacts with HOH to form HF or F^- and attacks glass (SiO_2) to yield the colorless gas SiF_4.

d. Hydrofluoric acid and the fluoride ion. The main aqueous species of F are HF, F^-, and HF_2^-. These are presented in Table 11.1 along with F_2, OF_2, and HOF, all three of these being thermodynamically unstable in HOH, the last two having a temporary existence due to slow reaction. Hydrogen fluoride HF is a colorless, highly corrosive gas which can be prepared by treating CaF_2 with H_2SO_4. It is very soluble in water to produce hydrofluoric acid, a weak, but dangerously corrosive, acid. Neutralizations of HF with oxides, hydroxides, and carbonates, or attacks by HF on metals give fluoride compounds. The fluorides of most alkali metals, Ag, Hg, Tl, and Be are soluble, whereas those of most alkaline earths, lanthanoids, actinoids, Li, Cu, Pb(II), Sn(II), Sb(III), Zn, and Fe are insoluble. The fluoride ion forms many complexes of the type MF_a^{n-a} where M is a simple or complex cation of formal charge n: Be^{+2}

Table 11.1
Fluorine Species

Name	Formula	State	Color	Solubility	$\Delta G°$ (kJ/mole)
Fluorine	F_2	g	Pale yellow	Decomp	0.0
Fluoride ion	F^-	aq	Colorless		-278.2
Hydrogen difluoride ion	HF_2^-	aq	Colorless		-578.2
Hydrofluoric acid	HF	aq	Colorless		-296.6
Oxygen difluoride	OF_2	aq	Colorless	Decomp	39.3

(a ≤ 4), Sc^{+3} (a ≤ 4), TiO^{+2} (a ≤ 4), VO^{+2} (a ≤ 4), VO_2^+ (a ≤ 4), UO_2^{+2} (a ≤ 4), Fe^{+3} (a ≤ 4), B^{+3} (a ≤ 4), and In^{+3} (a ≤ 4), Al^{+3} (a ≤ 6), Zr^{+4} (a ≤ 6), Hf^{+4} (a ≤ 6), Ta^{+5} (a ≤ 6), and Si^{+4} (a ≤ 6). Values of β_n for metal-fluoride complexes are presented under the respective metals.

e. Analyses. An electrometric method using a specific fluoride electrode will measure the ion down to about 50 ppb. The use of ion chromatography extends this limit to about 1 ppb.

f. Health aspects. Both F_2 and its inorganic compounds are highly toxic. For F_2 the lethal concentration in an hour for 50% of a population of rats, mice, or guinea pigs is about 150 ppm, and that for gaseous HF is about 500 ppm. The LD50 (oral rat) for NaF in rats is 52 mg/kg. The very small amount of F^- added to water supplies (1 ppm or less) is known to protect teeth from caries, and no adverse effects have been substantiated over many years of usage.

3. Chlorine (Cl) $3s^2 3p^5$

a. E–pH diagram. Figure 11.2 is the E–pH diagram for Cl with soluble species (except H^+) at $10^{-1.0}$ M. The diagram assumes that no insoluble Cl compound nor any Cl-containing soluble complex is involved. The species which have been entered into the construction of the diagram are Cl_2, HCl, Cl^-, HClO, ClO^-, $HClO_2$, ClO_2^-, $HClO_3$, ClO_3^-, $HClO_4$, ClO_4^-, and H^+. It will be noted that the final diagram includes only Cl^-, Cl_2, and ClO_4^-. This indicates that all the other species are thermodynamically unstable with respect to the remaining three. However, all the missing species are very slow to react which means that they can exist temporarily in HOH solutions. Further, HCl, $HClO_3$, and $HClO_4$ do not appear on the E–pH diagram because they are strong acids. Equations describing the lines between species are presented in the legend of Figure 11.2.

b. Discovery and occurrence. In 1774 Carl Wilhelm Scheele prepared a yellow-green gas by the oxidation of HCl with MnO_2. He believed that he had prepared a compound rather than an element. In 1810, Humphry Davy insisted that it was an element and named it chlorine after the Greek word chloros which means greenish-yellow. Chlorine is found in nature only in the combined state, most frequently as salt NaCl, sylvite KCl, and carnallite $KMgCl_3$. As NaCl, it occurs in sea water, salt lakes, brine wells, and mines.

c. The element. Cl_2 is a pungent, poisonous, yellow-green gas which is prepared commercially by the electrolysis of solutions of NaCl. Its dissolution in HOH is described by this equation: $Cl_2 + HOH \rightarrow HClO + Cl^- + H^+$ which has a log K value of -3.4. The position of Cl_2 on the E–pH diagram indicates that it is a very good oxidizing agent, although not nearly as strong as F_2. Cl forms several unstable oxides: Cl_2O (brownish-yellow gas), Cl_2O_3 (brown solid), ClO_2 (yellow-green gas), Cl_2O_4 (light yellow liquid), Cl_2O_6 (red liquid), and Cl_2O_7 (colorless liquid). Cl_2O is the anhydride of HClO, hypochlorous acid; and Cl_2O_7 is the anhydride of perchloric acid, $HClO_4$. Table 11.2 presents the predominant chlorine species which occur in aqueous solution.

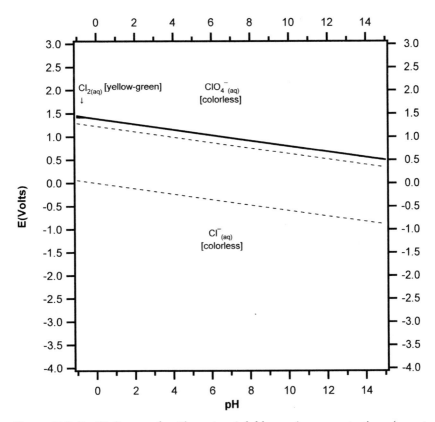

Figure 11.2 E–pH diagram for Cl species. Soluble species concentrations (except H^+) $= 10^{-1.0}$ M. Soluble species and most solids are hydrated. No agents producing complexes or insoluble compounds are present other than HOH and OH^-.

Species ($\Delta G°$ in kJ/mol): Cl_2 (7.1), $HClO_4$ (34.3), ClO_4^- (−8.4), $HClO_3$ (7.5), ClO_3^- (−7.9), $HClO_2$ (5.9), ClO_2^- (17.2), $HClO$ (−79.9), ClO^- (−36.3), HCl (−91.6) Cl^- (−131.4), HOH (−237.2), H^+ (0.0), and OH^- (−157.3)

Equations for the lines:

Cl_2/Cl^- $E = 1.40 - 0.059 \log [Cl^-] + 0.030 \log [Cl_2]$

ClO_4^-/Cl^- $E = 1.39 - 0.059 \, pH$

ClO_4^-/Cl_2 $E = 1.39 - 0.067 \, pH + 0.008 \log [ClO_4^-] - 0.004 \log [Cl_2]$

Table 11.2
Chlorine Species

Name	Formula	State	Color	Solubility	$\Delta G°$ (kJ/mole)
Chlorine	Cl_2	g	Yellow-green	Decomp	0.0
Chlorine	Cl_2	aq	Yellow-green		7.1
Chloride ion	Cl^-	aq	Colorless		−131.4
Hydrochloric acid	HCl	aq	Colorless		−91.6
Hypochlorite ion	ClO^-	aq	Colorless	Slow decomp	−36.3
Hypochlorous acid	HClO	aq	Colorless	Slow decomp	−79.9
Chlorite ion	ClO_2^-	aq	Colorless	Slow decomp	17.2
Chlorous acid	$HClO_2$	aq	Colorless	Slow decomp	5.9
Chlorate ion	ClO_3^-	aq	Colorless	Slow decomp	−7.9
Chloric acid	$HClO_3$	aq	Colorless	Slow decomp	7.5
Perchlorate ion	ClO_4^-	aq	Colorless	Slow decomp	−8.4
Perchloric acid	$HClO_4$	aq	Colorless	Slow decomp	34.3

d. Hydrochloric acid and chlorides. The action of H_2SO_4 on NaCl produces HCl, a pungent, toxic, colorless gas. The gas dissolves extensively in HOH to give the strong acid, hydrochloric acid, HCl. Neutralizations of metal oxides, hydroxides, and carbonates, as well as attacks on metals with appropriate E values yield chloride salts. The salts of Ag, Pb, Hg(I), Cu(I), Tl(I), BiO^+, SbO^+, Au(I), Pt(II), and Pd(II) are insoluble; most others are soluble. Chloride salts are white, except for those that have colored cations. The chloride ion Cl^- is a moderately strong complexing agent, forming complexes of the form MCl_a^{n-a} where M is a cation of formal charge n: Zr^{+4} ($a \leq 4$), Fe^{+3} ($a \leq 3$), Rh^{+3} ($a \leq 6$), Pd^{+2} ($a \leq 4$), Pt^{+2} ($a \leq 4$), Cu^+ ($a \leq 4$), Au^{+3} ($a \leq 4$), Cd^{+2} ($a \leq 4$), Hg^{+2} ($a \leq 4$), In^{+3} ($a \leq 3$), Tl^{+3} ($a \leq 4$), Sn^{+2} ($a \leq 4$), Pb^{+2} ($a \leq 4$), Sb^{+3} ($a \leq 6$), Bi^{+3} ($a \leq 6$).

e. Oxyacids of chlorine. Cl forms hypochlorous acid, HClO, chlorous acid, $HClO_2$, chloric acid, $HClO_3$, and perchloric acid, $HClO_4$. The first three of these have not been isolated, existing only in solution. All of them are strong oxidizing agents, E° for $HClO/Cl_2$ being 1.61 v, for $HClO_2/Cl_2$ 1.64 v, for $HClO_3/Cl_2$ 1.47 v, and for $HClO_4/Cl_2$ 1.39 v, for $HClO/Cl^-$ being 1.61 v, for $HClO_2/Cl^-$ 1.64 v, for $HClO_3/Cl^-$ 1.47 v, and for $HClO_4/Cl^-$ 1.39 v.

HClO is obtained in solution by $Cl_2 + HgO + HOH \rightarrow HgCl_2 + 2HClO$, the HgO and $HgCl_2$ being insoluble. The log K_p for the ClO^- anion is 7.5. Hypochlorite salts are prepared by reactions of the type $Cl_2 + 2Li^+ + 2OH^- \rightarrow 2Li^+ + ClO^- + Cl^- + HOH$. Among the more stable of the salts are those of Li, Na, Ca, Sr, and Ba.

$HClO_2$ is prepared by making $Ba(ClO_2)_2$ then treating it with H_2SO_4 to give a solution of $HClO_2$ and insoluble $BaSO_4$. $Ba(ClO_2)_2$ is made by treating BaO_2 with ClO_2. The anion ClO_2^- has a log K_p value of 2.0 indicating that the acid is moderately strong. Chlorites are usually made by reducing an HOH solution of ClO_2 and metal hydroxide with H_2O_2. The chlorite salts of alkali metals and the alkaline earths are white or colorless, and are soluble. Those of many of the heavier metals tend to explode upon being struck.

$HClO_3$ is obtained in solution by the following reaction: $Ba(ClO_3)_2 + H_2SO_4 \rightarrow 2H^+ + 2ClO_3^- + BaSO_4$, the $BaSO_4$ being insoluble. The barium chlorate for this reaction is made by $Ba^{+2} + 12OH^- + 6Cl_2 \rightarrow Ba(ClO_3)_2 + 10Cl^- + 6HOH$. Chlorates are commercially generated by the electrolysis of NaCl solutions which are rapidly stirred; the overall reaction is $Cl^- + 3HOH \rightarrow ClO_3^- + 3H_2$ which results from the attack of Cl_2 produced at the anode with the OH^- produced at the cathode plus further anodic oxidation. The chlorate anion has a log K_p of -2.7 which indicates a strong acid. Most metal chlorates are soluble.

f. Perchloric acid and perchlorates.

f. Perchloric acid and perchlorates. The electrolytic oxidation of $NaClO_3$ gives $NaClO_4$. This salt is treated with HCl to give a solution of ClO_4^-, Cl^-, H^+, and Na^+. The $HClO_4$ can be boiled off as a 72% azeotrope. The anhydrous liquid acid may be obtained by distillation at low pressure in the presence of fuming H_2SO_4. The log K_p of ClO_4^- is estimated to be -7.5, making $HClO_4$ an extremely strong acid. At room temperature, $HClO_4$ solutions and perchlorate solutions exhibit little oxidative behavior, but when warmed they function as very vigorous oxidants. Perchlorate salts are usually soluble except for those of K, Rb, and Cs, which are slightly soluble. The perchlorate ion is generally a very weak coordinating agent, although some examples of its coordination are known.

g. Analysis. The most sensitive methods for the determination of Cl as the chloride ion are differential pulse polarography and ion chromatography, both of which go down into the 1 ppb range. The lower limit for an ion-sensitive electrode is about 1 ppm.

h. Health aspects. The chloride ion is generally considered to be non-poisonous. The LD50(oral rat) for NaCl is 3 g/kg. In small quantities NaCl functions as an essential human nutrient factor. The LD50(oral rat) for $Ca(ClO)_2$ is 850 mg/kg, that for $NaClO_2$ is 165 mg/kg, that for $NaClO_3$ is 1.2 g/kg, and that for $NaClO_4$ is 2.1 g/kg. A 1-h inhalation of 293 ppm Cl_2 gas by rats gives 50% deaths. A concentration of 3.5 ppm in air produces a detectable odor, and immediate irritation occurs with 15 ppm.

4. Bromine (Br) $4s^2 4p^5$

a. E–pH diagram. Figure 11.3 is the E–pH plot which results for $10^{-1.0}$ M Br when these species are considered: Br_2, Br^-, HBr, BrO^-, HBrO, BrO_3^-, $HBrO_3$, BrO_4^-, $HBrO_4$. The assumption has been made that no insoluble compound and no complex compound is involved. The species BrO^- and HBrO do not appear on the diagram because they are thermodynamically unstable with respect to the species appearing on the diagram, however, they exist in solution because their decomposition is usually slow. HBr, $HBrO_3$,

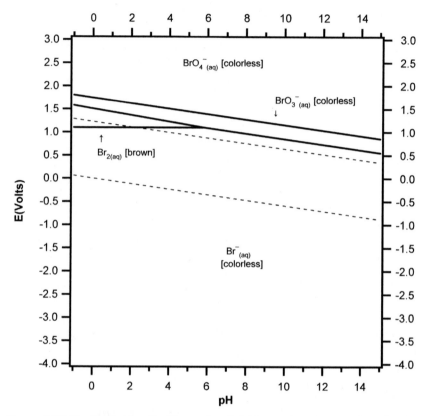

Figure 11.3 E–pH diagram for Br species. Soluble species concentrations (except H^+) = $10^{-1.0}$ M. Soluble species and most solids are hydrated. No agents producing complexes or insoluble compounds are present other than HOH and OH^-.

Species ($\Delta G°$ in kJ/mol): Br_2 (4.2), $HBrO_4$ (158.0), BrO_4^- (118.1), $HBrO_3$ (30.2), BrO_3^- (18.6), HBrO (−82.4), BrO^- (−33.5), HBr (−52.8), Br^- (−104.0), HOH (−237.2), H^+ (0.0), and OH^- (−157.3).

Equations for the lines:

Br_2/Br^-	$E = 1.10 - 0.059 \log[Br^-] + 0.030 \log[Br_2]$
BrO_3^-/Br^-	$E = 1.44 - 0.059\,pH$
BrO_3^-/Br_2	$E = 1.51 - 0.071\,pH + 0.012 \log[BrO_3^-] - 0.006 \log[Br_2]$
BrO_4^-/BrO_3^-	$E = 1.74 - 0.059\,pH$

and $HBrO_4$ are absent from the diagram because they are strong acids. Both BrO_4^- and BrO_3^- have the potential to react with HOH, but again, they are often slow to do so. Equations describing the lines between species are presented in the legend of Figure 11.3.

b. Discovery and occurrence. Antoine-Jerome Balard in 1828 treated water from a salt marsh with Cl_2 and extracted the yellow-colored layer with ether and KOH. He treated the product KBr with H_2SO_4 and MnO_2 to give a red liquid Br_2. The French Academy named the new element bromine from the Greek bromos which means stink. The largest source of Br is sea water where it exists as bromide ion Br^-. Br^- also exists in brine wells and salt lakes, where it is often more concentrated than in sea water.

c. The element. Bromine is made industrially by the oxidation of Br^- with Cl_2. It dissolves in HOH to about 0.2 M and undergoes this reaction: $Br_2 + HOH \rightarrow HBrO + Br^- + H^+$ (log K $= 10^{-8.1}$). Examination of the E–pH diagram shows that the oxidizing power of Br_2 is considerably less than that of Cl_2. There are only three well-characterized oxides of Br, all of them being highly unstable: brown Br_2O, yellow Br_2O_4, and orange Br_2O_3 (all solids at low temperatures). Table 11.3 presents the predominant bromine species which occur in aqueous solution.

d. Hydrobromic acid and bromides. Colorless HBr gas is generated industrially by the catalyzed reaction of H_2 and Br_2 at an elevated temperature. Smaller quantities can be prepared by the following reaction: $2P + 8HOH + 5Br_2 \rightarrow 10HBr + 2H_3PO_4$. The disagreeably smelling gas is highly soluble in HOH to yield HBr hydrobromic acid, a strong acid with an anion showing an approximate log K_p value of -9.0. Bromides can be made by reacting HBr with

Table 11.3
Bromine Species

Name	Formula	State	Color	Solubility	$\Delta G°$ (kJ/mole)
Bromine	Br_2	l	Brown		0.0
Bromine	Br_2	aq	Brown	Decomp	4.2
Bromide ion	Br^-	aq	Colorless		−104.0
Hydrobromic acid	HBr	aq	Colorless		−52.8
Hypobromite ion	BrO^-	aq	Colorless	Slow decomp	−33.5
Hypobromous acid	HBrO	aq	Colorless	Slow decomp	−82.4
Bromate ion	BrO_3^-	aq	Colorless	Slow decomp	18.6
Bromic acid	$HBrO_3$	aq	Colorless	Slow decomp	30.2
Perbromate ion	BrO_4^-	aq	Colorless	Slow decomp	118.1
Perbromic acid	$HBrO_4$	aq	Colorless	Slow decomp	158.0

metal oxides, hydroxides, or carbonates, and in many cases, with the metal itself. The solubilities of bromides are quite similar to those of chlorides. The bromide ion Br^- is a moderate complexing agent, forming complexes chiefly with heavy metal cations of the form MBr_a^{n-a} where M is a cation of formal charge n: Pd^{+2} (a ≤ 4), Pt^{+2} (a ≤ 4), Cu^+ (a ≤ 4), Ag^+ (a ≤ 4), Cd^{+2} (a ≤ 4), Hg^{+2} (a ≤ 4), In^{+3} (a ≤ 4), Tl^{+3} (a ≤ 4), Sn^{+2} (a ≤ 4), Pb^{+2} (a ≤ 5), Bi^{+3} (a ≤ 6).

e. Oxyacids of bromine. Br forms colorless hypobromous acid HBrO, colorless bromic acid $HBrO_3$, and colorless perbromic acid $HBrO_4$, these leading to hypobromite, bromate, and perbromate salts. The acids are stable only in HOH solution, none of them having been isolated. $HBrO_2$ is apparently very unstable, but two bromite salts have been prepared: $Sr(BrO_2)_2$ and $Ba(BrO_2)_2$. HBrO can be prepared in solution by the hydrolysis of Br as indicated above. The log K_p value of BrO^- is 8.6. Na and K hypobromites have been prepared in the cold, but when they warm up, they decompose. The perbromate ion BrO_4^- is prepared by the oxidation of BrO_3^- in strong alkali with F_2: $BrO_3^- + 2OH^- + F_2 \rightarrow BrO_4^- + 2F^- + HOH$. A cation exchange resin is used to produce a solution of $HBrO_4$. The acid and its anion thermodynamically have a powerful oxidizing potential, but a dilute solution is essentially inert at room temperature. Concentrated solutions are not so inert and act as potent oxidizing agents. Perbromate salts resemble perchlorate salts in their solubilities.

f. Bromic acid and bromates. The most important oxyacid of Br is bromic acid $HBrO_3$. It is prepared by the following reaction: $3Br_2 + 6OH^- \rightarrow BrO_3^- + 5Br^- + 3HOH$. As mentioned above, the acid has not been isolated, but exists only in solution. $HBrO_3$ is a strong acid, its anion having a log K_p value of -0.7, and it functions as a strong oxidant. Bromate salts are generally soluble, exceptions being those of Ag, Ba, Tl(I), Hg(II), and Pb.

g. Analysis. The most sensitive methods for the determination of Br as the bromide ion are differential pulse polarography and ion chromatography, the first going down into the 1 ppb range and the second to the 10 ppb range. An ion-sensitive electrode exhibits a detection limit around 100 ppb.

h. Health aspects. The LD50(oral rat) for NaBr is 3.5 g/kg and that for $KBrO_3$ is 320 mg/kg. For Br_2 the LD50(oral rat) is 2.6 mg/kg.

5. Iodine (I) $5s^2 5p^5$

a. E–pH diagram. Table 11.4 presents the predominant iodine species which occur in aqueous solution. When all these except I_3^- are employed to construct an E–pH diagram, Figure 11.4 results. As in other halide E–pH

Table 11.4
Iodine Species

Name	Formula	State	Color	Solubility	ΔG° (kJ/mole)
Iodine	I_2	s	Blue-black	Sl sol	0.0
Iodine	I_2	aq	Brown		16.3
Iodide ion	I^-	aq	Colorless		-51.9
Triiodide ion	I_3^-	aq	Colorless		-51.7
Hydroiodic acid	HI	aq	Colorless		3.1
Hypoiodite ion	IO^-	aq	Colorless	Decomp	-38.4
Hypoiodous acid	HIO	aq	Colorless	Decomp	-99.1
Iodate ion	IO_3^-	aq	Colorless	Slow decomp	-127.6
Iodic acid	HIO_3	aq	Colorless	Slow decomp	-132.5
Trihydrogenperiodate ion	$H_3IO_6^{-2}$	aq	Colorless		-485.6
Tetrahydrogenperiodate ion	$H_4IO_6^-$	aq	Colorless		-524.4
Periodic acid, ortho	H_5IO_6	aq	Colorless		-541.4

diagrams, the assumption has been made that no insoluble compound or any complex species is involved. The species HIO and IO^- do not appear because they are thermodynamically unstable with respect to other species in the diagram. HI does not show because of its acid strength. I_3^- (complex of I_2 and I^-) has not been included because it occupies a very narrow band parallel to and just beneath the lower I_2 line. Equations describing the lines between species are presented in the legend of Figure 11.4.

b. Discovery and occurrence. Iodine was first obtained in 1811 by Bernard Courtois who treated seaweed ash with concentrated H_2SO_4 and then condensed the purple vapors which came off. It was named shortly thereafter by Louis-Joseph Gay-Lussac after the Greek iodes which means violet-colored. Iodine of commercial importance occurs in subterranean brines as I^- and in iodate minerals such as lautarite $Ca(IO_3)_2$.

c. The element. The I^- in brines is oxidized with Cl_2 to yield I_2 which can be blown out with air or taken out as I_3^- on an ion exchange resin. The iodates are treated with $NaHSO_3$ to reduce the element to I^- which can then be reacted with fresh IO_3^- to precipitate I_2: $5I^- + IO_3^- + 6H^+ \rightarrow 3I_2 + 3HOH$. Oxides of I include the white solid I_2O_5 (the anhydride of HIO_3), and the much less stable yellow solids I_2O_4 and I_4O_9.

d. Hydroiodic acid and iodides. HI is readily produced by the action of concentrated H_3PO_4 on a metal iodide. The use of H_2SO_4 instead of H_3PO_4 is unsatisfactory because the former causes oxidation of the HI.

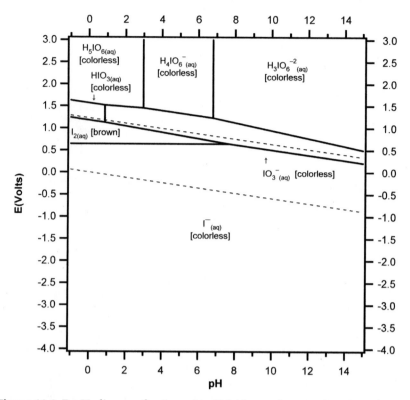

Figure 11.4 E–pH diagram for I species. Soluble species concentrations (except H^+) = $10^{-1.0}$ M. Soluble species and most solids are hydrated. No agents producing complexes or insoluble compounds are present other than HOH and OH^-.

Species ($\Delta G°$ in kJ/mol): I_2 (16.3), H_5IO_6 (−541.4), $H_4IO_6^-$ (−524.4), $H_3IO_6^{-2}$ (−485.6), HIO_3 (−132.5), IO_3^- (−127.6), HIO (−99.1), IO^- (−38.4), HI (3.1), I^- (−51.9), HOH (−237.2), H^+ (0.0), and OH^- (−157.3)

Equations for the lines:

I_2/I^-	$E = 0.62 - 0.059 \log [I^-] + 0.030 \log [I_2]$
IO_3^-/I^-	$E = 1.10 - 0.059 \, pH$
HIO_3/I_2	$E = 1.18 - 0.059 \, pH + 0.012 \log [HIO_3] - 0.006 \log [I_2]$
IO_3^-/I_2	$E = 1.19 - 0.071 \, pH + 0.012 \log [IO_3^-] - 0.006 \log [I_2]$
H_5IO_6/HIO_3	$E = 1.57 - 0.059 \, pH$
H_5IO_6/IO_3^-	$E = 1.54 - 0.030 \, pH$
$H_4IO_6^-/IO_3^-$	$E = 1.63 - 0.059 \, pH$
$H_3IO_6^{-2}/IO_3^-$	$E = 1.83 - 0.089 \, pH$
$H_3IO_6^{-2}/H_4IO_6^-$	$pH = 6.8$
$H_4IO_6^-/H_5IO_6$	$pH = 3.0$
IO_3^-/HIO_3	$pH = 0.9$

The odoriferous gas HI has a high solubility in HOH producing hydroiodic acid HI, a strong acid whose anion has an estimated log K_p value of -9.5. Iodide salts can be prepared by treatment of appropriate metal oxides, hydroxides, or carbonates with HI. For active metals, direct attack of HI will yield the salts. The solubilities of iodides compare favorably with those of bromides and chlorides. Iodides of Ag, Pb, Hg(I), Hg(II), Cu(I), Tl, BiO^+, SbO^+, Au(I), Pt(II), and Pd(II) are insoluble, with most others being soluble. In general, insoluble iodides are more insoluble than the corresponding bromides. Iodide forms complexes with heavy metal ions, particularly ones in the Periodic Table region of Hg and Pb. Some examples are of the form $MI_a{}^{n-a}$ where M is a cation of formal charge n: Fe^{+3} ($a \le 2$), Pd^{+2} ($a \le 4$), Pt^{+2} ($a \le 4$), Cu^+ ($a \le 4$), Ag^+ ($a \le 4$), Cd^{+2} ($a \le 4$), Hg^{+2} ($a \le 4$), Tl^{+3} ($a \le 4$), Sn^{+2} ($a \le 6$), Pb^{+2} ($a \le 4$), Bi^{+3} ($a \le 6$). Solutions of iodide show markedly increased solubility of I_2, the reason being the formation of the complex $I_3{}^-$.

e. Oxyacids of iodine.

I forms the colorless hypoiodous acid HIO, the white solid iodic acid HIO_3, and the white solid periodic acid H_5IO_6 (also known as orthoperiodic acid). All of these are strong oxidizing agents, the $E°$ value for HIO/I_2 being 1.44 v, that for $IO_3{}^-/I_2$ 1.19 v, and that for H_5IO_6/I_2 1.31 v. HIO is a weak unstable acid which exists only in solution. The acid anion IO^- has a log K_p value of about 10.6, and decomposes readily into iodide and iodate. No solid hypoiodites have been prepared, although the anion in basic solution is somewhat more stable than in acidic solution. Neither HIO_2 nor $IO_2{}^-$ has been prepared. Iodic acid HIO_3 is made by oxidizing a suspension of I_2 in HOH electrolytically or with very concentrated HNO_3. The white solid acid may be crystallized from the resulting solution. The acid is fairly strong, the log K_p value of $IO_3{}^-$ being 0.9. Iodates can be made by the standard method of the use of the acid with hydroxides, oxides, or carbonates. The iodates of the alkali metals, Mg^{+2}, Ni^{+2}, Cd^{+2}, Zn^{+2}, and Cr^{+3} are soluble, those of Ca^{+2}, Sr^{+2}, Co^{+2}, Cu^{+2}, and Fe^{+3} are slightly soluble, and most others are insoluble. H_5IO_6 is treated in the next section.

f. Periodic acid and periodates.

The three species across the top of the I E–pH diagram are (ortho)periodic acid and its two deprotonated anions. The three formulas represent what may be considered the parent species, since there also exist in solution many derivative species. These other species consist of dehydrated and polymeric forms of the parents. For examples, in equilibrium with H_5IO_6 are the species H_3IO_5, HIO_4, and some polymers; and in equilibrium with $H_4IO_6{}^-$ are the species $H_2IO_5{}^-$, $IO_4{}^-$, and some polymers. The oxidation of iodate with Cl_2 is described by this equation

$$IO_3{}^- + 7Na^+ + 4OH^- + Cl_2 \rightarrow H_2IO_6{}^{-3} + 2Cl^- + HOH + 7Na^+.$$

The solution can then be made acidic, and white solid H_5IO_6 can be isolated from it. Log K_p values for $H_3IO_6{}^{-2}$, $H_4IO_6{}^-$ are respectively 6.8, and 3.0, values which can be seen in the E–pH diagram. Periodic salts are known which contain the IO_6^{-5}, IO_5^{-3}, and IO_4^- ions, as well as some polymeric oxyanions having an iodine with an oxidation state of VII (such as $I_2O_9^{-4}$). The alkali metal periodates are soluble, but many others are slightly soluble or insoluble. Periodates complex strongly with a number of cations in their higher oxidation states: Mn(IV), Ni(IV), Fe(III), Pd(IV), Ce(IV), Co(III), Cu(III), Ag(III).

g. Analysis. An ion-sensitive electrode can determine the iodide ion down into the region of 50 ppb. Differential pulse polarography and ICPMS extend this limit to about 0.05 ppb.

h. Health aspects. LD50(oral rat) for NaI is 1.8 g/kg, LD50(oral mouse) for $NaIO_3$ is 500 mg/kg. The human body needs 75–150 μg iodine daily for optimal thyroid hormone production. There is a known case in which the consumption of 2–3 g of I_2 led to the death of a person.

6. Astatine (At) $6s^26p^5$

In 1940, Dale R. Corson, K. R. Mackenzie, and Emilo Gino Segrè bombarded Bi with alpha particles. Tracer studies indicated that the product element had an atomic number of 85. When the end of World War II afforded them the time, these researchers did further work which confirmed their initial discovery. They named the element after the Greek word astatos which means unstable. The isotope which they had produced was At-211 which has a half life of 7.2 h. Since then, numerous other isotopes have been discovered, and except for At-210 (half life 8.3 h), none has a half life greater than that of At-211, and those that occur in nature are present in extremely small amounts. These short half lives indicate that weighable amounts of At cannot be isolated. Most studies have therefore been conducted with very dilute solutions 10^{-11} to 10^{-15} M. These investigations have involved following the At as it precipitates with I compounds, as it extracts into organic solvents with I in various forms, as it migrates in electrophoretic and chromatographic investigations, and as its volatility is observed. Table 11.5 presents the predominant astatine species which are believed to occur in aqueous solution from which the E-pH diagram for At (Figure 11.5) is constructed.

The At-211 made from Bi metal can be boiled off the target and condensed onto a cool glass surface. It can be taken into solution by washing the surface with various reagents. Five oxidation states have been established:−I, 0, I, V, and VII. When the At on the glass surface is treated with a strong reductant, the resulting species precipitates with AgI, indicating At^-. A weak oxidant returns it to At(0) (extracts into organic solvents and is volatile), a somewhat stronger one produces HAtO, a still stronger one gives AtO_3^- (coprecipitates with $AgIO_3$), and a very powerful oxidant yields $H_4AtO_6^-$ (or perhaps AtO_4^-).

Table 11.5
Astatine Species

Name	Formula	State	Color	Solubility	ΔG° (kJ/mole)
Astatine	At or At$_2$	s	Unknown	Sol	0.0
Astatine	At or At$_2$	aq	Unknown		0.0
Astinide ion	At$^-$	aq	Unknown		-28.9
Hypoastinite ion	AtO$^-$	aq	Unknown		-79.9
Hypoastinous acid	HAtO	aq	Unknown		-140.6
Astinate ion	AtO$_3^-$	aq	Unknown		-36.0
Astinic acid	HAtO$_3$	aq	Unknown		-74.9
Trihydrogen perastinate ion	H$_3$AtO$_6^{-2}$	aq	Unknown	Decomp	-419.7
Tetrahydrogen perastinate ion	H$_4$AtO$_6^-$	aq	Unknown	Decomp	-443.5
Perastinic acid, ortho	H$_5$AtO$_6$	aq	Unknown		-372.8

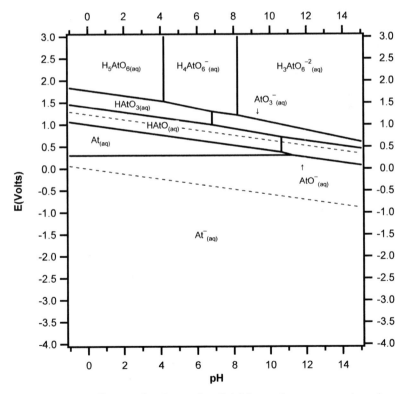

Figure 11.5 E–pH diagram for At species. Soluble species concentrations (except H$^+$) $= 10^{-13.0}$ M. Soluble species and most solids are hydrated. No agents producing complexes or insoluble compounds are present other than HOH and OH$^-$.

Figure 11.5 (Continued)

Species ($\Delta G°$ in kJ/mol): At (0.0), H_5AtO_6 (−443.5), $H_4AtO_6{}^-$ (−419.7), $H_3AtO_6{}^{-2}$ (−372.8), $HAtO_3$ (−74.9), $AtO_3{}^-$ (−36.0), $HAtO$ (−140.6), AtO^- (−79.9), At^- (−28.9), HOH (−237.2), H^+ (0.0), and OH^- (−157.3)

Equations for the lines:

$H_5AtO_6/HAtO_3$	$E = 1.78 - 0.059\ pH$
$H_4AtO_6{}^-/HAtO_3$	$E = 1.90 - 0.089\ pH$
$H_4AtO_6{}^-/AtO_3{}^-$	$E = 1.70 - 0.059\ pH$
$H_3AtO_6{}^{-2}/AtO_3{}^-$	$E = 1.94 - 0.089\ pH$
$HAtO_3/HAtO$	$E = 1.40 - 0.059\ pH$
$AtO_3{}^-/HAtO$	$E = 1.50 - 0.074\ pH$
$AtO_3{}^-/AtO^-$	$E = 1.34 - 0.059\ pH$
$HAtO/At$	$E = 1.00 - 0.059\ pH$
AtO^-/At	$E = 1.63 - 0.118\ pH$
AtO^-/At^-	$E = 0.96 - 0.059\ pH$
At/At^-	$E = 0.30$
$H_4AtO_6{}^-/H_5AtO_6$	$pH = 4.2$
$H_3AtO_6{}^{-2}/H_4AtO_6{}^-$	$pH = 8.2$
$AtO_3{}^-/HAtO_3$	$pH = 6.8$
$AtO^-/HAtO$	$pH = 10.6$

This latter species decomposes to $AtO_3{}^-$ fairly rapidly. Estimated $E°$ values are 0.30 v for At/At^-, 1.00 v for $HAtO/At$, 1.50 v for $AtO_3{}^-/HAtO$, 1.40 v for $HAtO_3/HAtO$, and 1.70 v for $AtO_4{}^-/AtO_3{}^-$. All of the above data are in accord with what one should expect by extrapolation from the properties of I.

Figure 11.5 is an E–pH diagram for At at $10^{-13.0}$ M. This diagram is quite speculative, with the free energy values estimated from the above $E°$ values, the reported chemical behavior, and assumed similarities with I. Because of the experimental difficulties, all these items are subject to question, and are open to amendment by further research. Equations describing the lines between species are presented in the legend of Figure 11.5.

12

The Scandium Group

1. Introduction

The members of the Sc Group are Sc scandium, Y yttrium, La lanthanum, Ce cerium, Pr praseodymium, Nd neodymium, Pm promethium, Sm samarium, Eu europium, Gd gadolinium, Tb terbium, Dy dysprosium, Ho holmium, Er erbium, Tm thulium, Yb ytterbium, Lu lutetium, and Ac actinium. All these elements resemble each other greatly, especially in the series La–Lu (called the lanthanoids). Their slight differences may be assigned largely to size similarities, but a few oxidation state changes give rise to marked differences. The predominant oxidation state is III, but the IV state for Ce, and the II state for Eu are also important in their aqueous chemistries. The electron structures of these elements along with some other of their pertinent properties are shown in Table 12.1. Note the progression in the sizes of M^{+3} rising from Sc to Ac, but decreasing from La to Lu. This behavior causes Y^{+3} to fall in between Dy^{+3} and Ho^{+3}, which results in yttrium's chemistry usually resembling the latter lanthanoids. For this reason, Y will be treated as a lanthanoid in succeeding sections. The successive filling of the 4f electron level from La through Lu should also be noted, as well as the interesting 5d occupancy for Gd.

Table 12.1
Electron Structure of Sc Group Elements

	Electron structure		Radius (pm)		Color of	Origin of
M	M	M^{+3}	M	M^{+3}	M^{+3}	element name
Sc	$4s^2 3d^1$	$4s^0 3d^0$	162	89	Colorless	Scandanavia
Y	$5s^2 4d^1$	$5s^0 4d^0$	180	104	Colorless	Ytterby, Sweden
La	$6s^2 5d^1$	$6s^0 5d^0$	183	117	Colorless	Greek lanthanein: to escape notice
Ac	$7s^2 6d^1 5f^0$	$7s^0 6d^0 5f^0$	188	126	Colorless	Greek aktinos: ray
La	$6s^2 5d^1$	$6s^0 5d^0$	183	117	Colorless	Greek lanthanein: to escape notice
Ce	$6s^2 5d^0 4f^2$	$6s^0 5d^0 4f^1$	182	115	Colorless	Asteroid Ceres discovered in 1801
Pr	$6s^2 5d^0 4f^3$	$6s^0 5d^0 4f^2$	182	113	Green	Greek praseos + didymos: green twin
Nd	$6s^2 5d^0 4f^4$	$6s^0 5d^0 4f^3$	181	112	Lilac	Greek neos + didymos: new twin
Pm	$6s^2 5d^0 4f^5$	$6s^0 5d^0 4f^4$	181	111	Pink	Prometheus stole fire from heaven
Sm	$6s^2 5d^0 4f^6$	$6s^0 5d^0 4f^5$	180	110	Yellow	Mineral samarskite
Eu	$6s^2 5d^0 4f^7$	$6s^0 5d^0 4f^6$	208	109	Pale pink	Europe
Gd	$6s^2 5d^1 4f^7$	$6s^0 5d^0 4f^7$	180	108	Colorless	Chemist Gadolin
Tb	$6s^2 5d^0 4f^9$	$6s^0 5d^0 4f^8$	177	106	Pale pink	Ytterby, Sweden
Dy	$6s^2 5d^0 4f^{10}$	$6s^0 5d^0 4f^9$	178	105	Yellow	Greek dysprositos: hard to get
Ho	$6s^2 5d^0 4f^{11}$	$6s^0 5d^0 4f^{10}$	176	104	Yellow	Latin for Stockholm: Holmia
Er	$6s^2 5d^0 4f^{12}$	$6s^0 5d^0 4f^{11}$	176	103	Rose	Ytterby, Sweden
Tm	$6s^2 5d^0 4f^{13}$	$6s^0 5d^0 4f^{12}$	176	102	Green	Latin Thule: northern land
Yb	$6s^2 5d^0 4f^{14}$	$6s^0 5d^0 4f^{13}$	193	101	Colorless	Ytterby, Sweden
Lu	$6s^2 5d^1 4f^{14}$	$6s^0 5d^0 4f^{14}$	174	100	Colorless	Latin Lutetia: Paris

2. Occurrences

The richest ore of Sc is the rare mineral thorveitite $Sc_2Si_2O_7$, but it also occurs in very small quantities in some lanthanoid, uranium, and tungsten ores. Yttrium and the lanthanoids (abbreviated Ln), except for Pm, occur in monazite $LnPO_4$ (mostly light lanthanoids), bastnaesite $LnCO_3F$ (mostly light lanthanoids), xenotime $LnPO_4$ (mostly heavy lanthanoids), loparite (mostly light lanthanoids), and lateritic clays (some with mostly light lanthanoids, others with mostly heavy lanthanoids). All isotopes of Pm are radioactive and it does not occur with the lanthanoids. Exceedingly small amounts are present in uranium ores where it has been produced by the spontaneous fission of U-238. Its major source is artificial production, the longest lived isotope being Pm-145 (half life of 17.7 years). Ac is also without a stable isotope, the radioactive element resulting from the decay of naturally occurring Th and U. The longest lived Ac species is Ac-227 which has a half life of 21.77 years.

3. Discoveries

In 1787, Carl Axel Arrhenius gave the name ytterbite to a new mineral he found in an Ytterby, Sweden quarry. This mineral was investigated in 1794 by Johan Gadolin who isolated from it a new oxide, which came to be called yttria. In 1804, Martin Heinrich Klaproth extracted from the mineral cerite another oxide which came to be called ceria. Ceria was independently discovered by Jöns Jacob Berzelius and Wilhelm Hisinger. Though the investigators thought at first that yttria and ceria were pure oxides, these two substances turned out to be mixtures of the lanthanoids. Over a period of 113 years, they were separated sufficiently to identify 15 lanthanoids (Y included, Pm excluded). The difficult separations were bought about by the tedious application of multiple stages of fractional precipitation and fractional crystallization. The separations were monitored by spectroscopic measurements and atomic weight determinations. The following chart indicates the progression of the separations and shows the dates and the investigators. Di stands for didymium, a mixture of oxides obtained by Mosander in 1841.

The discovery of Sc occurred separately from the above work. In 1840, C. J. A. Scheerer reported a new mineral euxenite which had been found near Jölster in Norway. An approximate analysis was made and many elements were detected including lanthanoids. Lars Fredrik Nilson in 1879 reported the isolation of 2 g of Sc_2O_3 from working up 10 kg of euxenite plus some residues from several other minerals.

After many false reports of its occurrence in nature, promethium was finally identified in 1947 in fission products by Jacob A. Marinsky, Lawrence E. Glendenin, and Charles D. Coryell. It was named after the Greek mythological character Prometheus who stole fire from the heavens. No stable isotopes of Pm exist, its longest lived isotope being Pm-145 (half life 17.7 years), but

1804^k Ce—Ce 1794^g Y——Y
      1839^m La—La 1842^m Tb
             1841^m Di—Di—Nd 1885^m 1842^m Er—Er—Er
                    | Pr 1885^m | Tm 1879^c
             1879^b Sm—Sm | Ho—Ho 1879^c
                    Gd 1886^b(1880ma) | Dy 1886^b
                    Eu 1901^d 1878ma Yb—Yb
                                                     Lu $1907^{u,w,j}$

b = Lecoq de Boisbaudran (France)   k = Martin Heinrich Klaproth
                                           (Germany)
c = Per Teodor Cleve (Sweden)   m = Carl Gustav Mosander (Sweden)
d = Eugène-Anatole Demarçay   ma = Jean Charles Galissard de
   (France)   Marignac (Switzerland)
g = Johann Gadolin (Finland)   u = Georges Urbain (France)
j = Charles James (United States)   w = Carl Auer von Welsbach
                                           (Austria)

most of the chemistry has been investigated with Pm-147 (half life 2.62 years). Recently, exceedingly tiny amounts of Pm have been found in U ores where it is a product of spontaneous fission of U-238.

In 1899, André Debierne added ammonium hydroxide to a solution of the U mineral pitchblende. When the lanthanoids precipitated as the hydroxides, a radioactive species was carried along. This element, which was a product of the radioactive decay of U-235 was named actinium. The species was Ac-227 (half life 21.77 years)

4. Scandium (Sc) $4s^2 3d^1$

a. E–pH diagram.   Figure 12.1 presents the E–pH diagram for Sc with soluble species (except H^+) at $10^{-1.0}$ M. The assumption is made that no complexes or any insoluble species other than the hydroxide are present. The metal is very active, the hydroxide precipitates at a relatively low pH, and the $Sc(OH)_4^-$ species indicates amphoterism. The Sc^{+3} is more accurately $Sc(HOH)_n^{+3}$ with n in the range of 4 through 6. At pH values just below the $Sc(OH)_3$ precipitation, the Sc^{+3} ion forms positively charged monomeric and polymeric hydroxy species. Negatively charged species with five or six hydroxides probably accompany the $Sc(OH)_4^-$ ion. The figure legend provides equations for the lines that separate the various species.

b. The element.   Scandium is generally recovered from residues resulting from the refining of uranium and tungsten ores. After preliminary treatment to isolate the element as the Sc^{+3} ion in solution, ScF_3 is precipitated, and then reduced to Sc with Ca. The element is a silvery-gray metal which oxidizes in

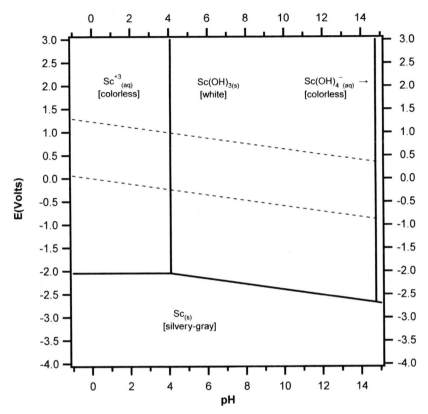

Figure 12.1 E–pH diagram for Sc species. Soluble species concentrations (except H^+) = $10^{-1.0}$ M. Soluble species and most solids are hydrated. No agents producing complexes or insoluble compounds are present other than HOH and OH^-.

Species ($\Delta G°$ in kJ/mol): Sc (0.0), Sc^{+3} (−586.2), $Sc(OH)_3$ (−1233.4), $Sc(OH)_4^-$ (−1380.7), HOH (−237.2), H^+ (0.0), and OH^- (−157.3)

Equations for the lines:

Sc^{+3}/Sc	$E = -2.03 + 0.020 \log [Sc^{+3}]$
$Sc(OH)_3/Sc$	$E = -1.80 - 0.059\, pH$
$Sc(OH)_4^-/Sc$	$E = -1.49 - 0.079\, pH + 0.020 \log [Sc(OH)_4^-]$
$Sc(OH)_3/Sc^{+3}$	$3\, pH = 11.3 - \log [Sc^{+3}]$
$Sc(OH)_4^-/Sc(OH)_3$	$pH = 15.8 + \log [Sc(OH)_4^-]$

air to Sc_2O_3. Sc attacks HOH and bases to liberate H_2 and produce $Sc(OH)_3$, and Sc reacts with acids to yield H_2 and Sc^{+3}.

c. Oxide and hydroxide. White Sc_2O_3 is produced by the oxidation of the metal in air. It dissolves in acids to yield Sc^{+3} and/or its partially hydrolyzed forms. The dissolution of Sc_2O_3 is usually slow, depending upon its method of preparation. The higher the temperature to which it is heated, the harder it is to put into solution. White $Sc(OH)_3$ is prepared by precipitating Sc^{+3} with OH^-. It is soluble both in acids and strong bases. Upon heating, the hydroxide forms Sc_2O_3.

d. Compounds and solubilities. Table 12.2 presents the most important compounds of Sc along with their solubilities. Most of these compounds can be

Table 12.2
Scandium Species

Name	Formula	State	Color	Solubility	$\Delta G°$ (kJ/mole)
Scandium	Sc	s	Silvery-gray	Insol	0.0
Scandium(III) ion	Sc^{+3}	aq	Colorless		−586.2
Hydroxoscandium(III)	$ScOH^{+2}$	aq	Colorless		−799.0
Dihydroxoscandium(III)	$Sc(OH)_2{}^+$	aq	Colorless		−1005.4
Scandium(III)					
acetate	$Sc(C_2H_3O_2)_3$	s	White	Sol	
arsenate	$ScAsO_4$	s	White	Insol	
bromate	$Sc(BrO_3)_3$	s	White	Sol	
bromide	$ScBr_3$	s	White	Sol	−717.6
carbonate	$Sc_2(CO_3)_2$	s	White	Insol	
chloride	$ScCl_3$	s	White	Sol	−846.0
chlorite	$Sc(ClO_2)_3$	s	White	Sol	
citrate	$ScC_6H_5O_7$	s	White	Insol	
dihydroxychloride	$Sc(OH)_2Cl$	s	White	Insol	
fluoride	ScF_3	s	White	Insol	−1573.0
hydroxide	$Sc(OH)_3$	s	White	Insol	−1233.4
iodate	$Sc(IO_3)_3$	s	White	Insol	
iodide	ScI_3	s	White	Sol	−561.9
nitrate	$Sc(NO_3)_3$	s	White	Sol	
oxalate	$Sc_2(C_2O_4)_3$	s	White	Insol	
oxide	Sc_2O_3	s	White	Insol	−1819.2
perchlorate	$Sc(ClO_4)_3$	s	White	Sol	
phosphate	$ScPO_4$	s	White	Insol	−1779.5
sulfate	$Sc_2(SO_4)_3$	s	White	Sol	
sulfide	Sc_2S_3	s		Insol	−1159.0
sulfite	$Sc_2(SO_3)_3$	s	White	Insol	

prepared by treatment of Sc, Sc(OH)$_3$, or Sc$_2$(CO$_3$)$_3$ with acid. Strong acids generally produce soluble salts (chloride, bromide, iodide, sulfate, nitrate, perchlorate), whereas weak acids often produce insoluble salts (fluoride, hydroxide, oxalate). A notable exception is the soluble acetate.

e. Redox reactions. The redox chemistry of Sc in aqueous solutions is ordinarily limited to the 0 and III oxidation states. The metal itself is a strong reducing agent.

f. Complexes. Notable complexes of Sc(III) include Sc(CO$_3$)$_2^-$, Sc(OH)$_4^-$, ScF$_4^-$, Sc(oxine)$_3$, Sc(ox)$_3^{-3}$, Sc(tart)$_2^-$, Sc(nta)$_2^{-3}$, Sc(edta)$^-$, and Sc(acac)$_3$. The oxine and acetylacetonate (acac) complexes are extractable into non-miscible organic media. Complexation constants for several species include (log β_1, log β_2, log β_3, log β_4): SCN$^-$ (0.2), NO$_3^-$(0.6, 0.1), OH$^-$(9.7, 18.3, 25.9, 30.0), SO$_4^{-2}$(4.2, 5.7), F$^-$(7.1, 12.9, 17.4, 20.3), Cl$^-$(1.9, 3.7), Br$^-$(−0.1), C$_2$O$_4^{-2}$(6.7, 11.3, 14.3, 16.7), malonate^{-2}(5.9, 10.1, 13.1), lactate$^-$(5.2), tartrate^{-2}(12.5), iminodiacetate^{-2}(9.8, 15.9), nta^{-3}(12.7, 24.1), edta^{-4}(23.1), and acac$^-$(8.0, 15.2).

g. Analysis. Colorimetric analysis for Sc using alizarin red permits determination down to about 1 ppm. AAS is slightly more competent, whereas ICPAES and ICPMS will go down to 0.1 ppb.

h. Health aspects. The LD50 (oral mouse) for ScCl$_3$ is 4 g/kg.

5. The Lanthanoids (La–Lu) 6s^{2}5d$^{0-1}$4f^n, Including Y, 5s^{2}4d^1 (All Referred to as Ln)

a. E–pH diagrams. The elements named in the title to this section are being called the lanthanoids (with a general abbreviation Ln). Notice that they include Y since its size and its chemical behavior usually locate it among the elements with atomic numbers of 65 through 68. Figures 12.2 through 12.17 display the E–pH diagrams for the lanthanoids. The figure legends provide equations for the lines that separate the species. The similarities of the diagrams reflect the fact that the lanthanoids resemble each other greatly, and that the small differences in their behavior often show a gradual trend through the series. The lanthanoids are highly active metals, reacting with HOH, acids, and bases. They show a predominant oxidation state of III, with Ce also showing IV, and Eu also showing II. Pr has a tendency to evidence an oxidation state of IV under extreme conditions, and Sm and Yb can be made to show a II oxidation state under extreme conditions. The aqueous lanthanoid(III) and Y(III) cations are believed to have HOH coordination numbers of 9 down to about the middle of the series at which point the number changes over to 8.

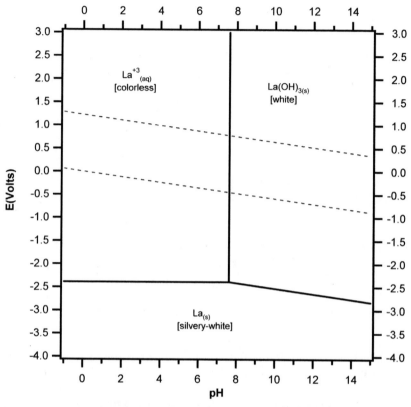

Figure 12.2 E–pH diagram for La species. Soluble species concentrations (except H^+) = $10^{-1.0}$ M. Soluble species and most solids are hydrated. No agents producing complexes or insoluble compounds are present other than HOH and OH^-.

Species ($\Delta G°$ in kJ/mol): La (0.0), La^{+3} (−686.1), $La(OH)_3$ (−1272.8), HOH (−237.2), H^+ (0.0), and OH^- (−157.3)

Equations for the lines:

$$La^{+3}/La \qquad E = -2.37 + 0.020 \log [La^{+3}]$$

$$La(OH)_3/La \qquad E = -1.94 - 0.059 \, pH$$

$$La(OH)_3/La^{+3} \qquad 3 \, pH = 21.9 - \log [La^{+3}]$$

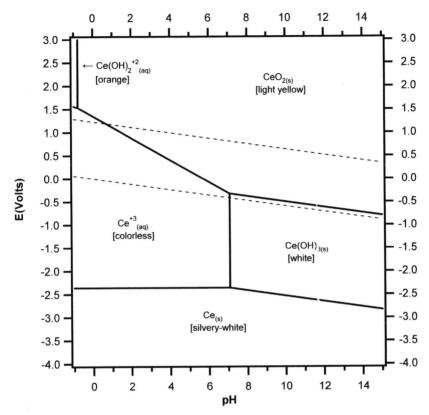

Figure 12.3 E–pH diagram for Ce species. Soluble species concentrations (except H^+) = $10^{-1.0}$ M. Soluble species and most solids are hydrated. No agents producing complexes or insoluble compounds are present other than HOH and OH^-.

Species ($\Delta G°$ in kJ/mol): Ce (0.0), Ce^{+3} (−677.4), $Ce(OH)_3$ (−1273.3), $Ce(OH)_2^{+2}$ (−1012.5), CeO_2 (−1027.1), HOH (−237.2), H^+ (0.0), and OH^- (--157.3)

Equations for the lines:

Ce^{+3}/Ce $\qquad$ $E = -2.34 + 0.020 \log [Ce^{+3}]$

$Ce(OH)_3/Ce$ $\qquad$ $E = -1.94 - 0.059 \, pH$

CeO_2/Ce^{+3} $\qquad$ $E = 1.29 - 0.236 \, pH - 0.059 \log [Ce^{+3}]$

$CeO_2/Ce(OH)_3$ $\qquad$ $E = 0.09 - 0.059 \, pH$

$Ce(OH)_2^{+2}/Ce^{+3}$ $\qquad$ $E = 1.44 - 0.118 \, pH$

$Ce(OH)_3/Ce^{+3}$ $\qquad$ $3 \, pH = 20.3 - \log [Ce^{+3}]$

$CeO_2/Ce(OH)_2^{+2}$ $\qquad$ $2 \, pH = -2.6 - \log [Ce(OH)_2^{+2}]$

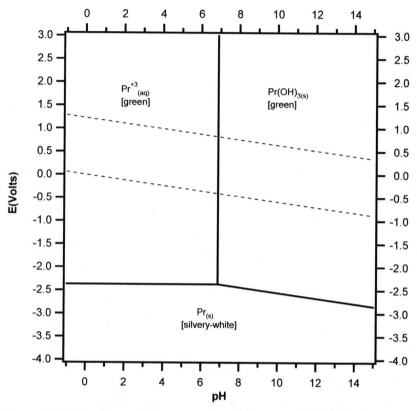

Figure 12.4 E–pH diagram for Pr species. Soluble species concentrations (except H^+) = $10^{-1.0}$ M. Soluble species and most solids are hydrated. No agents producing complexes or insoluble compounds are present other than HOH and OH^-.

Species ($\Delta G°$ in kJ/mol): Pr (0.0), Pr^{+3} (−680.3), $Pr(OH)_3$ (−1279.6), HOH (−237.2), H^+ (0.0), and OH^- (−157.3)

Equations for the lines:

$$Pr^{+3}/Pr \qquad E = -2.35 + 0.020 \log [Pr^{+3}]$$

$$Pr(OH)_3/Pr \qquad E = -1.96 - 0.059 \text{ pH}$$

$$Pr(OH)_3/Pr^{+3} \qquad 3 \text{ pH} = 19.7 - \log [Pr^{+3}]$$

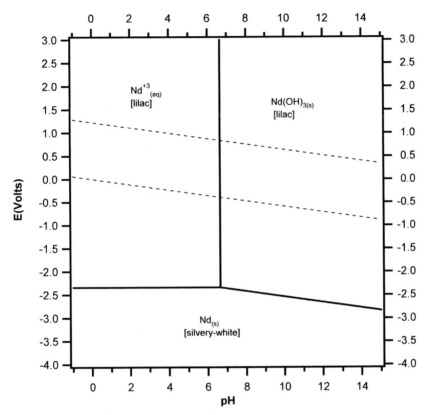

Figure 12.5 E–pH diagram for Nd species. Soluble species concentrations (except H^+) = $10^{-1.0}$ M. Soluble species and most solids are hydrated. No agents producing complexes or insoluble compounds are present other than HOH and OH^-.

Species ($\Delta G°$ in kJ/mol): Nd (0.0), Nd^{+3} (−671.6), $Nd(OH)_3$ (−1274.9), HOH (−237.2), H^+ (0.0), and OH^- (−157.3)

Equations for the lines:

$$Nd^{+3}/Nd \qquad E = -2.32 + 0.020 \log [Nd^{+3}]$$

$$Nd(OH)_3/Nd \qquad E = -1.95 - 0.059 \, pH$$

$$Nd(OH)_3/Nd^{+3} \qquad 3 \, pH = 19.0 - \log [Nd^{+3}]$$

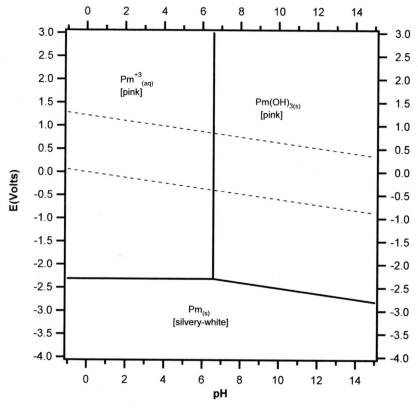

Figure 12.6 E–pH diagram for Pm species. Soluble species concentrations (except H^+) = $10^{-1.0}$ M. Soluble species and most solids are hydrated. No agents producing complexes or insoluble compounds are present other than HOH and OH^-.

Species ($\Delta G°$ in kJ/mol): Pm (0.0), Pm^{+3} (−663.0), $Pm(OH)_3$ (−1267.4), HOH (−237.2), H^+ (0.0), and OH^- (−157.3)

Equations for the lines:

$$Pm^{+3}/Pm \qquad E = -2.29 + 0.020 \log [Pm^{+3}]$$

$$Pm(OH)_3/Pm \qquad E = -1.92 - 0.059 \, pH$$

$$Pm(OH)_3/Pm^{+3} \qquad 3 \, pH = 18.8 - \log [Pm^{+3}]$$

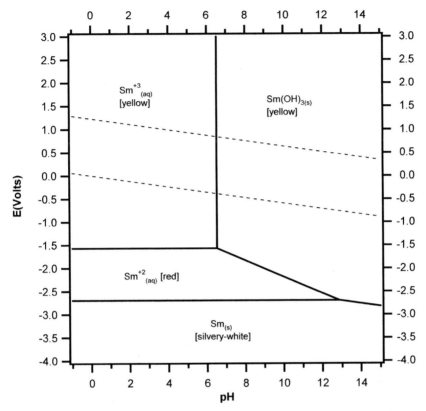

Figure 12.7 E–pH diagram for Sm species. Soluble species concentrations (except H$^+$) = 10$^{-1.0}$ M. Soluble species and most solids are hydrated. No agents producing complexes or insoluble compounds are present other than HOH and OH$^-$.

Species ($\Delta G°$ in kJ/mol): Sm (0.0), Sm^{+2} (−514.6), Sm^{+3} (−665.9), Sm(OH)$_3$ (−1271.5), HOH (−237.2), H$^+$ (0.0), and OH$^-$ (−157.3)

Equations for the lines:

$$Sm^{+2}/Sm \qquad E = -2.67 + 0.030 \log [Sm^{+2}]$$

$$Sm^{+3}/Sm^{+2} \qquad E = -1.57$$

$$Sm(OH)_3/Sm \qquad E = -1.93 - 0.059 \, pH$$

$$Sm(OH)_3/Sm^{+2} \qquad E = -0.47 - 0.177 \, pH - 0.059 \log [Sm^{+2}]$$

$$Sm(OH)_3/Sm^{+3} \qquad 3 \, pH = 18.6 - \log [Sm^{+3}]$$

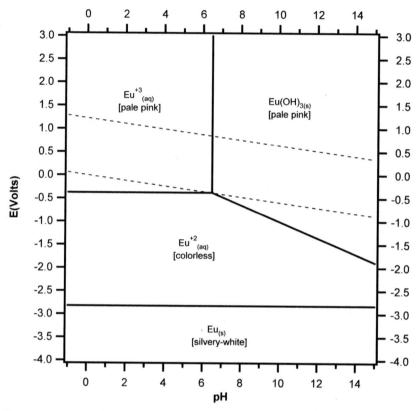

Figure 12.8 E–pH diagram for Eu species. Soluble species concentrations (except H^+) = $10^{-1.0}$ M. Soluble species and most solids are hydrated. No agents producing complexes or insoluble compounds are present other than HOH and OH^-.

Species ($\Delta G°$ in kJ/mol): Eu (0.0), Eu^{+2} (−543.9), Eu^{+3} (−576.1), $Eu(OH)_3$ (−1182.2), HOH (−237.2), H^+ (0.0), and OH^- (−157.3)

Equations for the lines:

$Eu^{+2}/Eu \qquad E = -2.82 + 0.030 \log [Eu^{+2}]$

$Eu^{+3}/Eu^{+2} \qquad E = -0.33$

$Eu(OH)_3/Eu^{+2} \quad E = 0.76 - 0.177\, pH - 0.059 \log [Eu^{+2}]$

$Eu(OH)_3/Eu^{+3} \quad 3\, pH = 18.5 - \log [Eu^{+3}]$

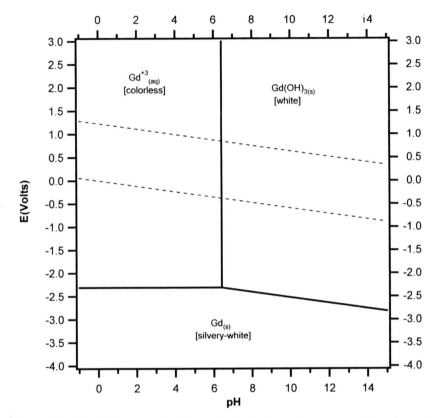

Figure 12.9 E–pH diagram for Gd species. Soluble species concentrations (except H^+) = $10^{-1.0}$ M. Soluble species and most solids are hydrated. No agents producing complexes or insoluble compounds are present other than HOH and OH^-.

Species (ΔG° in kJ/mol): Gd (0.0), Gd^{+3} (−663.0), $Gd(OH)_3$ (−1270.3), HOH (−237.2), H^+ (0.0), and OH^- (−157.3)

Equations for the lines:

$$Gd^{+3}/Gd \qquad E = -2.29 + 0.020 \log [Gd^{+3}]$$

$$Gd(OH)_3/Gd \qquad E = -1.93 - 0.059 \text{ pH}$$

$$Gd(OH)_3/Gd^{+3} \qquad 3 \text{ pH} = 18.3 - \log [Gd^{+3}]$$

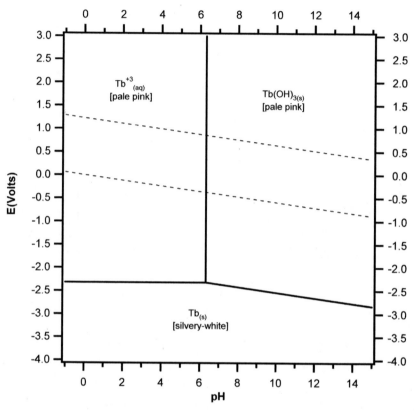

Figure 12.10 E–pH diagram for Tb species. Soluble species concentrations (except H^+) = $10^{-1.0}$ M. Soluble species and most solids are hydrated. No agents producing complexes or insoluble compounds are present other than HOH and OH^-.

Species ($\Delta G°$ in kJ/mol): Tb (0.0), Tb^{+3} (−665.9), $Tb(OH)_3$ (−1274.9), HOH (−237.2), H^+ (0.0), and OH^- (−157.3)

Equations for the lines:

$$Tb^{+3}/Tb \qquad E = -2.30 + 0.020 \log [Tb^{+3}]$$

$$Tb(OH)_3/Tb \qquad E = -1.95 - 0.059 \text{ pH}$$

$$Tb(OH)_3/Tb^{+3} \qquad 3 \text{ pH} = 18.0 - \log [Tb^{+3}]$$

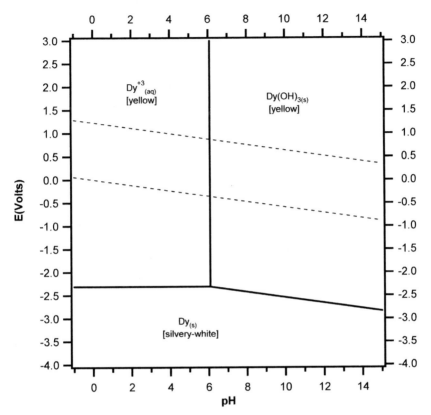

Figure 12.11 E–pH diagram for Dy species. Soluble species concentrations (except H^+) = $10^{-1.0}$ M. Soluble species and most solids are hydrated. No agents producing complexes or insoluble compounds are present other than HOH and OH^-.

Species ($\Delta G°$ in kJ/mol): Dy (0.0), Dy^{+3} (−663.0), $Dy(OH)_3$ (−1276.0), HOH (−237.2), H^+ (0.0), and OH^- (−157.3).

Equations for the lines:

$$Dy^{+3}/Dy \qquad E = -2.29 + 0.020 \log [Dy^{+3}]$$

$$Dy(OH)_3/Dy \qquad E = -1.95 - 0.059 \, pH$$

$$Dy(OH)_3/Dy^{+3} \qquad 3 \, pH = 17.3 - \log [Dy^{+3}]$$

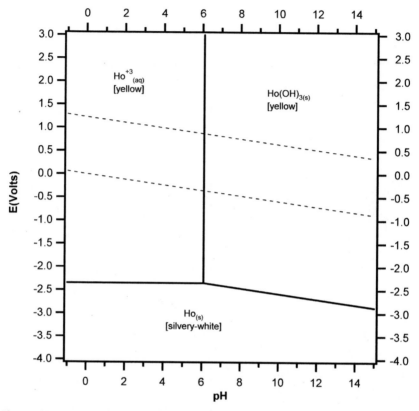

Figure 12.12 E–pH diagram for Ho species. Soluble species concentrations (except H^+) = $10^{-1.0}$ M. Soluble species and most solids are hydrated. No agents producing complexes or insoluble compounds are present other than HOH and OH^-.

Species ($\Delta G°$ in kJ/mol): Ho (0.0), Ho^{+3} (−674.5), $Ho(OH)_3$ (−1287.5), HOH (−237.2), H^+ (0.0), and OH^- (−157.3)

Equations for the lines:

Ho^{+3}/Ho $\qquad E = -2.33 + 0.020 \log [Ho^{+3}]$

$Ho(OH)_3/Ho$ $\qquad E = -1.99 - 0.059 \text{ pH}$

$Ho(OH)_3/Ho^{+3}$ $\quad 3 \text{ pH} = 17.3 - \log [Ho^{+3}]$

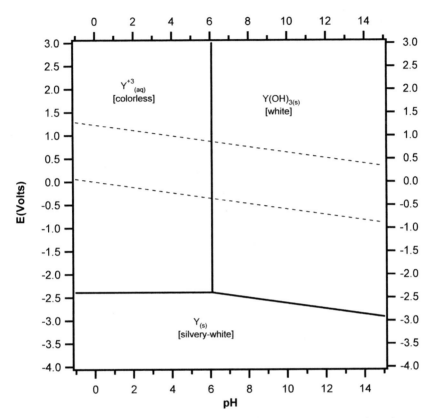

Figure 12.13 E–pH diagram for Y species. Soluble species concentrations (except H^+) = $10^{-1.0}$ M. Soluble species and most solids are hydrated. No agents producing complexes or insoluble compounds are present other than HOH and OH^-.

Species ($\Delta G°$ in kJ/mol): Y (0.0), Y^{+3} (−686.1), $Y(OH)_3$ (−1299.0), HOH (−237.2), H^+ (0.0), and OH^- (−157.3)

Equations for the lines:

$$Y^{+3}/Y \qquad E = -2.37 + 0.020 \log [Y^{+3}]$$

$$Y(OH)_3/Y \qquad E = -2.03 - 0.059 \, pH$$

$$Y(OH)_3/Y^{+3} \qquad 3 \, pH = 17.3 - \log [Y^{+3}]$$

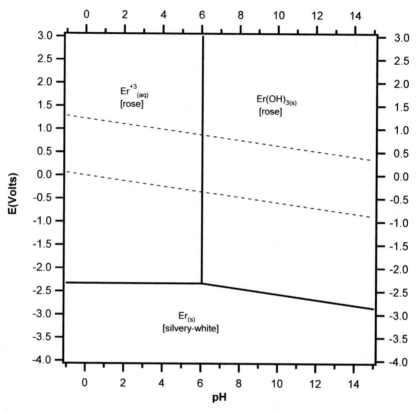

Figure 12.14 E–pH diagram for Er species. Soluble species concentrations (except H^+) = $10^{-1.0}$ M. Soluble species and most solids are hydrated. No agents producing complexes or insoluble compounds are present other than HOH and OH^-.

Species ($\Delta G°$ in kJ/mol): Er (0.0), Er^{+3} (−668.7), $Er(OH)_3$ (−1282.3), HOH (−237.2), H^+ (0.0), and OH^- (−157.3)

Equations for the lines:

Er^{+3}/Er $E = -2.31 + 0.020 \log [Er^{+3}]$

$Er(OH)_3/Er$ $E = -1.97 - 0.059$ pH

$Er(OH)_3/Er^{+3}$ 3 pH = $17.2 - \log [Er^{+3}]$

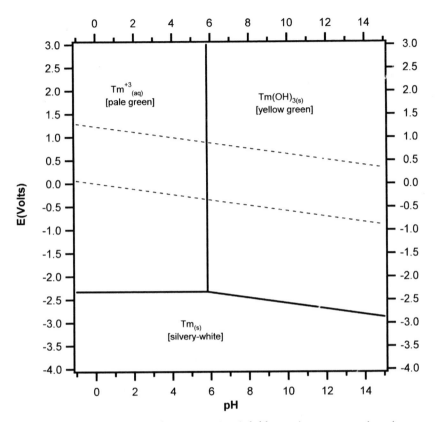

Figure 12.15 E–pH diagram for Tm species. Soluble species concentrations (except H^+) = $10^{-1.0}$ M. Soluble species and most solids are hydrated. No agents producing complexes or insoluble compounds are present other than HOH and OH^-.

Species (ΔG° in kJ/mol): Tm (0.0), Tm^{+3} (−668.7), $Tm(OH)_3$ (−1286.2), HOH (−237.2), H^+ (0.0), and OH^- (−157.3).

Equations for the lines:

$$Tm^{+3}/Tm \qquad E = -2.31 + 0.020 \log [Tm^{+3}]$$

$$Tm(OH)_3/Tm \qquad E = -1.99 - 0.059\, pH$$

$$Tm(OH)_3/Tm^{+3} \qquad 3\, pH = 16.5 - \log [Tm^{+3}]$$

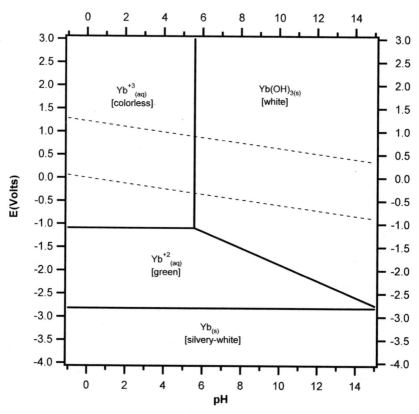

Figure 12.16 E–pH diagram for Yb species. Soluble species concentrations (except H^+) $= 10^{-1.0}$ M. Soluble species and most solids are hydrated. No agents producing complexes or insoluble compounds are present other than HOH and OH^-.

Species ($\Delta G°$ in kJ/mol): Yb (0.0), Yb^{+2} (−537.7), Yb^{+3} (−642.7), $Yb(OH)_3$ (−1264.2), HOH (−237.2), H^+ (0.0), and OH^- (−157.3)

Equations for the lines:

$$Yb^{+2}/Yb \qquad E = -2.79 + 0.030 \log [Yb^{+2}]$$

$$Yb^{+3}/Yb^{+2} \qquad E = -1.09$$

$$Yb(OH)_3/Yb^{+2} \qquad E = -0.15 - 0.177 \, pH - 0.020 \log [Yb^{+2}]$$

$$Yb(OH)_3/Yb^{+3} \qquad 3 \, pH = 15.8 - \log [Yb^{+3}]$$

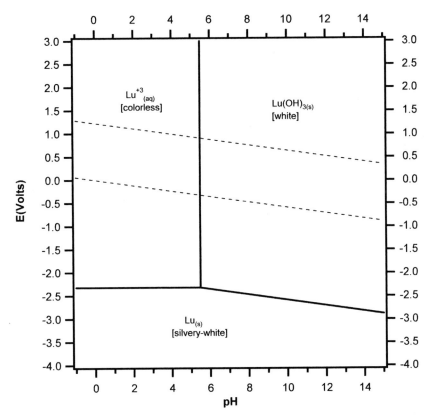

Figure 12.17 E–pH diagram for Lu species. Soluble species concentrations (except H^+) = $10^{-1.0}$ M. Soluble species and most solids are hydrated. No agents producing complexes or insoluble compounds are present other than HOH and OH^-.

Species ($\Delta G°$ in kJ/mol): Lu (0.0), Lu^{+3} (−665.9), $Lu(OH)_3$ (−1289.1), HOH (−237.2), H^+ (0.0), and OH^- (−157.3)

Equations for the lines:

$$Lu^{+3}/Lu \qquad E = -2.30 + 0.020 \log [Lu^{+3}]$$

$$Lu(OH)_3/Lu \qquad E = -1.99 - 0.059\, pH$$

$$Lu(OH)_3/Lu^{+3} \qquad 3\, pH = 15.5 - \log [Lu^{+3}]$$

b. Separation.　The lanthanoid minerals are cracked using strong acid or strong base. Then they are converted, if necessary, to a solution of chlorides. The older separation techniques of many stages of fractional crystallization or precipitation did not yield pure lanthanoids. However, in the late 1940s, the ion exchange method in conjunction with selective complexation proved to do so. This approach was superseded industrially in the 1960s by the use of multi-staged counter-current solvent extraction. Purities of 99.99% and better are achieved.

c. The elements.

Most of the metals are obtained by the reduction of chlorides or fluorides with Ca. For Eu, Sm, and Yb, Ca yields only $LnCl_2$, so the metals are prepared by reduction of the oxides with La. The Ln metals are silvery-white, very active, oxidize in air to Ln_2O_3, and attack HOH, acids, and bases to liberate H_2.

d. Oxides and hydroxides.

The sesquioxides Ln_2O_3 are prepared by burning the metals in air. These oxides are insoluble in HOH and bases, but will dissolve slowly in strong acids to give Ln^{+3}. The hydroxides $Ln(OH)_3$ are prepared by adding OH^- to Ln^{+3} solutions. Approximate experimental pH values at which 0.1 M perchlorate solutions precipitate are La(7.6)–Ce(7.1)–Pr(6.9)-Nd(6.7)–Pm(6.6)–Sm(6.5)–Eu(6.5)–Gd(6.4)–Tb(6.3)–Dy(6.1)– Ho(6.1)–Y(6.1)–Er(6.1)–Tm(5.8)–Yb(5.6)–Lu(5.5). The hydroxides are insoluble in HOH and bases, but dissolve readily in acids. When the hydroxides are heated, they first give LnO(OH), then Ln_2O_3. As the latter compound is heated to a higher temperature, the more difficult it is to dissolve in acids.

e. Compounds and solubilities.

Table 12.3 shows the solubilities in HOH of a number of trivalent Ln compounds along with other properties. Hydrated Ln(III) salts are prepared by the treatment of Ln, $Ln(OH)_3$, or $Ln_2(CO_3)_3$ with appropriate acids. Strong acids tend to give soluble salts, whereas weak acids usually produce insoluble salts, the acetate being an exception. Lanthanoid species in the II and IV oxidation states are presented in Table 12.4.

f. Redox reactions.

The E–pH diagrams of the lanthanoids show the redox characteristics of the elements. All the metals but Ce, Sm, Eu, and Yb transform into the triply charged ions Ln^{+3} at potentials in the region of −2.0 to −2.4 v. They persist well above the upper HOH line. Ce(IV) appears in ionic form well above the upper HOH line, and in the insoluble oxide form both within the HOH region and above it. They are stabilized respectively by kinetics and insolubility. The divalent ions Eu^{+2}, Yb^{+2}, and Sm^{+2} appear beneath the water lines. Kinetic retardation permits Eu^{+2} to persist in aqueous solution for hours, Yb^{+2} for much shorter times, and Sm^{+2} for even shorter times.

g. Complexes.

Notable complexes of Ln(III) include $Ln(CO_3)_2^-$, $Ln(P_2O_7)^-$, $Ln(SO_4)_2^-$, $Ln(S_2O_3)_2^-$, LnF^{+2}, $Ln(oxine)_3$, $Ln(C_2H_3O_2)_4^-$, $Ln(ox)_3^{-3}$, $Ln(tart)_2^-$, $Ln(cit)_2^{-3}$, $Ln(nta)_2^{-3}$, $Ln(edta)^-$, and $Ln(acac)_3$. The oxine and acetylacetonate (acac) complexes are extractable into non-miscible organic media. La^{+3} complexation constants for several species

Table 12.3
Lanthanoid(III) Species

Name	Formula	State	Color	Solubility	$\Delta G°$ (kJ/mole)
Lanthanoid	Ln	s	Silvery-white	Decomp	0.0
Lanthanoid(III) ion	Ln^{+3}	aq	Varies		$-664 \pm 7\%$
Lanthanoid(III)					
acetate	$Ln(C_2H_3O_2)_3$	s	Varies	Sol	
arsenate	$LnAsO_4$	s	Varies	Insol	
bromate	$Ln(BrO_3)_3$	s	Varies	Sol	
bromide	$LnBr_3$	s	Varies	Sol	
carbonate	$Ln_2(CO_3)_2$	s	Varies	Insol	
chloride	$LnCl_3$	s	Varies	Sol	$-927 \pm 8\%$
chlorite	$Ln(ClO_2)_3$	s	Varies	Sol	
citrate	$LnC_6H_5O_7$	s	Varies	Insol	
dihydroxychloride	$Ln(OH)_2Cl$	s	Varies	Insol	
fluoride	LnF_3	s	Varies	Insol	
hydroxide	$Ln(OH)_3$	s	Varies	Insol	$-1272 \pm 3\%$
iodate	$Ln(IO_3)_3$	s	Varies	Sl sol	
iodide	LnI_3	s	Varies	Sol	
molybdate	$Ln_2(MoO_4)_3$	s	Varies	Insol	
nitrate	$Ln(NO_3)_3$	s	Varies	Sol	
oxalate	$Ln_2(C_2O_4)_3$	s	Varies	Insol	
oxide	Ln_2O_3	s	Varies	Insol	$-1691 \pm 8\%$
perchlorate	$Ln(ClO_4)_3$	s	Varies	Sol	
phosphate	$LnPO_4$	s	Varies	Insol	
sulfate	$Ln_2(SO_4)_3$	s	Varies	Sol	
sulfide	Ln_2S_3	s	Varies	Decomp	
sulfite	$Ln_2(SO_3)_3$	s	Varies	Insol	

Colors of Ln^{+3} ions: La^{+3} colorless, Ce^{+3} colorless, Pr^{+3} green, Nd^{+3} lilac, Pm^{+3} pink, Sm^{+3} yellow, Eu^{+3} pink, Gd^{+3} colorless, Tb^{+3} pink, Dy^{+3} yellow, Ho^{+3} yellow, Y^{+3} colorless, Er^{+3} rose, Tm^{+3} green, Yb^{+3} colorless, Lu^{+3} colorless.

$\Delta G°$ (kJ/mole) of Ln^{+3} ions: $La^{+3} - 686$, $Ce^{+3} - 677$, $Pr^{+3} - 680$, $Nd^{+3} - 672$, $Pm^{+3} - 663$, $Sm^{+3} - 666$, $Eu^{+3} - 576$, $Gd^{+3} - 663$, $Tb^{+3} - 666$, $Dy^{+3} - 663$, $Ho^{+3} - 675$, $Y^{+3} - 686$, $Er^{+3} - 669$, $Tm^{+3} - 669$, $Yb^{+3} - 643$, $Lu^{+3} - 666$.

$\Delta G°$ (kJ/mole) of $Ln(OH)_3$: La -1273, Ce -1273, Pr -1280, Nd -1275, Pm -1267, Sm -1272, Eu -1182, Gd -1270, Tb -1275, Dy -1276, Ho -1288, Y -1299, Er -1282, Tm -1286, Yb -1264, Lu -1289.

$\Delta G°$ (kJ/mole) of Ln_2O_3: La -1706, Ce -1707, Pr -1734, Nd -1721, Pm -1723, Sm -1737, Eu -1565, Gd -1740, Tb -1777, Dy -1772, Ho -1792, Y -1817, Er -1809, Tm -1797, Yb -1729, Lu -1789.

$\Delta G°$ (kJ/mole) of $LnCl_3$: La -998, Ce -982, Pr -981, Nd -966, Pm -951, Sm -950, Eu -856, Gd -933, Tb -924, Dy -912, Ho -932, Y -950, Er -917, Tm -908, Yb -886, Lu -915.

Table 12.4
Lanthanoid(II and IV) Species

Name	Formula	State	Color	Solubility	$\Delta G°$ (kJ/mole)
Lanthanoid	Ln	s	Silvery-white	Decomp	0.0
Cerium(IV) ion	Ce^{+4}	aq	Orange		-508
Dihydroxocerium(IV) ion	$Ce(OH)_2{}^{+2}$	aq	Orange		-1013
Cerium(IV)					
fluoride	CeF_4	s	White	Insol	
iodate	$Ce(IO_3)_4$	s	Yellow	Insol	
oxalate	$Ce(C_2O_4)_2$	s	White	Insol	
oxide	CeO_2	s	Light yellow	Insol	-1027
nitrate	$Ce(NO_3)_4$	s	White	Decomp	
phosphate	$Ce_3(PO_4)_4$	s	White	Insol	
sulfate	$Ce(SO_4)_2$	s	Yellow	Decomp	
Praseodymium(IV) ion	Pr^{+4}	aq	Colorless	Decomp	-371
Praseodymium(IV)					
fluoride	PrF_4	s	White	Insol	
oxide	PrO_2	s	Black	Insol	-921
Hexapraseodymium					
undecoxide	Pr_6O_{11}	s	Black	Insol	-5314
Samarium(II) ion	Sm^{+2}	aq	Red	Decomp	-515
Samarium(II)					
bromide	$SmBr_2$	s	Brown	Decomp	
chloride	$SmCl_2$	s	Red-brown	Decomp	
fluoride	SmF_2	s	Purple	Insol	
iodide	SmI_2	s	Green	Decomp	
Europium(II) ion	Eu^{+2}	aq	Colorless		-544
Europium(II)					
bromide	$EuBr_2$	s	Brown	Sol	
carbonate	$EuCO_3$	s	White	Insol	
chloride	$EuCl_2$	s	White	Sol	
fluoride	EuF_2	s	Yellow	Insol	
iodide	EuI_2	s	Brown-green	Sol	
oxide	EuO	s	White	Sol	-560
sulfate	$EuSO_4$	s	White	Insol	
Terbium(IV) ion	Tb^{+4}	aq	Colorless	Decomp	
Terbium(IV)					
fluoride	TbF_4	s	White	Insol	
oxide	TbO_2	s	Brown		-913
Hexaterbium					
undecoxide	Tb_6O_{11}	s	Brown		-5396
Ytterbium(II) ion	Yb^{+2}	aq	Green		-538
Ytterbium(II)					
bromide	$YbBr_2$	s	Yellow	Sol	
chloride	$YbCl_2$	s	Green	Sol	
fluoride	YbF_2	s	Gray	Insol	
iodide	YbI_2	s	Yellow	Sol	
sulfate	$YbSO_4$	s	Green	Insol	

include $(\log \beta_1, \log \beta_2, \log \beta_3, \log \beta_4)$: SCN^- (0.1), NO_3^- (0.2), OH^- (5.5, 10.0, 15.0, 18.0), SO_4^{-2} (3.6, 5.2), $S_2O_3^{-2}$ (3.0), F^- (3.6, 7.3, 11.0), Cl^- (−0.1, −0.7), oxine$^-$ (5.9, 11.5, 17.0), $HCOO^-$ (1.1, 2.0), $C_2O_4^{-2}$ (4.3, 7.9, 10.3), acetate (1.6, 2.5, 3.0, 2.9), glycolate$^-$ (2.2, 3.8, 4.8, 5.1, 4.8), malonate^{-2} (3.7, 6.3), lactate$^-$ (2.6, 4.3, 5.7), maleate^{-2} (3.5, 5.4), malate^{-2} (4.3, 7.2), tartrate^{-2} (3.7, 6.1), citrate^{-3} (7.2, 10.2), mandelate$^-$ (2.3, 3.8), iminodiacetate^{-2} (5.9, 11.0), nta^{-3} (10.5, 17.8), edta^{-4} (16.0), and acac$^-$ (5.9, 10.3, 13.6). In general, $\log \beta$ values become slightly more positive from La through Lu, as they follow the trend in ionic radii. Observation of the above complexing agents indicates that the lanthanoids have a tendency to form strong complexes with oxygen as the donor atom.

h. Analysis. Analyses of the lanthanoids are usually carried out by ICPAES (limits 0.5 to 5 ppb), ICPMS (limits 0.01 to 0.05 ppb), and IC (limit 100 ppb). Colorimetric agents are ordinarily useful only after pre-separation, one of the better ones being 4-(2-pyridylazo)resorcinol.

i. Health aspects. The LD50 values (oral rat) for the lanthanum(III) chlorides are in the 2–4 g/kg range, as are those for the nitrates.

6. Actinium (Ac) $6s^2 5d^1$

The last chapter in the book treats the actinoid elements. The chemistry of Ac will be discussed there.

13

The Ti Group and the 5B, 6B, 7B, and
8B Heavy Elements

1. General Introduction

The elements to be treated in this chapter may be considered to be of three
types. All of them show one species which dominates the water domain in the
E–pH diagram. The dominant species in the E–pH diagrams and the elements
which display it are as follows:

  (1) an insoluble oxide: Ti, Zr, Hf (Group 4B) and Nb, Ta (Group 5B),

  (2) a high-oxidation-state anion: Mo, W (Group 6B) and Tc, Re
      (Group 7B),

  (3) a noble metal: Ru, Rh, Pd, Os, Ir, Pt (Group 8B).

2. Introduction to the Insoluble-oxide Elements: Ti, Zr, Hf, Nb, Ta

These five elements all show highly stable inert oxides which occupy the
majority of the water domain in their E–pH diagrams. This can be seen
in Figures 13.1 through 13.5. The three 4B oxides (TiO_2, ZrO_2, HfO_2) are
insoluble in HOH, dilute acids, dilute bases, and concentrated bases, but

are soluble in strong concentrated acids to give TiO^{+2}, ZrO^{+2}, and HfO^{+2}. The two 5B oxides (Nb_2O_5, Ta_2O_5) are insoluble in HOH, dilute acids, and dilute bases, but Nb_2O_5 dissolves in concentrated bases whereas Ta_2O_5 does not. All the elements in their highest oxidation state are hard cations and therefore will be particularly attracted to the hard atoms F and O.

3. Titanium (Ti) $4s^2 3d^2$

a. E–pH diagram.

The E–pH diagram in Figure 13.1 shows Ti in oxidation states of 0, II, III, and IV. In the legend of the diagram, equations for the lines between the species are presented. Table 13.1 displays ions and

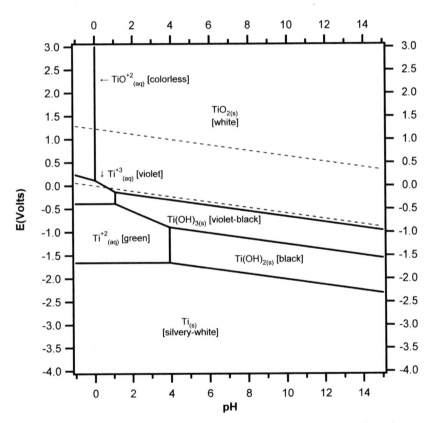

Figure 13.1 E–pH diagram for Ti species. Soluble species concentrations (except H^+) $= 10^{-1.0}$ M. Soluble species and most solids are hydrated. No agents producing complexes or insoluble compounds are present other than HOH and OH^-.

Species ($\Delta G°$ in kJ/mol): Ti (0.0), $Ti(OH)_2$ (-750.6), TiO_2 (-821.3), $Ti(OH)_3$ (-1049.8), Ti^{+2} (-314.2), Ti^{+3} (-349.8), TiO^{+2} (-577.4), HOH (-237.2), H^+ (0.0), and OH^- (-157.3).

Figure 13.1 (Continued)

Equations for the lines:

Ti^{+2}/Ti	$E = -1.63 + 0.030 \log[Ti^{+2}]$
$Ti(OH)_2/Ti$	$E = -1.43 - 0.059 \, pH$
Ti^{+3}/Ti^{+2}	$E = -0.37$
$Ti(OH)_3/Ti^{+2}$	$E = -0.25 - 0.177 \, pH - 0.059 \log[Ti^{+2}]$
$Ti(OH)_3/Ti(OH)_2$	$E = -0.64 - 0.059 \, pH$
TiO^{+2}/Ti^{+3}	$E = 0.10 - 0.118 \, pH$
TiO_2/Ti^{+3}	$E = 0.03 - 0.236 \, pH - 0.059 \log[Ti^{+3}]$
$TiO_2/Ti(OH)_3$	$E = -0.09 - 0.059 \, pH$
$Ti(OH)_2/Ti^{+2}$	$2 \, pH = 6.7 - \log[Ti^{+2}]$
$Ti(OH)_3/Ti^{+3}$	$3 \, pH = 2.0 - \log[Ti^{+3}]$
TiO_2/TiO^{+2}	$2 \, pH = -1.2 - \log[TiO^{+2}]$

compounds of Ti. The metal appears to be very active, but a thin refractory oxide coating renders it inactive to all but extreme treatment. Ions and compounds in oxidation states of II and III are unstable with regard to atmospheric O_2 and also with regard to HOH except for Ti^{+3} in strongly acidic solution.

b. Discovery, occurrence, and extraction. Ti, named after the Titans, the mythological first sons of the earth, was discovered by Gregor in 1791 in the mineral menachanite, a variety of ilmenite. The major sources of Ti are the minerals rutile TiO_2 and ilmenite $FeTiO_3$. They are treated with Cl_2 and C at elevated temperatures to generate gaseous $TiCl_4$ which condenses to a colorless liquid at 136°C. $TiCl_4$ vapor is treated with Mg at a high temperature to produce metallic Ti.

c. The element. Ti is a silvery-white ductile metal which reacts with air (O_2), HOH, cold acids, and cold bases to coat itself with a thin layer of oxide which renders it inert to most reagents. The coated metal can be dissolved in HF or in hot concentrated strong acids such as HCl. The product of dissolution in excess HF is probably the complex $TiF_6{}^{-2}$, and HCl yields Ti^{+3} with Ti in excess or TiO^{+2} with acid in excess.

d. Compounds and complexes. Ti shows three main oxides: TiO_2, Ti_2O_3, and TiO. The white solid TiO_2 is prepared by heating Ti in air or by adding OH^- to a TiO^{+2} solution. That prepared in air can be

Table 13.1
Titanium Species

Name	Formula	State	Color	Solubility	$\Delta G°$ (kJ/mole)
Titanium	Ti	s	Silvery-white	Insol	0.0
Oxotitanium(IV) ion	TiO^{+2}	aq	Colorless		−577.4
Titanium (III) ion	Ti^{+3}	aq	Violet		−349.8
Titanium(II) ion	Ti^{+2}	aq	Green		−314.2
Titanium(II)					
bromide	$TiBr_2$	s	Black	Decomp	−382.4
chloride	$TiCl_2$	s	Black	Decomp	−401.7
fluoride	TiF_2	s	Black	Decomp	−782.8
hydroxide	$Ti(OH)_2$	s	Black	Decomp	−750.6
iodide	TiI_2	s	Black	Decomp	−257.7
oxide	TiO	s	Black	Decomp	−489.2
sulfide	TiS	s	Red-black	Decomp	
Titanium(III)					
bromide	$TiBr_3$	s	Violet-black	Decomp	−521.3
chloride	$TiCl_3$	s	Dark violet	Decomp	−619.2
fluoride	TiF_3	s	Violet-red	Decomp	−1217.1
hydroxide	$Ti(OH)_3$	s	Violet-black	Decomp	−1049.8
iodide	TiI_3	s	Violet-black	Decomp	−320.1
oxide	Ti_2O_3	s	Violet-black	Decomp	−1432.2
sulfate	$Ti_2(SO_4)_3$	s	Green		
Titanium(IV)					
bromide	$TiBr_4$	s	Orange-yellow	Decomp	−610.9
chloride	$TiCl_4$	l	Colorless	Decomp	−674.5
fluoride	TiF_4	s	White	Decomp	−1448,9
iodide	TiI_4	s	Red	Decomp	−426.8
nitrate	$Ti(NO_3)_4$	s	White	Decomp	
oxide(hydrated)	TiO_2	s	White	Insol	−821.3
oxide	TiO_2	s	White	Insol	−888.4
oxoacetate	$TiO(C_2H_3O_2)_2$	s	White	Sol, decomp	
oxochloride	$TiOCl_2$	s	White	Sol, decomp	
oxosulfate	$TiO(SO_4)$	s	White	Sol, decomp	

difficult to dissolve, but TiO_2 prepared from TiO^{+2} is readily soluble in acids to regenerate TiO^{+2}. HF will dissolve TiO_2 prepared by both methods to give TiF_6^{-2}, but dissolution is slower for the air-made compound. A number of salts of the TiO^{+2} ion can be prepared, but they are subject to hydrolysis to TiO_2 even at low pH values. Important complexes of Ti(IV) include the fluoride complex mentioned above, $TiOCl_4^{-2}$, and those with acetylacetonate$^-$, catecholate^{-2}, and citrate^{-3}. Log β_n values for some complexes with TiO^{+2} are: SO_4^{-2} (2.5), F$^-$ (6.7, 11.7, 16.3, 20.4), Cl$^-$ (0.5, 0.2), edta^{-4} (17.5). When added to HOH, all four tetrahalides of Ti are hydrolyzed. When treated with H_2O_2, Ti(IV) in acid gives an orange peroxy ion $Ti(O_2)OH^+$.

Violet-black Ti_2O_3 can be made by reducing hot TiO_2 with H_2 and dark violet $TiCl_3$ results when $TiCl_4$ vapor is reduced with H_2. Dissolution of Ti_2O_3 in acid yields violet Ti^{+3}. Ti^{+3} can also be prepared by reducing TiO^{+2} with Zn. The Ti^{+3} ion reduces HOH to give TiO^{+2} or TiO_2 depending upon the pH. However, the reaction is slow which permits Ti(III) salts to be prepared. Complexation constants log β_n values for Ti(III) complexes include OH^- (11.8), Cl^- (0.3), and $edta^{-4}$ (21.3). As in the case of the halides of Ti(IV), those of Ti(III) are subject to hydrolysis, as are those of Ti(II). The black TiO is made by heating TiO_2 with Ti. This oxide dissolves in acid to give Ti^{+2} which immediately reduces HOH.

e. Analysis. Ti is best determined by ICPAES or ICPMS, both of which have a lower detection limit of about 0.1 ppb. Colorimetric methods permit 50 ppb to be measured.

f. Health aspects. Oral ingestion of TiO_2 is not a threat, but inhalation of its dust is to be avoided. Both inhalation and ingestion of $TiCl_4$ are to be avoided, mainly because of its hydrolysis to produce HCl.

4. Zirconium (Zr) $5s^2 4d^2$ and Hafnium (Hf) $6s^2 5d^2$

a. Similarity and E–pH diagrams. The two elements Zr and Hf are treated together because their properties are exceedingly similar. This is due to the identical values of their atomic and ionic radii. The identity stems from the decrease in atomic radii from La through Lu (lanthanide contraction) which drops the Hf radius to the Zr value. The E–pH diagrams in Figures 13.2 and 13.3 exemplify the similarity in the chemistries of the two elements. In the legends of the diagrams, equations for the lines between the species are presented. Tables 13.2 and 13.3 show some differences in the values of the standard free energies, but the differences between species are almost the same. Both elements show an oxidation state of IV in aqueous systems.

b. Discovery, occurrence, and extraction. Klaproth in 1789 analyzed a mineral known as jargon and isolated Zr. Hf, however, because of its similarity to Zr, and because its abundance is relatively small, was not definitively identified until 1923, by Coster and von Hevesy. The name zirconium comes from the Syrian zargun (color of gold) and the name hafnium is the Latin word for Copenhagen. The major sources of Zr are baddeleyite ZrO_2 and zircon $ZrSiO_4$. Hf occurs to the extent of 1–2% in these Zr minerals. The heating of ZrO_2 with C and Cl_2 results in $ZrCl_4$ vapor, which is reduced to the metal with Mg. Zr is a silvery-white metal which coats with a highly inactive film of ZrO_2 when exposed to air or HOH. The coating resists HOH, dilute acid, and base, but hot concentrated strong acids break it and dissolve the Zr to form ZrO^{+2}. (This formula is a simplification of the tetrameric species that is believed to predominate, namely $Zr_4(OH)_4^{+8}$.) The best solvent for Zr is HF which yields colorless ZrF_6^{-2}.

c. Compounds and complexes. The white ZrO_2, which results from the hydrolysis of most other Zr(IV) compounds, is a very stable solid. It may

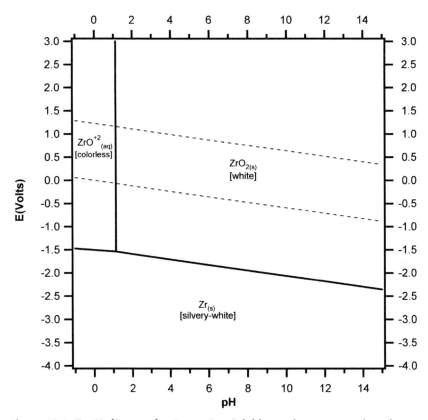

Figure 13.2 E–pH diagram for Zr species. Soluble species concentrations (except H^+) = $10^{-1.0}$ M. Soluble species and most solids are hydrated. No agents producing complexes or insoluble compounds are present other than HOH and OH^-.

Species ($\Delta G°$ in kJ/mol): Zr (0.0), ZrO_2 (−1041.8), ZrO^{+2} (−811.7), HOH (−237.2), H^+ (0.0), and OH^- (−157.3)

Equations for the lines:

$$ZrO^{+2}/Zr \qquad E = -1.49 - 0.030\ pH + 0.015\ \log[ZrO^{+2}]$$

$$ZrO_2/Zr \qquad E = -1.47 - 0.059\ pH$$

$$ZrO_2/ZrO^{+2} \qquad 2\ pH = 1.3 - \log[ZrO^{+2}]$$

also be produced by heating Zr in air. When ZrO_2 is prepared from ZrO^{+2} solutions, it is insoluble in dilute acid, HOH, and bases, but goes back into solution with concentrated acid. If this ZrO_2 is heated or if ZrO_2 is prepared by heating in air, it tends to resist solubility even in concentrated acid. Salts of ZrO^{+2} and a few salts of Zr^{+4} can be prepared, but they are stable only in acid solution. Zr(IV) complexes strongly with F^-, $C_2O_4^{-2}$, tartrate^{-2}, SO_4^{-2}, acac$^-$, citrate^{-3}, nta^{-3}, and edta^{-4}. Some log β_n values for Zr(IV) are as

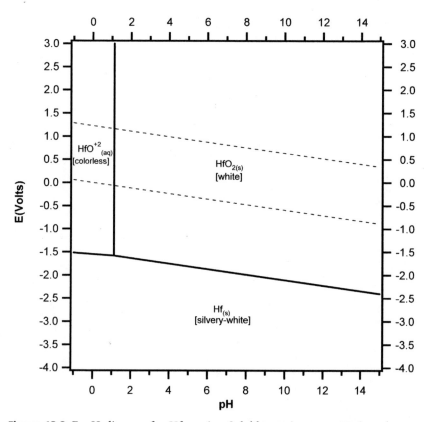

Figure 13.3 E–pH diagram for Hf species. Soluble species concentrations (except H^+) = $10^{-1.0}$ M. Soluble species and most solids are hydrated. No agents producing complexes or insoluble compounds are present other than HOH and OH$^-$.

Species ($\Delta G°$ in kJ/mol): Hf (0.0), HfO_2 (−1058.6), HfO^{+2} (−828.4), HOH (−237.2), H^+ (0.0), and OH$^-$ (−157.3)

Equations for the lines:

$$HfO^{+2}/Hf \qquad E = -1.53 - 0.030\,pH + 0.015\,\log[HfO^{+2}]$$

$$HfO_2/Hf \qquad E = -1.51 - 0.059\,pH$$

$$HfO_2/HfO^{+2} \qquad 2\,pH = 1.2 - \log[HfO^{+2}]$$

Table 13.2
Zirconium Species

Name	Formula	State	Color	Solubility	$\Delta G°$ (kJ/mole)
Zirconium	Zr	s	Silvery-white	Insol	0.0
Oxozirconium(IV) ion	ZrO^{+2}	aq	Colorless		−811.7
Zirconium(IV)					
bromide	$ZrBr_4$	s	White	Decomp	−724.0
chloride	$ZrCl_4$	s	White	Decomp	−889.0
fluoride	ZrF_4	s	White	Decomp	−1813.0
iodide	ZrI_4	s	Orange-yellow	Decomp	−480.6
nitrate	$Zr(NO_3)_4$	s	White	Decomp	
oxide(hydrated)	ZrO_2	s	White	Insol	−1041.8
oxide	ZrO_2	s	White	Insol	
oxoacetate	$ZrO(C_2H_3O_3)_2$	s	White	Sol, decomp	
oxochloride	$ZrOCl_2$	s	White	Sol, decomp	
oxonitrate	$ZrO(NO_3)_2$	s	White	Sol, decomp	
oxosulfate	$ZrOSO_4$	s	White	Sol, decomp	
sulfate	$Zr(SO_4)_2$	s	White	Decomp	

Table 13.3
Hafnium Species

Name	Formula	State	Color	Solubility	$\Delta G°$ (kJ/mole)
Hafnium	Hf	s	Silvery-white	Insol	0.0
Oxohafnium(IV) ion	HfO^{+2}	aq	Colorless		−828.4
Hafnium(IV)					
bromide	$HfBr_4$	s	White	Decomp	
chloride	$HfCl_4$	s	White	Decomp	
fluoride	HfF_4	s	White	Decomp	
iodide	HfI_4	s	Orange-yellow	Decomp	
nitrate	$Hf(NO_3)_4$	s	White	Decomp	
oxide(hydrated)	HfO_2	s	White	Insol	−1058.6
oxide	HfO_2	s	White	Insol	
oxoacetate	$HfO(C_2H_3O_3)_2$	s	White	Sol, decomp	
oxochloride	$HfOCl_2$	s	White	Sol, decomp	
oxonitrate	$HfO(NO_3)_2$	s	White	Sol, decomp	
oxosulfate	$HfOSO_4$	s	White	Sol, decomp	
sulfate	$Hf(SO_4)_2$	s	White	Decomp	

follows: NO_3^- (0.3, 0.1), OH^- (14,3, 26.3, 36.9, 46.3, 54.0), SO_4^{-2} (3.7, 6.4, 7.4), F^- (9.0, 16.5, 23.1, 28.8, 34.0, 38.0), nta^{-3} (20.8), $edta^{-4}$ (29.5).

d. Separation and analyses. Hf accompanies Zr in the extraction procedures discussed above, since their compounds resemble each other very closely. Industrially, Hf is separated from Zr by ion exchange and solvent extraction, the latter technique employing the differential solubilities of the metal thiocyanates in methyl isobutyl ketone. The two elements are measurable down to about 1 ppb by ICPAES and 0.1 ppb by ICPMS. The colorimetric agent SCN^- makes determinations down to 0.5 ppm possible, but provides no discrimination between the two elements.

e. Health aspects. The LD50 (oral) of $ZrCl_4$ in mice is 0.7 g/kg, that for ZrF_4 is 0.1 g/kg. The LD50 for $Zr(SO_4)_2$ in rats is 3.5 g/kg and that for $ZrO(C_2H_3O_2)_2$ is 4.1 g/kg. These figures suggest that the liberation of HCl or HF in the first two cases is largely responsible for the poisonous effects.

5. Niobium (Nb) $5s^24d^3$

a. E–pH diagram. The fourth member of the group of five elements whose insoluble oxides dominate the E–pH diagram is Nb. Its E–pH diagram is shown in Figure 13.4 which indicates that the major oxidation state in aqueous systems is V and that Nb_2O_5 is its chief compound. Other oxidation states which appear on the diagram are 0, II, and IV with all resting below the lower HOH stability line. In the legend of the diagram, equations for the lines between the species are presented.

b. Discovery, occurrence, and extraction. Nb was discovered in 1801 by Hatchett, and its congener Ta in 1802 by Ekeberg. However, most chemists viewed them as the same element until 1844 when Marignac showed them to be different. Tantalum takes its name from the Greek mythological god Tantalos whose son was named Niobe, which led to niobium. Nb and Ta almost always occur together in nature, the major ores being niobite $Fe(NbO_3)_2$, pyrochlore $NaCa(NbO_3)_2F$, tantalite $Fe(TaO_3)_2$, and microlite $NaCa(TaO_3)_2F$, the first two being richer in Nb and the latter ones being richer in Ta. The mineral is dissolved in HF, Ta is extracted into methyl isobutyl ketone at lower HF concentrations, and Nb is extracted at higher HF concentrations. The Nb is back-extracted into HOH, NaOH is added to precipitate Nb_2O_5 which is reduced to Nb with Na or C.

c. The element. Nb is a silvery-white metal which coats with inactive Nb_2O_5 when exposed to air. This coating resists HOH, bases, and acids,

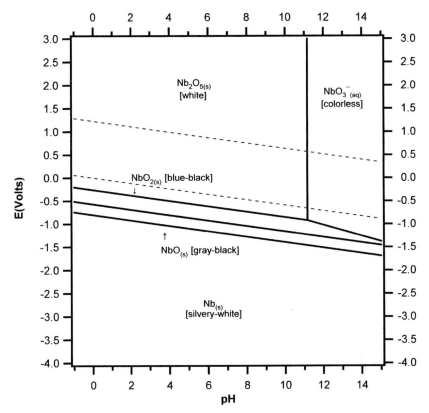

Figure 13.4 E–pH diagram for Nb species. Soluble species concentrations (except H^+) = $10^{-1.0}$ M. Soluble species and most solids are hydrated. No agents producing complexes or insoluble compounds are present other than HOH and OH^-.

Species ($\Delta G°$ in kJ/mol): Nb (0.0), NbO (−392.0), NbO$_2$ (−739.3), Nb$_2$O$_5$ (−1766.1), NbO$_3^-$ (−932.2), HOH (−237.2), H$^+$ (0.0), and OH$^-$ (−157.3).

Equations for the lines:

NbO/Nb	$E = -0.80 - 0.059 \text{ pH}$
NbO$_2$/NbO	$E = -0.57 - 0.059 \text{ pH}$
Nb$_2$O$_5$/NbO$_2$	$E = -0.26 - 0.059 \text{ pH}$
NbO$_3^-$/NbO$_2$	$E = 0.46 - 0.118 \text{ pH} + 0.059 \log[\text{NbO}_3^-]$
NbO$_3^-$/Nb$_2$O$_5$	$2 \text{ pH} = 24.4 + 2 \log[\text{NbO}_3^-]$

Table 13.4
Niobium Species

Name	Formula	State	Color	Solubility	$\Delta G°$ (kJ/mole)
Niobium	Nb	s	Silvery-white	Insol	0.0
Niobate(V)	NbO_3^-	aq	Colorless		−932.2
Niobium(II) oxide	NbO	s	Gray-black	Insol, decomp	−392.0
Niobium(IV) oxide	NbO_2	s	Blue-black	Insol, decomp	−739.3
Niobium(V)					
bromide	$NbBr_5$	s	Yellow-red	Decomp	−510.4
chloride	$NbCl_5$	s	Yellow	Decomp	−682.8
fluoride	NbF_5	s	White	Sol, decomp	−1699.1
iodide	NbI_5	s	Yellow-brown	Decomp	
oxide	Nb_2O_5	s	White	Insol	−1766.1
oxobromide	$NbOBr_3$	s	Yellow	Decomp	
oxochloride	$NbOCl_3$	s	White	Sol, decomp	
Sodium niobate	$NaNbO_3$	s	White	Decomp	

except hot concentrated HF or HF/HNO_3. The species formed are $NbOF_5^{-2}$ and NbF_6^-.

d. Compounds. Table 13.4 shows compounds of Nb which are pertinent to the aqueous chemistry. The oxide Nb_2O_5 can be prepared by heating Nb powder in air or by the hydrolysis of the flouride complexes mentioned above. The compound is insoluble in HOH, bases, and acids, except hot concentrated H_2SO_4, HF, or HF/HNO_3. A niobate, with the anion over simply represented as NbO_3^- (the actual species is probably $Nb_6O_{19}^{-8}$), is made by fusing Nb_2O_5 with NaOH or Na_2CO_3. It is water soluble but its solutions are stable only at higher pH values. If the oxide is fused with $NaHSO_4$, the melt can be dissolved in oxalate or tartrate or citrate solution due to the formation of complexes.

e. Analysis. Analysis for Nb can be performed down to about 1 ppb by ICPAES and down to about 0.1 ppb by ICPMS. For colorimetric determination down to about 0.5 ppm, SCN^- can be employed.

f. Health aspects. Nb and its compounds are considered to be relatively non-toxic due to the non-reactivity of the metal and the oxide in addition to the propensity of most of its compounds to hydrolyze to Nb_2O_5. However, direct ingestion of the pentahalides can be dangerous, probably due to the liberation of the hydrohalic acids. The LD50 (oral mice) for $NbCl_5$ is 0.8 g/kg.

6. Tantalum (Ta) $6s^2 5d^3$

a. E–pH diagram. Figure 13.5 is the E–pH diagram for Ta, and Table 13.5 is a listing of the metal and the main compounds which are related to its aqueous chemistry. The E–pH diagram indicates that the sole aqueous oxidation state is V, and that the diagram is occupied only by the metal and Ta_2O_5. In the legend of the diagram, an equation for the line between the species is presented.

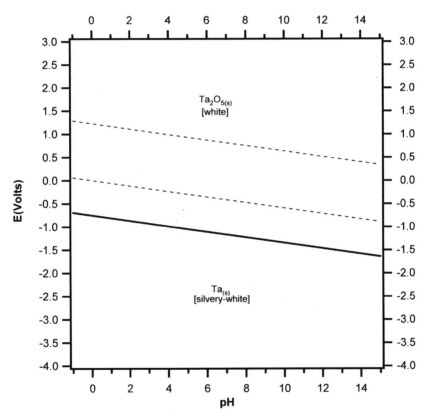

Figure 13.5 E–pH diagram for Ta species. Soluble species concentrations (except H^+) $= 10^{-1.0}$ M. Soluble species and most solids are hydrated. No agents producing complexes or insoluble compounds are present other than HOH and OH^-.

Species ($\Delta G°$ in kJ/mol): Ta (0.0), Ta_2O_5 (−1910.8), HOH (−237.2), H^+ (0.0), and OH^- (−157.3)

Equations for the lines:

$$Ta_2O_5/Ta \qquad E = -0.75 - 0.059\ pH$$

Table 13.5
Tantalum Species

Name	Formula	State	Color	Solubility	$\Delta G°$ (kJ/mole)
Tantalum	Ta	s	Silvery-white	Insol	0.0
Tantalum(V)					
bromide	$TaBr_5$	s	Yellow-red	Decomp	−552.3
chloride	$TaCl_5$	s	Yellow	Decomp	−707.1
fluoride	TaF_5	s	White	Sol, decomp	−1790.8
iodide	TaI_5	s	Yellow-brown	Decomp	
oxide	Ta_2O_5	s	White	Insol	−1910.8
Sodium tantalate	$NaTaO_3$	s	White	Decomp	

b. Discovery, occurrence, and extraction. As referred to in the previous section, the chief minerals of Ta are tantalite $Fe(TaO_3)_2$ and microlite $NaCa(TaO_3)_2F$. The minerals are put into solution with HF, and the tantalum–fluoride complex extracted into methyl isobutyl ketone. Back extraction into HOH is carried out, and the extractant is treated with NaOH to precipitate Ta_2O_5. The oxide is then reduced to the metal with Na or C.

c. The element and its compounds. Ta is a silvery-white metal which should dissolve in HOH. Ta metal, as do Ti, Zr, Hf, and Nb, coats in air with a thin film of oxide which makes it highly resistant to HOH, and most acids and bases. The most effective solvents are hot concentrated HF and HF/HNO_3 which convert Ta_2O_5 to fluoride complexes such as TaF_6^- and TaF_7^{-2}. The white oxide Ta_2O_5 is produced by heating of Ta in air or oxygen or by adding base to a solution of the tantalum–fluoride complexes. It is insoluble in HOH, and most acids and bases, except for hot, concentrated HF, HF/HNO_3, and H_2SO_4. Small amounts also dissolve in hot concentrated base. Oxalate, tartrate, and citrate complexes may be prepared in the same way as those for Nb. Tantalates are usually prepared by fusion methods, fusion with excess alkali hydroxides or carbonates gives tantalates of very complex kinds which are soluble in HOH.

d. Analysis. Analysis for Ta can be performed down to about 1 ppb by ICPAES and down to about 0.1 ppb by ICPMS. For colorimetric determination down to about 2.0 ppm, hydroquinone can be employed.

e. Health aspects. As is the case for Nb, Ta metal and Ta_2O_5 are relatively non-toxic, but the pentahalides can be dangerous due to their liberation of hydrohalic acids upon hydrolysis. The LD50 (oral rat) for Ta_2O_5 is 8 g/kg, that for $TaCl_5$ is 1.9 g/kg, and that for K_2TaF_7 is 2.5 g/kg.

7. Introduction to the High Oxidation State Oxyanion Elements: Mo, W, Tc, Re

The elements which make up this type are those in which a high-oxidation state oxyanion occupies most of the water domain in their E–pH diagrams. They are Mo and W in Group 6B and Tc and Re in Group 7B, with the corresponding simple oxyanions being MoO_4^{-2}, WO_4^{-2}, TcO_4^-, and ReO_4^-. In the cases of Mo and W, the formulas are grossly oversimplified. These elements form very complex oxyanionic species known as isopolymetallates. As the pH is lowered from the species MoO_4^{-2} or WO_4^{-2}, anionic clusters begin to form. The number of metal atoms in the clusters can rise to high numbers, and the metal:oxygen ratio goes from 4:1 to 3:1 at which point the charge is lost and the trioxide MO_3 precipitates. Examples of these species are $Mo_7O_{24}^{-6}$ and $W_{10}O_{32}^{-4}$, but species with many more atoms have been identified. Indications are that concentrations, temperature, rates of change of pH, and other ions in the solution all affect the compositions of the species and exactly where they change. Usually, a low pH is required for the precipitation of the trioxides.

8. Molybdenum (Mo) $5s^1 4d^5$

a. E–pH diagram.   The E–pH diagram for Mo is given in Figure 13.6. Mo and its major compounds are described in Table 13.6. The main oxidation state is VI, and IV, III and 0 are also seen in the diagram. Further to be noted is an intermediate region in which complex mixtures of molybdenum(VI)–oxygen anionic clusters occur. This area is labeled isopolyanions, examples being $Mo_2O_7^{-2}$, $Mo_6O_{19}^{-2}$, $Mo_7O_{24}^{-6}$, and $Mo_8O_{26}^{-4}$. Along the line which separates the Mo(VI) and the Mo(IV) species, mild reduction of Mo(VI) gives mixtures of high molecular weight ions containing Mo, O, and H which are called molybdenum blue. They evidence an average Mo oxidation state which varies between V and VI. As will be seen later, some less stable complexes with Mo in oxidation states of II, III, IV, and V can be prepared in solution.

A special note is required regarding the equation for the line which divides the MoO_4^{-2} region and the isopolyanion region on the Mo E–pH diagram. In setting up the diagram, the region of isopolyanions has been simply represented by the typical species $Mo_7O_{24}^{-6}$. When an equation for this line is derived, it takes the form: $8pH = 60.4 + 7\ log[MoO_4^{-2}] - log[isopolyanions] = 60.4 + 7\ log[MoO_4^{-2}] - log[Mo_7O_{24}^{-6}]$. Since the E–pH diagram operates assuming equal moles of Mo atoms on both sides of the line, if $[MoO_4^{-2}]$ is entered as 0.10 M, then $[Mo_7O_{24}^{-6}]$ must be entered as 0.10/7 or 0.014 M to make the moles of Mo atoms on both sides equal. Similar considerations apply to all lines involving solution species which contain more than one atom of the principal element, such as the Mo atom above.

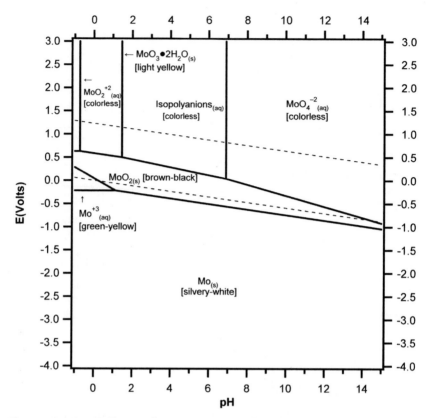

Figure 13.6 E–pH diagram for Mo species. Soluble species concentrations (except H^+) $= 10^{-1.0}$ M. Soluble species and most solids are hydrated. No agents producing complexes or insoluble compounds are present other than HOH and OH^-.

Species ($\Delta G°$ in kJ/mol): Mo (0.0), Mo^{+3} (−57.7), MoO_2 (−533.0), MoO_4^{-2} (−836.4), $MoO_3 \cdot 2H_2O$ (−1131.4), MoO_2^{+2} (−405.8), Isopolyanions (−5250.5), HOH (−237.2), H^+ (0.0), and OH^- (−157.3)

Equations for the lines:

Mo^{+3}/Mo	$E = -0.20 + \log[Mo^{+3}]$
MoO_2/Mo	$E = -0.15 - 0.059\,pH$
MoO_2/Mo^{+3}	$E = -0.01 - 0.236\,pH - 0.059\,\log[Mo^{+3}]$
MoO_2^{+2}/MoO_2	$E = 0.66 + 0.030\,\log[MoO_2^{+2}]$
$MoO_3 \cdot 2H_2O/MoO_2$	$E = 0.59 - 0.059\,pH$
Isopolyanions/MoO_2	$E = 0.63 - 0.084\,pH + 0.004\,\log[\text{Isopolyanions}]$
MoO_4^{-2}/MoO_2	$E = 0.89 - 0.118\,pH + 0.030\,\log[MoO_4^{-2}]$
$MoO_4^{-2}/\text{Isopolyanions}$	$8\,pH = 60.4 + 7\,\log[MoO_4^{-2}] - \log[\text{Isopolyanions}]$
Isopolyanions/$MoO_3 \cdot 2H_2O$	$6\,pH = 10.5 + \log[\text{Isopolyanions}]$
$MoO_3 \cdot 2H_2O/MoO_2^{+2}$	$2\,pH = -2.5 - \log[MoO_2^{+2}]$

Table 13.6
Molybdenum Species

Name	Formula	State	Color	Solubility	$\Delta G°$ (kJ/mole)
Molybdenum	Mo	s	Silvery-white	Insol	0.0
Molybdenum(III) ion	Mo^{+3}	aq	Green-yellow		−57.7
Dioxomolydenum(VI)	MoO_2^{+2}	aq	Colorless		−405.8
Molybdate(VI)	MoO_4^{-2}	aq	Colorless		−836.4
Polymolybdate ion	$Mo_7O_{24}^{-6}$	aq	Colorless		−5250.5
Molybdenum(IV)					
bromide	$MoBr_4$	s	Black	Decomp	
chloride	$MoCl_4$	s	Black	Decomp	−574.0
fluoride	MoF_4	s	Green	Decomp	−920.1
oxide	MoO_2	s	Brown-black	Insol	−533.0
Molybdenum(V)					
chloride	$MoCl_5$	s	Black	Decomp	−423.0
fluoride	MoF_5	s	Yellow	Decomp	
Molybdenum(VI)					
dioxydifluoride	MoO_2F_2	s	White	Decomp to MoO_3	
oxide	MoO_3	s	White	Insol	−668.0
oxide dihydrate	$MoO_3 \cdot 2HOH$	s	Light yellow	Insol	−1131.4
fluoride	MoF_6	l	Colorless	Decomp to MoO_3	−1473.0
oxytetrafluoride	$MoOF_4$	s	White	Decomp to MoO_3	

In the legend of the diagram, equations for the lines between the species are presented.

b. Discovery, occurrence, and extraction. In 1778, Scheele identified the mineral molybdenite as a compound of a new element rather than as a variation of a lead mineral or graphite. The name of the element derives from this early confusion, since molybdos is the Greek word for lead. Mo occurs as the mineral molybdenite MoS_2, and in the same compound in the refining of Cu, which can be roasted in air to MoO_3. The heated oxide is then reduced to the metal using H_2.

c. The element and its compounds. Mo is a silvery-white metal which is insoluble in HOH, dilute acids, and dilute bases. It dissolves readily in hot concentrated HNO_3, HF/HNO_3, or H_2SO_4, and goes into solution slowly with concentrated base. Mo is oxidized to white MoO_3 when heated in air. This main oxide does not dissolve in acid, but strong bases (such as NaOH) convert it to MoO_4^{-2} and the weak base NH_4OH converts it to the isopolyanion $Mo_7O_{24}^{-6}$. In the mid-pH range, many polyanions occur which cause the precipitation of yellow $MoO_3 \cdot 2HOH$ to take place at much lower pH values than might be predicted assuming a simple MoO_4^{-2} anion.

This yellow dihydrate is often referred to as molybdic acid. Salts of the molybdate anion may be made, most of them being insoluble, except for those of the alkali metal ions. Mo(VI) not only forms isopolyanions (such as $Mo_7O_{24}^{-6}$), but also heteropolyanions (such as $PMo_{12}O_{40}^{-3}$), as well as complexes such as $MoO_2Cl_4^{-2}$, $MoO_2(acac)_2$, $MoO_2(tart)_2^{-2}$, $MoO_2(cit)_2^{-4}$, and $MoO_2(ox)_2^{-2}$.

d. Redox reactions. Very carefully controlled reduction reactions of MoO_3 in acid can give a number of lower oxidation-state species, including Mo_2^{+4} (dark red), Mo^{+3} (green-yellow), MoF_6^{-3}, $MoCl_6^{-3}$ (red), $Mo(CN)_6^{-3}$, $Mo_3O_4^{+4}$, $MoCl_6^{-2}$, MoF_6^{-2}, $MoOCl_4^{-2}$, $Mo(CN)_8^{-4}$, and $Mo_2O_4^{+2}$ (yellow-orange). Note that oxidation states of II, III, IV, and V are represented. The brown-black oxide MoO_2 can be prepared by selective reduction of $MoO_3 \cdot 2HOH$. It is insoluble in acids and bases, but is readily oxidized to $MoO_3 \cdot 2HOH$, isopolyanions of Mo, or MoO_4^{-2}.

e. Analysis. The best methods of analysis for Mo are ICPAES (limit 1 ppb) and ICPMS (limit 0.1 ppb). Molecular absorption after reduction to Mo(V), then formation of a complex with thiocyanate can be employed down to about 0.1 ppm .

f. Health aspects. In general, Mo compounds show low toxicity. The LD50 (oral rat) for a single dose of MoO_3 is 3.0 g/kg.

9. Tungsten (W) $6s^1 5d^5$

a. E–pH diagram. The E–pH diagram for W is given in Figure 13.7. In the legend of the diagram, equations for the lines between the species are presented. W and its major compounds are described in Table 13.7. The main oxidation state is VI, with IV and 0 also appearing on the diagram. It is also to be noted that there is a sizable intermediate region in which complex mixtures of tungsten(VI)–oxygen anionic clusters occur. This area is labeled isopolyanions, examples being $W_7O_{24}^{-6}$, $H_2W_{12}O_{42}^{-10}$, $H_2W_{12}O_{40}^{-6}$, $W_{10}O_{32}^{-4}$. These species are listed in order of decreasing pH. Notice that the O:W ratio moves from 4.0 to 3.0 as the pH decreases. At very low pH values $WO_3 \cdot 2H_2O$, generally referred to as tungstic acid, precipitates out.

b. Discovery, occurrence, and extraction. In 1779, Woulfe concluded that the mineral now known as wolframite contained a new element. Then in 1782 Scheele isolated a new acid from a mineral known in Sweden as tung sten (heavy stone), today called scheelite. And in the following year, the two brothers de Elhuyar isolated the same acid from wolframite. Succeeding arguments over the name and the symbol ended up assigning the symbol W to the element tungsten. W occurs in wolframite (Fe, Mn)WO_4, scheelite $CaWO_4$, huebnerite $MnWO_4$, and ferberite $FeWO_4$. Wolframite is fused with

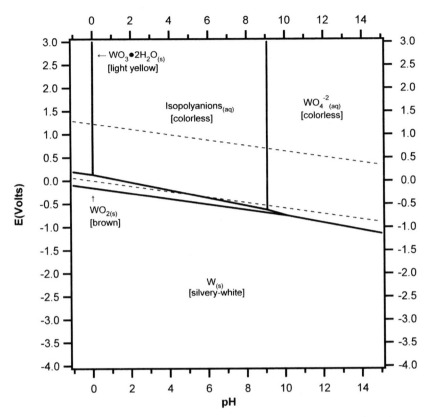

Figure 13.7 E–pH diagram for W species. Soluble species concentrations (except H^+) = $10^{-1.0}$ M. Soluble species and most solids are hydrated. No agents producing complexes or insoluble compounds are present other than HOH and OH^-.

Species ($\Delta G°$ in kJ/mol): W (0.0), WO_2 (−534.1), WO_4^{-2} (−916.7), $WO_3 \cdot 2H_2O$ (−1220.0), Isopolyanions (−5175.6), HOH (−237.2), H^+ (0.0), and OH^- (−157.3)

Equations for the lines:

WO_2/W	$E = -0.15 - 0.059 \, pH$
WO_4^{-2}/W	$E = 0.06 - 0.079 \, pH - 0.010 \, \log[WO_4^{-2}]$
WO_4^{-2}/WO_2	$E = 0.48 - 0.118 \, pH + 0.030 \, \log[WO_4^{-2}]$
Isopolyanions/WO_2	$E = 0.14 - 0.084 \, pH + 0.005 \, \log[\text{Isopolyanions}]$
$WO_3 \cdot 2H_2O/WO_2$	$E = 0.13 - 0.059 \, pH$
$WO_4^{-2}/\text{Isopolyanions}$	$7 \, pH = 67.9 + 6 \log[WO_4^{-2}] - \log[\text{Isopolyanions}]$
Isopolyanions/$WO_3 \cdot 2H_2O$	$5 \, pH = 1.7 + \log[\text{Isopolyanions}]$

Table 13.7
Tungsten Species

Name	Formula	State	Color	Solubility	$\Delta G°$ (kJ/mole)
Tungsten	W	s	Silvery-white	Insol	0.0
Tungsten(III) ion	W^{+3}	aq	Green-yellow		
Dioxotungsten(VI)	WO_2^{+2}	aq	Colorless		−502.1
Tungstate(VI)	WO_4^{-2}	aq	Colorless		−916.7
Hydrogentungstate	HWO_4^-	aq	Colorless		−938.1
Polytungstate ion	$HW_6O_{21}^{-5}$	aq	Colorless		−5175.6
Tungsten(IV)					
bromide	WBr_4	s	Black	Decomp	
chloride	WCl_4	s	Black	Decomp	−362.4
fluoride	WF_4	s	Red-brown	Decomp	
oxide	WO_2	s	Brown	Insol	−534.1
Tungsten(V)					
bromide	WBr_5	s	Brown-black	Decomp	−271.3
chloride	WCl_5	s	Green	Decomp	−410.2
fluoride	WF_5	s	Yellow	Decomp	−1329.7
Tungsten(VI)					
bromide	WBr_6	s	Blue-black		−292.4
chloride	WCl_6	s	Blue		−469.0
dioxydifluoride	WO_2F_2	s	White	Decomp to WO_3	−1057.5
oxide-hydrated	WO_3	s	Yellow	Insol	−740.0
oxide	WO_3 (cryst)	s	Yellow	Insol	−763.9
oxide dihydrate	$WO_3 \cdot 2HOH$	s	Light yellow	Insol	−1220.0
fluoride	WF_6	g	Colorless	Decomp to WO_3	−1635.9
oxytetrafluoride	WOF_4	s	White	Decomp to WO_3	−1298.1

NaOH, then acidified with HCl, whereas scheelite is dissolved in strong HCl, the end product in both cases being the insoluble $WO_3 \cdot 2HOH$, which is heated to form WO_3. This oxide is then reduced at high temperatures with H_2.

c. The element and its compounds. The silvery-white metal forms a coat of inactive WO_3 in air, and is insoluble in HOH, acids, and dilute bases. It slowly dissolves in concentrated HF/HNO_3. W burns in air to give yellow WO_3 which dissolves in strong base to produce WO_4^{-2}. As indicated by the E–pH diagram, as the pH drops, isopolyanions are formed, and finally $WO_3 \cdot 2H_2O$ precipitates when the acidity becomes high. Tungstates of many metal ions are known, those of the alkali metal ions, NH_4^+, Mg^{+2}, and Tl^+ being soluble, with most others insoluble. In addition to isopolyanions, W(VI) also forms heteropoly anions (such as $PW_{12}O_{40}^{-3}$ and $H_2Co_2W_{11}O_{40}^{-8}$), as well as complexes with concentrated HF and HCl, examples being WOF_5^-, $WO_2F_4^{-2}$, $WO_3F_3^{-3}$, and $WO_2Cl_4^{-2}$.

d. Redox reactions. Mild reduction in acid medium of $WO_3 \cdot 2H_2O$ or W isopolyanions gives intensely colored tungsten blue, this being a general term which is applied to complex mixtures of W(VI) and W(V) oxides and hydroxides, both neutral and anionic. Stronger reductions under controlled conditions can result in a number of lower oxidation number W species. Examples of some of these are WX_8^{-3} (X = F, Cl, Br), $W(CN)_8^{-3}$, WX_6^{-2} (X = Cl, Br), $WOCl_5^{-2}$, $W(OH)Cl_5^{-2}$, $W(CN)_8^{-4}$, $WO_2(CN)_4^{-4}$, and $W_2Cl_9^{-3}$.

e. Analysis. Analytical methods for W include ICPAES (limit 1 ppb) and ICPMS (limit 0.1 ppb). Molecular absorption using SCN^- shows a limit of 1.0 ppm.

f. Health aspects. As in the case of Mo, W and its compounds show low toxicity. The LD50 (oral rat) for WO_3 is reported as 1–2 g/kg.

10. Rhenium (Re) $6s^25d^5$

a. E–pH diagram. Figure 13.8 is a representation of the E–pH diagram for Re, and Table 13.8 sets out the compounds of this element which are pertinent to its aqueous chemistry. Notice that the predominant species is the anionic ReO_4^- and that a (IV) species, namely insoluble ReO_2, appears. In the legend of the diagram, equations for the lines between the species are presented.

b. Discovery, occurrence, and extraction. In 1925, Noddack, Tacke, and Berg identified Re in platinum ores, columbite, and gadolinite. The major source of the element now is the flue dusts of Mo and Cu refineries. From them, the volatile Re_2O_7 (boiling point 362°C) comes off with the dust. Treatment with HOH converts the oxide into the soluble ReO_4^-. Addition of NH_4^+ precipitates NH_4ReO_4, which is dried, then subjected to reduction with H_2 at elevated temperatures to give Re metal.

c. The element and its compounds. Silvery-white Re is insoluble in HOH, bases, and non-oxidizing acids, but Br_2 water, Cl_2 water, H_2O_2, and hot concentrated oxidizing acids (HNO_3, H_2SO_4) put it into solution as ReO_4^-. Neutralization of acidic solutions of ReO_4^- leads to perrhenate salts, those of Ag(I), Tl(I), Rb(I), and Cs(I) being slightly soluble, most others being soluble. $HReO_4$ is a strong acid (log K_p of $ReO_4^- = -1.3$), and the ReO_4^- is a weak complexing agent.

d. Redox reactions. Reduction of ReO_4^- with Zn leads to black ReO_2, reductions with I^- in the presence of HX (X = Cl, Br, I) give MX_6^{-2}, and controlled reduction in the presence of other complexing agents leads to complexes such as monovalent $Re(CN)_5^{-5}$, trivalent $Re(CN)_7^{-4}$,

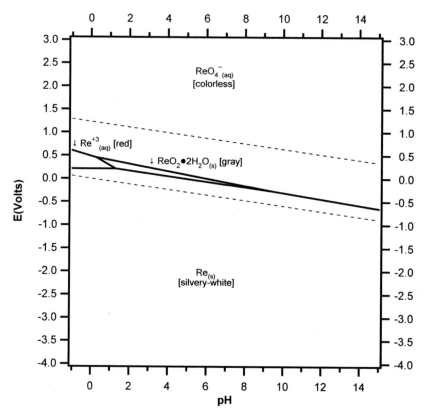

Figure 13.8 E–pH diagram for Re species. Soluble species concentrations (except H^+) = $10^{-1.0}$ M. Soluble species and most solids are hydrated. No agents producing complexes or insoluble compounds are present other than HOH and OH^-.

Species (ΔG° in kJ/mol): Re (0.0), $ReO_2 \cdot 2H_2O$ (−837.6), ReO_4^- (−694.7), Re^{+3} (66.9), HOH (−237.2), H^+ (0.0), and OH^- (−157.3)

Equations for the lines:

ReO_4^-/Re^{+3} $\qquad$ $E = 0.49 - 0.118 \text{ pH}$

$ReO_4^-/ReO_2 \cdot 2H_2O$ $\qquad$ $E = 0.49 - 0.079 \text{ pH} + 0.020 \log[ReO_4^-]$

ReO_4^-/Re $\qquad$ $E = 0.38 - 0.067 \text{ pH} + 0.008 \log[ReO_4^-]$

Re^{+3}/Re $\qquad$ $E = 0.23 + 0.020 \log[Re^{+3}]$

$ReO_2 \cdot 2H_2O/Re$ $\qquad$ $E = 0.29 - 0.059 \text{ pH}$

$ReO_2 \cdot 2H_2O/Re^{+3}$ $\qquad$ $E = 0.46 - 0.236 \text{ pH} - 0.059 \log[Re^{+3}]$

Table 13.8
Rhenium Species

Name	Formula	State	Color	Solubility	$\Delta G°$ (kJ/mole)
Rhenium	Re	s	Silvery-white	Insol	0.0
Rhenium(III) ion	Re^{+3}	aq	Red		66.9
Hexachlororhenate(IV)	$ReCl_6^{-2}$	aq	Green		−589.1
Perrhenate(VII) ion	ReO_4^-	aq	Colorless		−694.7
Rhenium(III)					
bromide	$ReBr_3$	s	Red-brown	Decomp	
chloride	$ReCl_3$	s	Red	Decomp	−188.0
hydroxide	$Re(OH)_3$	s	Brown-black	Decomp	−661.1
iodide	ReI_3	s	Black	Decomp	
oxide	Re_2O_3	s	Brown-black	Decomp	−425.1
Rhenium(IV)					
bromide	$ReBr_4$	s	Red	Decomp	
chloride	$ReCl_4$	s	Purple-black	Decomp	−169.5
fluoride	ReF_4	s	Blue	Decomp	
iodide	ReI_4	s	Black	Decomp	
oxide (cryst)	ReO_2	s	Black	Insol	−391.2
oxide-dihydrate	$ReO_2 \cdot 2HOH$	s	Gray	Insol	−837.6
Rhenium(V)					
bromide	$ReBr_5$	s	Brown	Decomp	
chloride	$ReCl_5$	s	Brown-black	Decomp	−264.0
fluoride	ReF_5	s	Yellow-green	Decomp	−831.8
oxide	Re_2O_5	s	Red	Decomp	
Rhenium(VI)					
chloride	$ReCl_6$	s	Red-green	Decomp	
fluoride	ReF_6	s	Yellow	Decomp	
oxide	ReO_3	s	Red	Insol	−507.1
Rhenium(VII)					
fluoride	ReF_7	s	Yellow	Decomp	
oxide	Re_2O_7	s	Yellow	Decomp	−1089.1

and $Re_2X_8^{-2}$. The latter compound is of interest since it has a Re-to-Re quadruple bond.

e. Analysis. With Sn(II) in sulfuric acid solution, ReO_4^- forms colored chelates with oximes and thiocyanate. These are extracted into organic media and determined spectrophotometrically down to about 30 ppb. Re can be detected down to about 3.0 ppb by ICPAE and down to about 0.1 ppb by ICPMS.

f. Health aspects. No non-ambiguous evidence of toxic effects of perrhenate have been reported, but only a few reports have been made.

11. Technetium (Tc) $5s^2 4d^5$

a. E–pH diagram. The E–pH diagram for Tc is presented as Figure 13.9, and a listing of some of its more important compounds is given in Table 13.9.

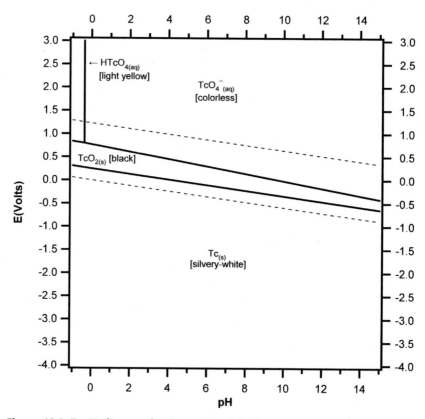

Figure 13.9 E–pH diagram for Tc species. Soluble species concentrations (except H^+) = $10^{-1.0}$ M. Soluble species and most solids are hydrated. No agents producing complexes or insoluble compounds are present other than HOH and OH^-.

Species ($\Delta G°$ in kJ/mol): Tc (0.0), TcO_2 (−378.3), TcO_4^- (−624.0), $HTcO_4$ (−621.9), HOH (−237.2), H^+ (0.0), and OH^- (−157.3).

Equations for the lines:

$HTcO_4/TcO_2$	$E = 0.80 - 0.059 \, pH + 0.020 \, \log[HTcO_4]$
TcO_4^-/TcO_2	$E = 0.79 - 0.079 \, pH + 0.020 \, \log[TcO_4^-]$
TcO_2/Tc	$E = 0.25 - 0.059 \, pH$
$TcO_4^-/HTcO_4$	$pH = -0.4$

Table 13.9
Technetium Species

Name	Formula	State	Color	Solubility	$\Delta G°$ (kJ/mole)
Technetium	Tc	s	Silvery-white	Insol	0.0
Technetium(II) ion	Tc^{+2}	aq		Decomp	77.2
Hexabromotechnetate(IV)	$TcBr_6^{-2}$	aq	Red-brown	Decomp	
Hexachlorotechnetate(IV)	$TcCl_6^{-2}$	aq	Yellow	Decomp	
Hexaiodotechnetate(IV)	TcI_4^{-2}	aq	Purple-black	Decomp	
Pertechnetate (VII) ion	TcO_4^-	aq	Colorless		−624.0
Pertechnetic acid	$HTcO_4$	aq	Light yellow		−621.9
Technetium(IV)					
chloride	$TcCl_4$	s	Purple-black	Decomp	
oxide	TcO_2	s	Black	Insol	−378.3
Technetium(V)					
fluoride	TcF_5	s	Yellow	Decomp	
Technetium(VI)					
fluoride	TcF_6	s	Yellow	Decomp	
oxide	TcO_3	s	Red	Insol	−461.1
Technetium(VII)					
fluoride	TcF_7	s	Yellow	Decomp	
oxide	Tc_2O_7	s	Yellow	Decomp	−938.1

The strong resemblance to the Re data is to be noted. In the legend of the diagram, equations for the lines between the species are presented.

b. Discovery and production. Technetium was the first element to be produced artificially. It was initially identified in an Mo target which had been bombarded with deuterons to produce the radioisotopes Tc-95m and Tc-97m, with respective half lives of 61 and 90 days. This was carried out by Perrier and Segré in 1937. Although Tc was believed to be absent in nature, it has been seen in the spectra of stars and traces of Tc-99m are present on the earth as a result of spontaneous fission of U. In line with its production, the element was named from the Greek word technikos, which means artificial.

Tc-99, which has a half life of 2.12×10^5 years, can be recovered from nuclear fission waste in kilogram quantities. Solvent extraction, ion exchange, and volatilization processes are employed to separate it from the numerous other fission products. Because of its long half life and its emission of a soft (low energy) beta particle, it can be safely handled in milligram quantities. Almost all chemical studies of the element have been carried out with this isotope.

c. The element and its compounds. From E–pH diagrams and other properties of elements, it can be seen that Tc resembles Re far more than it does Mn. Most of the remarks in the previous section apply to Tc as well as Re. Both metals are silvery-white, both have similar solubility characteristics (oxidizing agents needed to solubilize them), both have similar MO_4^- salts, and both can be easily reduced to MO_2.

d. Redox reactions. Carefully controlled reductions of TcO_4^- lead to a variety of species including $TcCl_6^{-2}$, $TcBr_6^{-2}$, $Tc_2Cl_8^{-3}$, and $Tc(NCS)_6^{-2}$. Reductions in the presence of organic ligands can result in the preparation of Tc(IV) complexes containing acetylacetonate, citrate, tartrate, nitrilotriacetate, and ethylenediaminetetraacetate.

e. Analysis. ICPAES is sensitive to 2 ppb of Tc, and ICPMS to 0.1 ppb. Use of colorimetry permits measurements down to about 20 ppb.

f. Health aspects. If ingested, Tc-99 concentrates in the thyroid gland and the gastrointestinal tract. The half time for excretion is 60 h. Adverse health effects may be assigned to its radioactivity.

12. Introduction to the Noble Metal Elements: Ru, Rh, Pd, Os, Ir, Pt

a. E–pH diagrams. Figures 13.10 through 13.15 present the E–pH diagrams for the noble metals in chloride media with metal concentrations of $10^{-1.0}$ M and chloride concentrations of 1.0 M, except for Pt ($10^{-3.0}$ M). The reason for this chloride medium is that the best known and the most useful solution chemistry of the noble metals is that of their chloride complexes. It is to be noted that Cl^- is oxidized in the upper portion of the diagrams, which eliminates the chloride complexes. However, there is usually some kinetic stabilization of the upper-valent complexes such that they do exist temporarily above the Cl^- oxidation line. In the legends of the diagrams, equations for the lines between the species are presented. Tables 13.10 through 13.15 list the ions and compounds of the metals along with some important properties relating to the solution chemistry.

b. Discoveries. Impure Pt, found as the element, was used both in ancient Egypt and in ancient Central America. The first European reference to the Central American substance was made in 1557 by Julius C. Scaliger. In 1748, don Antonio de Ulloa described small grains of a silvery unworkable metal which he had seen in Columbia in 1735. They were called platina del Pinto, which means small silver from the Pinto (Mountains). Shortly thereafter, Charles Wood and William Brownrigg investigated some platina from Jamaica, and in 1750 Brownrigg's account of this "new metal" was presented to the Royal Society of London. The process to purify platinum was developed by William H. Wollaston. The crude platinum ore was selectively dissolved into aqua regia , precipitated by ammonium chloride, and heated to give residual platinum powder. The resulting solution was initially thrown away. However, Wollaston studied this solution leading to the discovery of palladium (1802), and of rhodium (1804). Smithson Tennant examined the insoluble residue and found osmium (1803) and iridium (1803). Forty years later ruthenium was found in this waste.

Pd, named after the asteroid Pallas and Pallas, the Greek god of wisdom, was first prepared in 1802 by Wollaston. He dissolved impure Pt in aqua regia, evaporated the acid off, then added $Hg(CN)_2$ to produce a yellow precipitate. Upon ignition, the silvery-white Pd remained. In 1803, Tennant dissolved impure Pt in aqua regia, and noticed that a metallic black powder was left. The powder was fused with NaOH, then treated with acid, and distilled. The acrid-smelling condensate contained a compound of a metal which was called osmium, and the residue contained a compound of a second metal which was called iridium. Os was named after the Greek word osme which means stench, and Ir was named after Iris, the Greek goddess of the rainbow.

In 1803–04, Wollaston dissolved impure Pt in aqua regia, precipitated Pt with NH_4Cl, precipitated Pd with $Hg(CN)_2$, decomposed the excess $Hg(CN)_2$, then evaporated the solution to obtain a red residue containing Rh. It was so named because the word comes from the Greek rhodon which translates as rose. In 1840 Karl K. Klaus began some investigations on Pt residues (that which was insoluble in aqua regia) in an effort to settle a controversy over whether there was another element in them. He fused the residue with KOH and KNO_3, took it up in water, and added HNO_3 to give a black precipitate. This precipitate was boiled with aqua regia, and a volatile Os compound came off. The remaining solution was treated with NH_4Cl which precipitated a salt of a new element, which had previously been called ruthenium, ruthenia being a latinized term for Russia.

c. Occurrence and separation.

The six noble metals (Ru, Rh, Pd, Os, Ir, Pt) ordinarily occur together, but the amounts of each vary widely, Pt most often being predominant. Analysis of a sample from South Africa shows 61% Pt, 26% Pd, 8% Ru, 3% Rh, 1% Ir, and 1% Os. The metals are often found as native metal, natural alloys, and as sulfides, tellurides, and arsenides in light silicate matrices containing ores of Ni, Cu, and Fe. The major sources are the Cu–Ni ores in South Africa and Russia, where the noble elements are present in the range of 10–50 ppm. For the Merensky Reef (South Africa) materials, the ore is pulverized, the metal-containing particles are separated by flotation and gravity concentration, and the concentrate is then oxidatively smelted with an Fe-removing flux to produce a deposit containing Cu, Ni, and the noble metals. The Cu and Ni are removed magnetically, and the resulting residue is separated into its constituent noble metals.

Flowchart 13.1 shows one of the several separation methods employed with the noble metals. The noble-metal concentrate is treated with a strongly oxidizing mixture of HCl and Cl_2 to put Ru and Os in solution as the

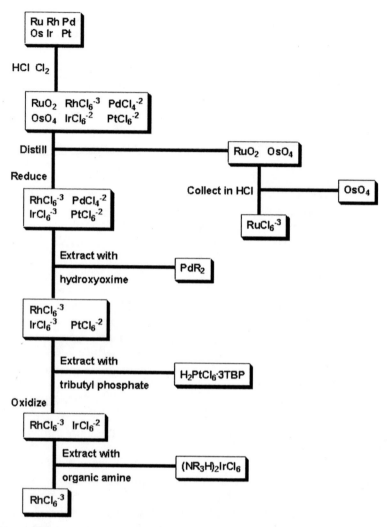

Flowchart 13.1 Noble-metals separation scheme.

tetraoxides and the others as chloride complexes. Distillation of the mixture takes off the volatile RuO_4 and OsO_4 which are collected in HCl. This gives $RuCl_6^{-3}$, leaving the OsO_4 unchanged, which can be distilled off. The solution of the chloride complexes of the other four elements is reduced with SO_2 which leaves the Rh, Pd, and Pt as they were, and puts the Ir in the (III) state. The solution is then contacted with an immiscible organic solvent containing a beta-hydroxyoxime which extracts the Pd into the organic phase. The remaining solution is then contacted with tributyl phosphate in an immiscible organic solvent to take out the Pt. Oxidation with Cl_2 puts the Ir into the (IV) state which permits its extraction with n-octylamine in an immiscible organic solvent. This leaves the Rh.

d. Analyses.   The most widely used methods for determining the noble metals are ICPAES and ICPMS. The detection limits for the former method are about 1 ppb and those for the latter are about 10 times more sensitive. A further advantage is that all six of the elements may be measured simultaneously.

13. Ruthenium (Ru) $5s^1 4d^7$

a. The element.   See Figure 13.10 and Table 13.10. Ru is a gray-white, lustrous metal which shows no reaction with air, HOH, acids, or bases. The oxidation states which it displays in aqueous systems are 0, II, III, IV, and VIII, with VI and VII in the strongly basic region. The most prevalent oxidation state is III. The powdered metal burns in O_2 or combines with Cl_2 at elevated temperatures to yield RuO_2 or $RuCl_3$. Ru also can be fused with NaOH or Na_2O_2 to produce material which gives RuO_4^{-2} when dissolved in HOH. If a strong oxidizing agent such as Cl_2 or $NaClO_3$ is combined with HCl, Ru can be put in solution as a chloride complex. The most important industrially prepared substance is represented as $RuCl_{3/4} \cdot 5HOH$, but is probably a variable mixture of Ru(III) and Ru(IV) chlorides, oxychlorides, hydroxychlorides, and aquates. It is made by passing Cl_2 plus CO over heated Ru, then dissolution in HOH and recrystallization. This substance serves as the starting point of much Ru chemistry. The lower-valent states of Ru are soft cations which gives them strong affinities for Cl, Br, I, S, and N in their covalent and ionic forms. Ignition of most Ru compounds gives the metal.

RuO_4 is an important compound in numerous reactions of the element. Just below room temperature, it is a yellow solid; just above it is a highly volatile liquid, soluble in HOH. RuO_4 can be produced from Ru and most Ru compounds by treatment with very strong oxidizing agents. Those often used are Cl_2, MnO_4^-, and IO_4^- in acid solution. As can be concluded from its position on the E–pH diagram, it is a powerful oxidizing agent. RuO_4 can be reduced to Ru(IV) chloride complexes by HCl and can be reduced to orange RuO_4^{-2} by NaOH.

b. Compounds and complexes of Ru(III).   The pink Ru^{+2} ion can be prepared by reducing a solution of RuO_4 in a non-complexing acid with Pb. Oxidation of the Ru^{+2} with air gives the yellow-red ion Ru^{+3}. Salts of the type RuX_3 (X = OH, F, Cl, Br, I, CN) can be prepared, the hydroxide being precipitated as an amorphous, hydrated material, which is soluble in acid. Excess ligand added to the above compounds or to Ru^{+3} gives complexes, some of the simpler being RuF_6^{-3}, red $RuCl_6^{-3}$, red-orange $Ru(NO_2)_6^{-3}$, $Ru(NH_3)_6^{+3}$, green $Ru(C_2O_4)_3^{-3}$, and red Ru(acetylacetone)$_3$. In these, and in numerous mixed complexes, Ru(III) functions with a coordination number of 6. Ru(III) species have a tendency to be inert (react slowly) which often gives them a temporary existence, even though they may be thermodynamically unstable.

c. Compounds and complexes of Ru(IV).
Controlled reduction of RuO_4 with H_2O_2 produces a Ru(IV) cation with the formula $Ru_4O_6^{+4}$. The anhydrous dioxide RuO_2 results from the burning of Ru in air, and the hydrous

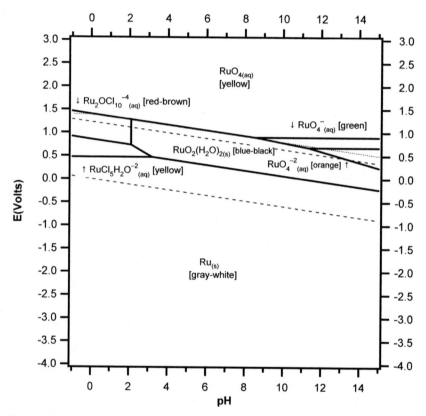

Figure 13.10 E–pH diagram for Ru/Cl species. Ru soluble species concentrations (except H^+) $= 10^{-1.0}$ M. Cl soluble species concentrations (except H^+) $= 10^{0.0}$ M. Soluble species and most solids are hydrated. No agents producing complexes or insoluble compounds are present other than HOH, OH^-, and Cl^-. The finely-dashed line (which appears at 0.5 v on the right then moves leftward along the top of the Ru(IV) lines) represents the oxidation of Cl^- to Cl_2 and/or ClO_4^- and hence the disappearance of all species containing Cl^-.

Species ($\Delta G°$ in kJ/mol): Ru (0.0), RuO_4 (−147.7), RuO_4^- (−234.3), RuO_4^{-2} (−299.6), $RuO_2(H_2O)_2$ (−694.5), $RuCl_5H_2O^{-2}$ (−752.3), $Ru_2OCl_{10}^{-4}$ (−1106.7), Cl_2 (7.1), ClO_4^- (−8.4), Cl^- (−131.4), HOH (−237.2), H^+ (0.0), and OH^- (−157.3)

Figure 13.10 (Continued)

Equations for the lines:

$RuO_4/Ru_2OCl_{10}{}^{-4}$	$E = 1.42 - 0.063 \text{ pH} - 0.001 \log[Ru_2OCl_{10}{}^{-4}] -$
	$0.001 \log[RuO_4] + 0.007 \log[ClO_4{}^-]$
$Ru_2OCl_{10}{}^{-4}/RuCl_5H_2O^{-2}$	$E = 0.83 - 0.059 \text{ pH} - 0.059 \log[RuCl_5H_2O^{-2}] +$
	$0.030 \log[Ru_2OCl_{10}{}^{-4}]$
$RuCl_5H_2O^{-2}/Ru$	$E = 0.49 - 0.098 \log[Cl^-] + 0.020 \log[RuCl_5H_2O^{-2}]$
$RuO_4/RuO_2(H_2O)_2$	$E = 1.42 - 0.059 \text{ pH} + 0.015 \log[RuO_4]$
$RuO_2(H_2O)_2/Ru$	$E = 0.66 - 0.059 \text{ pH}$
$RuO_4/RuO_4{}^-$	$E = 0.90$
$RuO_4{}^-/RuO_4{}^{-2}$	$E = 0.68$
$RuO_4{}^-/RuO_2(H_2O)_2$	$E = 1.59 - 0.079 \text{ pH} + 0.020 \log[RuO_4{}^-]$
$RuO_4{}^{-2}/RuO_2(H_2O)_2$	$E = 2.05 - 0.118 \text{ pH} + 0.030 \log[RuO_4{}^{-2}]$
$RuO_2(H_2O)_2/RuCl_5H_2O^{-2}$	$E = 1.16 - 0.236 \text{ pH} - 0.059 \log[RuCl_5H_2O^{-2}] +$
	$0.295 \log[Cl^-]$
$RuO_2(H_2O)_2/Ru_2OCl_{10}{}^{-4}$	$6 \text{ pH} = 11.2 - \log[Ru_2OCl_{10}{}^{-4}] + 10 \log[Cl^-]$

dioxide is obtained by reduction of aqueous RuO_4 with H_2 or from treatment of Ru(IV) compounds with base. The anhydrous compound is resistant to dissolution in acids. Some common complexes of Ru(IV) include $RuX_6{}^{-2}$ (X = F, Cl, Br). As in the above case, Ru(IV) complexes have a tendency to be inert.

d. Other species. Ru shows a few other upper-valent species of importance, chiefly $RuO_4{}^-$ and $RuO_4{}^{-2}$. The dissolution of RuO_4 in cold dilute KOH yields the green $RuO_4{}^-$, which goes over to the orange $RuO_4{}^{-2}$ upon increasing the concentration of KOH and/or heating. $RuO_4{}^-$ may also be prepared by the alkaline oxidation of $RuCl_3$ with $NaIO_4$, and $RuO_4{}^{-2}$ by the alkaline oxidation of $RuCl_3$ with $K_2S_2O_8$. Salts of both anions can be readily prepared. Special attention is to be given to the $RuNO^{+3}$ group which is formed in many systems involving Ru and HNO_3 or HNO_2. The ion readily complexes with many other ligands to form a wide variety of ions and compounds.

e. Health aspects. Most Ru compounds are only moderately toxic; for example, the LD50 for $RuCl_3$ in guinea pigs is 210 mg/kg. However, RuO_4

Table 13.10
Ruthenium Species

Name	Formula	State	Color	Solubility	$\Delta G°$ (kJ/mole)
Ruthenium	Ru	s	Gray-white	Insol	0.0
Ruthenium (II) ion	Ru^{+2}	aq	Pink		
Ruthenium (III) ion	Ru^{+3}	aq	Yellow-red		
Ruthenium (IV) ion	$Ru_4O_6^{+4}$	aq	Red		
Hexacyanoruthenate(II)	$Ru(CN)_6^{-4}$	aq	Colorless		
Hexammineruthenium(II)	$Ru(NH_3)_6^{+3}$	aq	Colorless		
Hexachlororuthenate(III)	$RuCl_6^{-3}$	aq	Red		
Hexacyanoruthenate(III)	$Ru(CN)_6^{-3}$	aq	Green		
Aquapentachlororuthenate(III)	$RuCl_5HOH^{-2}$	aq	Yellow		−752.3
Hexafluorruthenate(III)	RuF_6^{-3}	aq	Pale yellow		
Hexabromoruthenate(IV)	$RuBr_6^{-2}$	aq	Dark brown		
Hexachlororuthenate(IV)	$RuCl_6^{-2}$	aq	Brown		
Hexafluororuthenate(IV)	RuF_6^{-2}	aq	Yellow		
μ-oxo-bis [pentachlororuthenate(IV)]	$Ru_2OCl_{10}^{-4}$	aq	Red-brown		−1106.7
Ruthenate(VI)	RuO_4^{-2}	aq	Orange		−299.6
Perruthenate(VII)	RuO_4^-	aq	Green		−234.3
Ruthenium(II)					
bromide	$RuBr_2$	s	Black	Decomp	
chloride	$RuCl_2$	s	Brown	Decomp	
iodide	RuI_2	s	Blue	Decomp	
Ruthenium(III)					
bromide	$RuBr_3$	s	Brown	Sol	
chloride-hydrated	$RuCl_3 \cdot HOH$	s	Brown	Sol	−395.8
cyanide	$Ru(CN)_3$	s	Blue	Insol	
fluoride	RuF_3	s	Brown	Insol	
hydroxide-hydrated	$Ru(OH)_3 \cdot HOH$	s	Black	Insol	−766.0
iodide	RuI_3	s	Black	Insol	
Ruthenium (IV)					
fluoride	RuF_4	s	Pink	Hydrol	
oxide	RuO_2	s	Blue-black	Insol	−252.7
oxide-hydrated	$RuO_2 \cdot 2HOH$	s	Blue-black	Insol	−694.5
sulfide	RuS_2	s	Gray	Insol	
Ruthenium(V) fluoride	RuF_5	s	Brown	Decomp	
Ruthenium(VI) fluoride	RuF_6	s	Green	Decomp	
Ruthenium(VIII) oxide	RuO_4	aq	Yellow		−147.7

is a very volatile, highly toxic substance, which is especially hazardous to the respiratory system and the eyes, which is due to its strong oxidizing character. RuO_4 is thermodynamically unstable with reference to RuO_2 and O_2, and as a consequence, can be explosive when heated.

14. Osmium (Os) $6s^2 5d^6$

a. The element. See Figure 13.11 and Table 13.11. The oxidation states that Os displays in its aqueous chemistry are 0, II, III, IV, VI, and VIII, with the IV state showing the largest number of stable ions and compounds. The chloride system is the most important one for this element because it has been the basis for most of the research on the element's aqueous chemistry. The pertinent aqueous ions are chiefly anions, only a few cations having been substantiated. Os is a gray-white, lustrous metal which does not react with air, HOH, acids, and bases at room temperature. Upon heating, Os powder is oxidized to the volatile yellow OsO_4, which is highly toxic.

b. Osmium(VIII) oxide. OsO_4 vapor can be passed into HOH in which it dissolves. Treatment of the solution with ethanol gives a black precipitate of OsO_2·2HOH. Treatment with ethanol and OH^- results in the purple $OsO_2(OH)_4^{-2}$ anion with the Os in the VI oxidation state. On carrying the reaction out in the presence of a high concentration of Cl^-, $OsO_2Cl_4^{-2}$ results. Similar anions can be made by substitution for Cl. Treatment of the OsO_4 solution with ethanol and HCl gives yellow $OsCl_6^{-2}$. Salts of the above anions have been prepared.

c. Hexachloroosmate(IV) anion. The anion $OsCl_6^{-2}$ can act as the starting point for a number of other compounds and ions. Treatment with CO_3^{-2} gives OsO_2·2HOH, with Br^- gives $OsBr_6^{-2}$, with I^- gives OsI_6^{-2}, with H_2S gives the persulfide $Os(S_2)$, and with Zn and NH_3 gives $Os(NH_3)_5Cl^{+2}$. Controlled reduction of $OsCl_6^{-2}$ yields the unstable $OsCl_6^{-3}$.

d. Health aspects. The exceptionally toxic, highly volatile OsO_4 can be produced in numerous ways: ignition in air of Os compounds and Os metal, and treatment of Os compounds and Os with strong oxidizing agents. The vapor pressure at 25°C is 10 mm, at 40°C is 25 mm, and at 70°C is 100 mm. The vapors are very poisonous in that they severely corrode the tissues of the eyes and the respiratory tract, and can temporarily affect the bone marrow, kidneys, and liver. The LD50 for OsO_4 (oral, rat) is 14 mg/kg and the lethal concentration (low) in air is 0.13 mg/m^3.

15. Rhodium (Rh) $5s^1 4d^8$

a. The element. See Figure 13.12 and Table 13.12. Rh is a silvery-white, lustrous metal which is resistant to attack by air, HOH, bases, and most acids. In its compounds, the predominant oxidation state is III, with oxidation to higher states being very difficult, and with reduction to the metal being

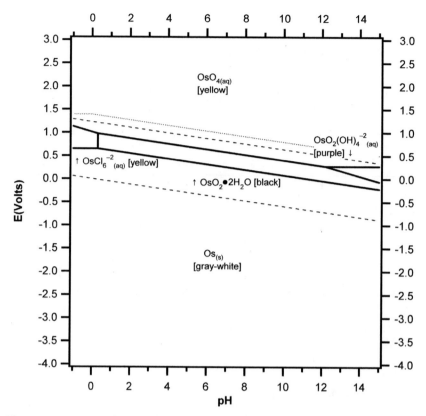

Figure 13.11 E–pH diagram for Os/Cl species. Os soluble species concentrations (except H+) = $10^{-1.0}$ M. Cl soluble species concentrations (except H+) = $10^{0.0}$ M. Soluble species and most solids are hydrated. No agents producing complexes or insoluble compounds are present other than HOH, OH⁻ and Cl⁻. The finely-dashed line represents the oxidation of Cl⁻ to Cl_2 and/or ClO_4^- and hence the disappearance of all species containing Cl⁻.

Species ($\Delta G°$ in kJ/mol): Os (0.0), $OsO_2(H_2O)_2$ (−690.4), OsO_4 (−301.7), $OsCl_6^{-2}$ (−531.4), $OsO_2(OH)_4^{-2}$ (−828.4), Cl_2 (7.1), ClO_4^- (−8.4), Cl⁻ (−131.4), HOH (−237.2), H+ (0.0), and OH⁻ (−157.3)

Equations for the lines:

$OsO_4/OsO_2(OH)_4^{-2}$	$E = 0.27$
$OsO_4/OsO_2\cdot2H_2O$	$E = 1.01 - 0.059\,pH + 0.015\log[OsO_4]$
$OsO_4/OsCl_6^{-2}$	$E = 1.01 - 0.118\,pH + 0.089\log[Cl^-]$
$OsO_2(OH)_4^{-2}/OsO_2\cdot2H_2O$	$E = 1.74 - 0.118\,pH + 0.030\log[OsO_2(OH)_4^{-2}]$
$OsO_2\cdot2H_2O/Os$	$E = 0.67 - 0.059\,pH$
$OsCl_6^{-2}/Os$	$E = 0.67 - 0.089\log[Cl^-] + 0.015\log[OsCl_6^{-2}]$
$OsO_2\cdot2H_2O/OsCl_6^{-2}$	$4\,pH = 0.25 - \log[OsCl_6^{-2}] + 6\log[Cl^-]$

Table 13.11
Osmium Species

Name	Formula	State	Color	Solubility	$\Delta G°$ (kJ/mole)
Osmium	Os	s	Gray-white	Insol	0.0
Hexacyanoosmate(II)	$Os(CN)_6{}^{-4}$	aq	Colorless		
Hexammineosmium(III)	$Os(NH_3)_6{}^{+3}$	aq	Colorless		
Hexachloroosmate(III)	$OsCl_6{}^{-3}$	aq	Red		
Hexachloroosmate(IV)	$OsCl_6{}^{-2}$	aq	Yellow		-531.4
Tetrachlorodioxoosmate(VI)	$OsO_2Cl_4{}^{-2}$	aq	Red		
Osmate(VI)	$OsO_2(OH)_4{}^{-2}$	aq	Purple		-828.4
Osmium(II)					
iodide	OsI_2	s	Black	Decomp	
Osmium(III)					
chloride	$OsCl_3$	s	Gray	Insol	-119.7
iodide	OsI_3	s	Black	Insol	
Osmium (IV)					
bromide	$OsBr_4$	s	Black	Insol	
chloride	$OsCl_4$	s	Red	Sol	-158.2
fluoride	OsF_4	s	Yellow	Hydrol	
oxide	OsO_2	s	Black	Insol	-239.7
oxide-hydrated	$OsO_2 \cdot 2HOH$	s	Black	Insol	-690.4
Osmium(V) fluoride	OsF_5	s	Blue	Decomp	
Osmium(VI) fluoride	OsF_6	s	Yellow	Hydrol	
Osmium(VIII) oxide	OsO_4	aq	Yellow		-301.7

very easy. Ignition and mild reducing agents will bring about the latter. One effective way to put Rh into solution is to powder it, then heat it in air to brown Rh_2O_3, then dissolve the oxide in strong acid. Another route is to heat Rh powder mixed with KCl in Cl_2, to dissolve the melt in HOH, to precipitate hydrated Rh_2O_3 by addition of OH^-, then to dissolve the hydrated solid in HCl. Both methods can lead to hydrated $RhCl_3$ which is usually employed as the starting material in the preparation of Rh compounds.

b. Simple compounds. Rh forms two oxides, brown Rh_2O_3, made as stated above, and black RhO_2. The hydrated form of the latter substance is produced by the treatment of Rh(III) compounds with very strong oxidizing agents, such as Cl_2, or by electrochemical methods. There is disagreement concerning the exact character and formula of the hydrated material, there being indications that it is a peroxide or a superoxide, or partly so. When hydrated Rh_2O_3 (yellow) is dissolved in $HClO_4$, the yellow Rh^{+3} is produced; this occurs because of the non-complexing character of $ClO_4{}^-$.

Rh forms all the trihalides (F, Cl, Br, I). The first three have interesting forms, one insoluble and one soluble. When the trihalides are prepared by combination of the elements, they are all insoluble. However, water-soluble forms of the fluoride, chloride, and bromide are produced by solution methods such as the treatment of hydrated Rh_2O_3 with an appropriate acid. These soluble salts are probably complex compounds such as $RhCl_3(HOH)_3$. Other salts of Rh(III) may be made by similar methods.

c. Complexes. Rh(III) shows a wide variety of complexes with a coordination number of 6 and with an inert character. These include RhX_6^{-3} (X = F, Cl, Br, I, CN, SCN, NO_2), $Rh(NH_3)_6^{+3}$, $Rh(oxalate)_3^{-3}$, $Rh(acetylacetone)_3$,

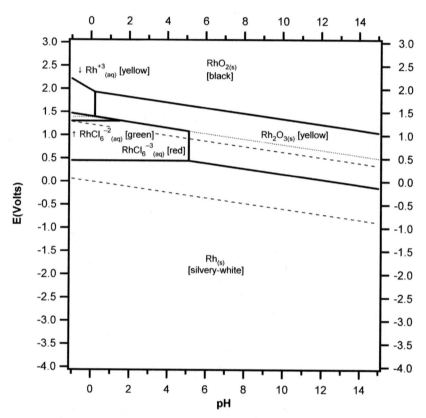

Figure 13.12 E–pH diagram for Rh/Cl species. Rh soluble species concentrations (except H^+) = $10^{-1.0}$ M. Cl soluble species concentrations (except H^+) = $10^{0.0}$ M. Soluble species and most solids are hydrated. No agents producing complexes or insoluble compounds are present other than HOH, OH^-, and Cl^-. The finely-dashed line represents the oxidation of Cl^- to Cl_2 and/or ClO_4^- and hence the disappearance of all species containing Cl^-.

Species ($\Delta G°$ in kJ/mol): Rh (0.0), Rh_2O_3 (−276.6), Rh^{+3} (219.7), RhO_2 (−69.5), $RhCl_6^{-3}$ (−652.7), $RhCl_6^{-2}$ (−527.2), Cl_2 (7.1), ClO_4^- (−8.4), Cl^- (−131.4), HOH (−237.2), H^+ (0.0), and OH^- (−157.3)

Figure 13.12 (Continued)

Equations for the lines:

RhO_2/Rh^{+3}	$E = 1.92 - 0.236 \, pH - 0.059 \log[Rh^{+3}]$
RhO_2/Rh_2O_3	$E = 1.94 - 0.059 \, pH$
$Rh^{+3}/RhCl_6{}^{-2}$	$E = 1.31 - 0.060 \, pH + 0.008 \log[ClO_4{}^-]$
$Rh_2O_3/RhCl_6{}^{-2}$	$E = 1.41 - 0.064 \, pH - 0.001 \log[RhCl_6{}^{-2}] + 0.008 \log[ClO_4{}^-]$
$Rh_2O_3/RhCl_6{}^{-3}$	$E = 1.41 - 0.063 \, pH - 0.001 \log[RhCl_6{}^{-3}] + 0.007 \log[ClO_4{}^-]$
$RhCl_6{}^{-3}/Rh$	$E = 0.47 + 0.020 \log[RhCl_6{}^{-3}] - 0.118 \log[Cl^-]$
Rh_2O_3/Rh	$E = 0.75 - 0.059 \, pH$
$RhCl_6{}^{-2}/RhCl_6{}^{-3}$	$E = 1.30$
Rh_2O_3/Rh^{+3}	$6 \, pH = -0.77 - 2 \log[Rh^{+3}]$
$Rh_2O_3/RhCl_6{}^{-3}$	$6 \, pH = 28.7 - 2 \log[RhCl_6{}^{-3}] + 12 \log[Cl^-]$

Table 13.12
Rhodium Species

Name	Formula	State	Color	Solubility	$\Delta G°$ (kJ/mole)
Rhodium	Rh	s	Silvery-white	Insol	0.0
Rhodium(III) ion	Rh^{+3}	aq	Yellow		219.7
Hexamminerhodium(III)	$Rh(NH_3)_6{}^{+3}$	aq	Colorless		
Aquapentamminerhodate(III)	$RhCl_5HOH^{+2}$	aq	Yellow		
Hexachlororhodate(III)	$RhCl_6{}^{-3}$	aq	Red		−652.7
Hexachlororhodate(IV)	$RhCl_6{}^{-2}$	aq	Green		−527.2
Rhodium(III)					
bromide	$RhBr_3$	s	Red-brown	Insol	
bromide-hydrated	$RhBr_3$-hyd	s	Red-brown	Sol	
chloride	$RhCl_3$	s	Red	Insol	−228.0
chloride-hydrated	$RhCl_3$-hyd	s	Red-brown	Sol	
cyanide	$Rh(CN)_6$	s	Brown	Insol	
fluoride	RhF_3	s	Yellow	Insol	
fluoride-hydrated	RhF_3-hyd	s	Red	Sol	
iodide	RhI_3	s	Black	Insol	
oxide	Rh_2O_3	s	Brown	Insol	
oxide-hydrated	Rh_2O_3-hyd	s	Yellow	Insol	−276.6
sulfate	$Rh_2(SO_4)_3$	s	Yellow	Sol	
Rhodium (IV)					
oxide	RhO_2	s	Black	Insol	
oxide-hydrated	RhO_2-hyd	s	Black	Insol	−69.5

$Rh(SO_3)_3^{-3}$, all of which are usually prepared by starting with hydrated $RhCl_3$. Controlled reactions will lead to complexes with mixed ligands such as $RhCl_5HOH^{-2}$ and $RhCl_2(NO_2)_4^{-3}$. In the IV oxidation state, among the few complexes which have been characterized are RhF_6^{-2} and $RhCl_6^{-2}$, both being strongly subject to hydrolysis and reduction.

d. Health aspects.   The LD50(oral, rat) for $RhCl_3$ is about 1.3 g/kg which indicates that the compound is moderately toxic. Similar values probably apply to other compounds. Contact with eyes and skin is to be carefully avoided.

16. Iridium (Ir) $6s^25d^7$

a. The element.   See Figure 13.13 and Table 13.13. Ir is a silvery-white metal which shows no reaction with acids, HOH, bases, and even aqua regia at room temperature. Its powdered form, when heated in air, gives the blue-black IrO_2. This oxide can be dissolved in aqua regia to give red-brown $IrCl_6^{-2}$. Another way to put Ir in solution is to fuse Ir with Na_2O_2 plus KNO_3, then to dissolve the melt in HCl. Still a further approach is to mix Ir, concentrated HCl, plus Cl_2 or $NaClO_3$ and heat. In all cases, the resulting $IrCl_6^{-2}$ can be reduced to yellow-green $IrCl_6^{-3}$ with agents such as $C_2O_4^{-2}$ or SO_3^{-2}. These two hexachloro complexes usually serve as starting ions for the derivation of other Ir species. As the E–pH diagram shows, the major oxidation states are III and IV. Reduction of Ir compounds to the metal is easily done, and ignition of such compounds also yields the metal.

b. Simple compounds.   Many simple anhydrous compounds of Ir can be made by direct combination of elements, usually at controlled high temperatures: Ir_2O_3, IrO_2, $IrCl_3$, $IrBr_3$, IrI_3. However, when the oxides are prepared by reactions of bases with $IrCl_6^{-2}$ or $IrCl_6^{-3}$, hydrated species are produced. Likewise, when the halides are obtained by treatment of the hydrated oxides or the hexachloride ions, hydrated halides result. The anhydrous halides are insoluble, but the hydrated ones are soluble. This probably indicates that the hydrated species are actually complex compounds $MX_3(HOH)_3$. Dissolution of hydrated Ir_2O_3 in $HClO_4$ produces the yellow Ir^{+3} ion.

c. Complex compounds.   Treatment of the basic complexes $IrCl_6^{-2}$ and $IrCl_6^{-3}$ with proper complexing agents or treatment of the hydrated oxides with appropriate reagents leads to complexes in the IV and III oxidation states. Ir typically shows a coordination number of 6 and its complexes tend to be quite inert (slow to react). Examples of species derived from $IrCl_6^{-2}$ are red-yellow IrF_6^{-2} and blue-black IrI_6^{-2}. Complexes prepared from $IrCl_6^{-3}$ include green $IrBr_6^{-3}$, green IrI_6^{-3}, colorless $Ir(CN)_6^{-3}$, orange-yellow $Ir(SCN)_6^{-3}$, colorless $Ir(NO_2)_6^{-3}$, yellow $Ir(OH)_6^{-3}$, red $Ir(C_2O_4)_3^{-3}$, yellow $Ir(acetylacetone)_3$, and colorless $Ir(NH_3)_6^{+3}$. Many of these are subject to hydrolysis, replacement of ligands by HOH, and replacement of ligands by other ligands.

d. Health aspects. The LD50 (oral, rat) for hydrated $IrCl_3$ is reported as 1.2 g/kg, indicating that the compound is moderately toxic. Precautions must be taken to avoid contact with eyes and skin.

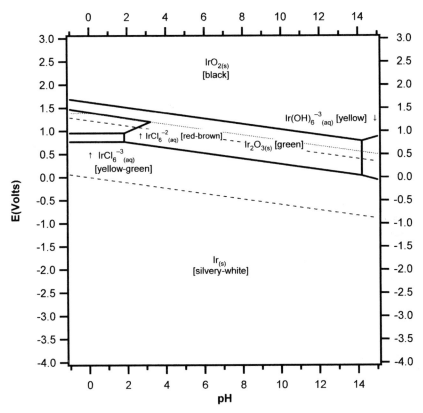

Figure 13.13 E–pH diagram for Ir/Cl species. Ir soluble species concentrations (except H^+) = $10^{-1.0}$ M. Cl soluble species concentrations (except H^+) = $10^{0.0}$ M. Soluble species and most solids are hydrated. No agents producing complexes or insoluble compounds are present other than HOH, OH^-, and Cl^-. The finely-dashed line represents the oxidation of Cl^- to Cl_2 and/or ClO_4^- and hence the disappearance of all species containing Cl^-.

Species ($\Delta G°$ in kJ/mol): Ir (0.0), Ir_2O_3 (−204.6), IrO_2 (−64.0), $Ir(OH)_6^{-3}$ (−920.5), $IrCl_6^{-2}$ (−468.6), $IrCl_6^{-3}$ (−560.7), Cl_2 (7.1), ClO_4^- (−8.4), Cl^- (−131.4), HOH (−237.2), H^+ (0.0), and OH^- (−157.3)

Figure 13.13 (Continued)

Equations for the lines:

IrO_2/Ir_2O_3	$E = 1.63 - 0.059\,pH$
$IrO_2/Ir(OH)_6^{-3}$	$E = -0.96 + 0.118\,pH - 0.059\,\log[Ir(OH)_6^{-3}]$
$Ir_2O_3/IrCl_6^{-2}$	$E = 1.40 - 0.064\,pH - 0.001\,\log[IrCl_6^{-2}] + 0.008\,\log[ClO_4^{-}]$
$IrCl_6^{-2}/Ir_2O_3$	$E = 0.69 + 0.177\,pH + 0.059\,\log[IrCl_6^{-2}] - 0.354\,\log[Cl^-]$
$IrCl_6^{-2}/IrCl_6^{-3}$	$E = 0.95$
$IrCl_6^{-3}/Ir$	$E = 0.79 + 0.020\,\log[IrCl_6^{-3}] - 0.118\,\log[Cl^-]$
Ir_2O_3/Ir	$E = 0.88 - 0.059\,pH$
$Ir(OH)_6^{-3}/Ir$	$E = 1.74 - 0.118\,pH + 0.020\,\log[Ir(OH)_6^{-3}]$
$Ir_2O_3/IrCl_6^{-3}$	$6\,pH = 9.1 - 2\,\log[IrCl_6^{-3}] + 12\,\log[Cl^-]$
$Ir(OH)_6^{-3}/Ir_2O_3$	$6\,pH = 87.4 + \log[Ir(OH)_6^{-3}]$

Table 13.13
Iridium Species

Name	Formula	State	Color	Solubility	$\Delta G°$ (kJ/mole)
Iridium	Ir	s	Silvery-white	Insol	0.0
Iridium(III) ion	Ir^{+3}	aq	Yellow		79.6
Hexammineiridium(III)	$Ir(NH_3)_6^{+3}$	aq	Colorless		
Hexachloroiridate(III)	$IrCl_6^{-3}$	aq	Yellow-green		-560.7
Hexahydroxoiridate(III)	$Ir(OH)_6^{-3}$	aq	Yellow		-920.5
Hexachloroiridate(IV)	$IrCl_6^{-2}$	aq	Red-brown		-468.6
Iridium(III)					
bromide	$IrBr_3$	s	Red-brown	Insol	
bromide-hydrated	$IrBr_3$-hyd	s	Green	Sol	
chloride	$IrCl_3$	s	Red	Insol	
chloride-hydrated	$IrCl_3$-hyd	s	Green	Sol	
fluoride	IrF_3	s	Brown	Insol	
iodide	IrI_3	s	Brown-black	Insol	
iodide-hydrated	IrI_3-hyd	s	Yellow-brown	Sol	
oxide	Ir_2O_3	s	Brown	Insol	
oxide-hydrated	Ir_2O_3-hyd	s	Green	Insol	-204.6
sulfate	$Ir_2(SO_4)_3$	s	Yellow	Insol	
Iridium (IV)					
fluoride	IrF_4	s	Red-brown	Insol	
oxide	IrO_2	s	Blue-black	Insol	
oxide-hydrated	IrO_2-hyd	s	Black	Insol	-64.0

17. Palladium (Pd) $5s^0 4d^{10}$

a. The element. See Figure 13.14 and Table 13.14. Pd is a gray-white lustrous metal which shows no reaction with air, HOH, bases, or non-oxidizing acids. However, it burns in air or O_2 to black PdO, dissolves in HNO_3 to red-brown Pd^{+2}, and dissolves in aqua regia. The predominant oxidation states are 0 and II, with a fairly unstable state of IV being evidenced

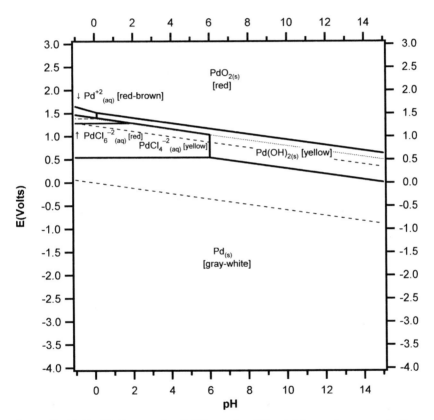

Figure 13.14 E–pH diagram for Pd/Cl species. Pd soluble species concentrations (except H^+) = $10^{-1.0}$ M. Cl soluble species concentrations (except H^+) = $10^{0.0}$ M. Soluble species and most solids are hydrated. No agents producing complexes or insoluble compounds are present other than HOH, OH^-, and Cl^-. The finely-dashed line represents the oxidation of Cl^- to Cl_2 and/or ClO_4^- and hence the disappearance of all species containing Cl^-.

Species ($\Delta G°$ in kJ/mol): Pd (0.0), $Pd(OH)_2$ (−300.8), PdO_2 (−7.1), Pd^{+2} (177.8), $PdCl_4^{-2}$ (−415.1), $PdCl_6^{-2}$ (−430.1), Cl_2 (7.1), ClO_4^- (−8.4), Cl^- (−131.4), HOH (−237.2), H^+ (0.0), and OH^- (−157.3)

Figure 13.14 (Continued)

Equations for the lines:

PdO_2/Pd^{+2}	$E = 1.50 - 0.118 \, pH - 0.030 \log[Pd^{+2}]$
$PdCl_6^{-2}/PdCl_4^{-2}$	$E = 1.28 - 0.059 \log[Cl^-]$
$PdCl_4^{-2}/Pd$	$E = 0.57 - 0.118 \log[Cl^-] + 0.030 \log[PdCl_4^{-2}]$
$PdO_2/Pd(OH)_2$	$E = 1.52 - 0.059 \, pH$
$Pd(OH)_2/Pd$	$E = 0.90 - 0.059 \, pH$
$Pd^{+2}/PdCl_6^{-2}$	$E = 1.41 - 0.062 \, pH + 0.008 \log[ClO_4^-]$
$Pd(OH)_2/PdCl_6^{-2}$	$E = 1.41 - 0.064 \, pH - 0.001 \log[PdCl_6^{-2}] + 0.008 \log[ClO_4^-]$
$Pd(OH)_2/PdCl_4^{-2}$	$E = 1.41 - 0.063 \, pH - 0.002 \log[PdCl_4^{-2}] + 0.007 \log[ClO_4^-]$
$Pd(OH)_2/PdCl_4^{-2}$	$2 \, pH = 11.1 - \log[PdCl_4^{-2}] + 4 \log[Cl^-]$
$Pd(OH)_2/Pd^{+2}$	$2 \, pH = -0.74 - \log[Pd^{+2}]$

in some compounds under strongly oxidizing conditions. Pd(II), as all of the low-valent noble-metal cations is a soft cation which shows little affinity for hard ions and atoms (F, O), but strong affinity for softer ions and atoms (Cl, Br, I, S, N) and for ligands that can pi bond. Pd compounds in general are readily reduced to Pd by reducing agents and by ignition.

b. Compounds. Pd, PdO, and $Pd(OH)_2$ all dissolve in hot concentrated HNO_3 or $HClO_4$ to give Pd^{+2}. This ion then is used to prepare numerous other compounds of Pd. Pd, PdO, and $Pd(OH)_2$ dissolve in aqua regia to give red $PdCl_6^{-2}$ which decomposes to yellow $PdCl_4^{-2}$ and Cl_2 upon heating the solution. $Pd(OH)_2$ is made by precipitation from Pd^{+2} or $PdCl_4^{-2}$, and the yellow, hydrated compound readily dissolves in acid. The halide compounds of Pd include PdX_2 (X = F, Cl, Br, I) and PdF_4. $PdCl_2$ is soluble, both of the fluorides decompose in HOH, and the other two PdX_2 are insoluble. Other simple compounds that have been prepared include $Pd(C_2H_3O_2)_2$, $Pd(CN)_2$, $Pd(SCN)_2$, $Pd(NO_3)_2$, and $PdSO_4$. Very few simple Pd(IV) compounds exist because of the strong tendency of such compounds to disproportionate.

c. Complexes. Numerous Pd(II) complexes with a coordination number of 4 can be prepared by treatment of Pd^{+2} or simple Pd compounds with appropriate complexing agents. Among the better known are PdX_4^{-2} (X = Cl, Br, I, CN, SCN, NO_2), PdZ_2^{-2} (Z = SO_3, S_2O_3, C_2O_4), PdE_2 (E = acetylacetonate, glycinate, dimethylglyoxime), and $Pd(NH_3)_4^{+2}$.

Table 13.14
Palladium Species

Name	Formula	State	Color	Solubility	$\Delta G°$ (kJ/mole)
Palladium	Pd	s	Gray-white	Insol	0.0
Palladium (II) ion	Pd^{+2}	aq	Red-brown		177.8
Tetramminepalladium(II)	$Pd(NH_3)_4^{+2}$	aq	Colorless		-75.0
Tetrabromopalladate(II)	$PdBr_4^{-2}$	aq	Red-brown		-318.0
Tetrachloropalladate(II)	$PdCl_4^{-2}$	aq	Yellow		-415.1
Tetracyanopalladate(II)	$Pd(CN)_4^{-2}$	aq	Yellow		628.0
Tetraiodopalladate(II)	PdI_4^{-2}	aq	Red-brown		-159.0
Tetranitritopalladate(II)	$Pd(NO_2)_4^{-2}$	aq	Yellow		-68.0
Tetrathiocyanatopalladate(II)	$Pd(CNS)_4^{-2}$	aq	Red		410.5
Hexabromopalladate(IV)	$PdBr_6^{-2}$	aq	Brown-black		-335.1
Hexachloropalladate(IV)	$PdCl_6^{-2}$	aq	Red		-430.1
Hexaiodopalladate(IV)	PdI_6^{-2}	aq	——		-170.3
Palladium(II)					
acetate	$Pd(C_2H_3O_2)_2$	s	Brown	Sol	
bromide	$PdBr_2$	s	Red-black	Insol	
chloride-hydrated	$PdCl_2$-hydr	s	Red	Sol	-128.8
cyanide	$Pd(CN)_2$	s	Yellow	Insol	
fluoride	PdF_2	s	Violet	Decomp	
hydroxide	$Pd(OH)_2$	s	Yellow	Insol	-300.8
iodide	PdI_2	s	Black	Insol	-11.5
nitrate	$Pd(NO_3)_2$	s	Yellow-brown	Sol	
oxide	PdO	s	Black	Insol	
perchlorate	$Pd(ClO_4)_2$	s	Yellow-brown	Sol	
sulfate	$PdSO_4$	s	Red-brown	Sol	
sulfide	PdS	s	Green-brown	Insol	-67.0
sufide(per-)	$Pd(S_2)$	s	Brown black	Insol	
thiocyanate	$Pd(CNS)_2$	s	Red-brown	Sol	243.3
Palladium(IV)					
fluoride	PdF_4	s	Red	Decomp	
oxide	PdO_2	s	Red	Insol	-7.1

Some log β_n values for Pd^{+2} are NH_3 (9.6, 18.5, 26.0, 36.3), OH^- (13.0, 25.8, 29.4, 42.2). By oxidation PdX_6^{-2} complexes ($X = Cl, Br, I$) can be formed from the PdX_4^{-2} complexes, but they are thermodynamically unstable. However, they have a tendency to be inert, which means their decompositions occur slowly.

d. Health aspects. The LD50 values for $PdCl_2$ (oral rat) has been reported as 200 mg/kg. A number of Pd compounds are also known to produce allergenic reactions in some persons.

18. Platinum (Pt) 6s¹5d⁹

a. The element. See Figure 13.15 and Table 13.15. Pt is a gray-white lustrous metal which shows no reaction with air, HOH, bases, and acids, but dissolves in aqua regia or HCl plus HClO$_3$, to give yellow-red PtCl$_6^{-2}$. The finely powdered element dissolves to some extent in a KCN solution in the presence of air. The predominant oxidation states are 0, II, and IV.

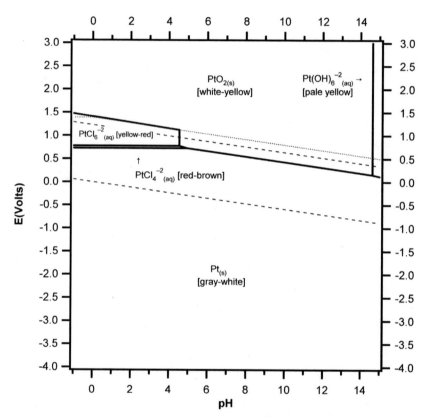

Figure 13.15 E–pH diagram for Pd/Cl species. Pt soluble species concentrations (except H$^+$) = $10^{-3.0}$ M. Cl soluble species concentrations (except H$^+$) = $10^{0.0}$ M. Soluble species and most solids are hydrated. No agents producing complexes or insoluble compounds are present other than HOH, OH$^-$, and Cl$^-$. The finely-dashed line represents the oxidation of Cl$^-$ to Cl$_2$ and/or ClO$_4^-$ and hence the disappearance of all species containing Cl$^-$.

Species ($\Delta G°$ in kJ/mol): Pt (0.0), PtO$_2$ (−80.8), PtCl$_4^{-2}$ (−368.2), PtCl$_6^{-2}$ (−481.2), Pt(OH)$_6^{-2}$ (−845.2), Cl$_2$ (7.1), ClO$_4^-$ (−8.4), Cl$^-$ (−131.4), HOH (−237.2), H$^+$ (0.0), and OH$^-$ (−157.3)

Figure 13.15 (Continued)

Equations for the lines:

PtO_2/Pt	$E = 1.02 - 0.059 \, pH$
$Pt(OH)_6^{-2}/Pt$	$E = 1.50 - 0.089 \, pH + 0.015 \log [Pt(OH)_6^{-2}]$
$PtCl_6^{-2}/PtCl_4^{-2}$	$E = 0.78 - 0.059 \log [Cl^-]$
$PtCl_4^{-2}/Pt$	$E = 0.82 + 0.030 \log [PtCl_4^{-2}] - 0.118[Cl^-]$
$PtO_2/PtCl_4^{-2}$	$E = 1.22 - 0.118 \, pH - 0.030 \log [PtCl_4^{-2}] + 0.118 \log [Cl^-]$
$PtO_2/PtCl_6^{-2}$	$E = 1.41 - 0.064 \, pH - 0.001 \log [PtCl_6^{-2}] + 0.007 \log [ClO_4^-]$
$PtO_2/PtCl_6^{-2}$	$4 \, pH = 15.2 - \log [PtCl_6^{-2}] + \log [Cl^-]$
$Pt(OH)_6^{-2}/PtO_2$	$2 \, pH = 32.4 + \log [Pt(OH)_6^{-2}]$

Table 13.15
Platinum Species

Name	Formula	State	Color	Solubility	$\Delta G°$ (kJ/mole)
Platinum	Pt	s	Gray-white	Insol	0.0
Platinum(II) ion	Pt^{+2}	aq	Yellow		254.8
Tetrabromoplatinate(II)	$PtBr_4^{-2}$	aq	Red-brown		−262.5
Tetrachloroplatinate(II)	$PtCl_4^{-2}$	aq	Red-brown		−368.2
Tetracyanoplatinate(II)	$Pt(CN)_4^{-2}$	aq	Yellow-brown		710.5
Tetrammineplatinum(II)	$Pt(NH_3)_4^{+2}$	aq	Colorless		−52.9
Hexabromoplatinate(IV)	$PtBr_6^{-2}$	aq	Red		−319.3
Hexachloroplatinate(IV)	$PtCl_6^{-2}$	aq	Yellow-red		−481.2
Hexachloroplatinic(IV) acid	H_2PtCl_6	s	Brown-red	Sol	
Hexahydroxoplatinate(IV)	$Pt(OH)_6^{-2}$	aq	Pale yellow		−845.2
Hexaiodoplatinate(IV)	PtI_6^{-2}	aq	Brown-black		−108.4
Platinum(II)					
bromide	$PtBr_2$	s	Brown		
chloride	$PtCl_2$	s	Green		−93.2
hydroxide	$Pt(OH)_2$	s	Black	Insol	−272.4
iodide	PtI_2	s	Black		−17.1
sulfide	PtS	s	Green	Insol	−334.1
Platinum(IV)					
bromide	$PtBr_4$	s	Brown-black	Insol	−104.6
chloride	$PtCl_4$	s	Red-brown	Sol	−163.7
iodide	PtI_4	s	Brown-black	Insol	−45.0
oxide-hydrated	$PtO_2(-hydr)$	s	White-yellow	Insol	−80.8
sulfide	PtS_2	s	Black	Insol	−99.6

Pt(II) and Pt(IV) are both soft cations, showing low affinities for hard ions and atoms (F, O) and high affinities for softer ions and atoms (Cl, Br, I, S, N) and for ligands that can pi bond. Practically all compounds of Pt can be reduced to Pt metal by ignition or moderately strong reducing agents.

b. Compounds and complexes of Pt(IV).

Most of the aqueous chemistry of Pt has been based upon the use of yellow-red $PtCl_6^{-2}$ as the starting material. Addition of base to a solution of $PtCl_6^{-2}$ gives a white-yellow hydrated PtO_2 which dissolves in acid. If treated with strong base, it dissolves to give pale yellow $Pt(OH)_6^{-2}$. Heating of the white-yellow solid gives a black material which not only has lost water but has also lost some oxygen, and is insoluble in acids. Treatment of a $PtCl_6^{-2}$ solution with H_2S or a soluble sulfide yields the black, insoluble PtS_2. Metathetical reactions with the proper ions produce $PtCl_6^{-2}$ salts, those of ammonium, potassium, rubidium, and cesium being insoluble. Platinum(IV) shows four tetrahalides ($X = F, Cl, Br, I$). The yellow-red fluoride is prepared by combination of the elements and is rapidly hydrolyzed by HOH. The red-brown chloride is prepared by heating H_2PtCl_6 and it dissolves readily in HOH. The brown-black bromide and brown-black iodide can be made by treatment of $PtCl_6^{-2}$ solution with Br^- and with I^-. They are both insoluble in HOH.

Pt(IV) forms a wide variety of complexes which have a coordination number of 6 and a strong tendency to be inert (react very slowly). Some of the simplest of these are PtX_6^{-2} ($X = F, Cl, Br, I, CN, SCN$), $Pt(NH_3)_6^{+4}$, and numerous ones with mixed ligands such as $Pt(NH_3)_4Cl_2^{+2}$. Numerous salts of the complex anions have been prepared.

c. Compounds and complexes of Pt(II).

The aqueous $PtCl_6^{-2}$ can be readily reduced with hydrazine or SO_2 to yield $PtCl_4^{-2}$, which is often the starting substance for Pt(II) compounds. Reaction of KOH with a solution of $PtCl_4^{-2}$ gives a white amphoteric hydrated $Pt(OH)_2$ which turns black upon aging. It is easily oxidized by air. Reaction of $AgClO_4$ with $PtCl_4^{-2}$ gives the Pt^{+2} cation. All three of the dihalide compounds (Cl, Br, I) are insoluble in HOH, but they are taken into solution by the appropriate hydrohalic acids. A soluble sulfide or H_2S precipitates from $PtCl_4^{-2}$ solution the dark green PtS.

Pt(II) forms many complexes, most of which have a coordination number of 4 and a strong tendency to be inert (react very slowly). Some notable ones are PtX_4^{-2} ($X = Cl, Br, I, OH, NO_2, NO_3, CN, SCN$), $Pt(NH_3)_4^{+2}$, $Pt(C_2O_4)_2^{-2}$, $Pt(acac)_2$, and $Pt(glycinate)_2$. Complexes of mixed ligands of many possible combinations are also well known, for examples, the isomers $Pt(NH_3)_2Cl_2$ and $[Pt(NH_3)_4][PtCl_4]$. As is the case for Pt(IV) complex anions, many salts of Pt(II) complex anions have been made. Log β_n values for some Pt(II) complexes are NH_3 (log $\beta_4 = 35.3$), Cl^- (7.0, 11.0, 14.0, 16.0).

d. Health aspects. The salts of $PtX_6{}^{-2}$ $(X = Cl, Br, I)$ have strong allergenic potential. Further, all compounds of Pt are toxic by oral ingestion, having LD50 values in the range of 10–50 mg/kg. No reports have appeared of toxic effects due to exposure arising from automotive catalysts which can contain Pt, Pd, and/or Rh.

14

The V–Cr–Mn Group

1. Introduction

The three elements to be treated in this chapter (V, Cr, Mn) are the third, fourth, and fifth members of the first transition series. The first two members (Sc, Ti) have been treated in previous chapters (Chapters 12 and 13). The ten elements of this first transition series (Sc through Zn) are characterized by electron activity in the 3d–4s levels. All elements in the 3d transition series are metals, and many of their compounds tend to be colored as a result of unpaired electrons. Most of the elements have a strong tendency to form complex ions due to participation of the d electrons in bonding. Since both the 4s and the 3d electrons are active, most of the elements show a considerable variety of oxidation states (Sc and Zn being exceptions). For the first five (Sc through Mn), the maximum oxidation number is the total number of electrons in the 4s and 3d levels. Complexing is often so strong that the most stable oxidation state for simple compounds may differ from that for complex compounds.

2. Vanadium (V) $4s^2 3d^3$

a. E–pH diagram. The E–pH diagram in Figure 14.1 shows V in oxidation states of 0, II, III, IV, and V. This diagram, which involves vanadium at $10^{-3.0}$ M is somewhat oversimplified in that there are some isopolyanions present in the 4–6 pH regions. The prevalence of isopolyanions increases as the V concentration increases. This is illustrated in Figure 14.2 which has V at $10^{-1.0}$ M. Further, the cations V^{+2}, V^{+3}, VO^{+2}, and VO_2^+ are probably aquated to satisfy a coordination number of six, and the $V(OH)_3$ may actually be hydrated V_2O_3. Note that the soluble solution chemistries of V(IV) and V(V) are dominated by the VO^{+2} and VO_2^+ complex ions. Three of these cations (III, IV, V) are subject to hydrolysis, the processes setting in around pH values of just under 3, 3, and 2. The E–pH diagram indicates that elemental V is very active, but a thin coat of oxide protects it from all except strong action. The II cation is unstable with regard to HOH, and the III and IV cations are unstable in the presence of air. The legends of the figures give equations for the lines between the species.

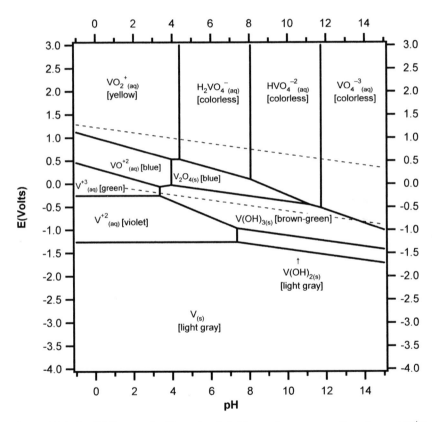

Figure 14.1 E–pH diagram for V species. Soluble species concentrations (except H^+) = $10^{-3.0}$ M. Soluble species and most solids are hydrated. No agents producing complexes or insoluble compounds are present other than HOH and OH^-.

Species ($\Delta G°$ in kJ/mol): V (0.0), V^{+3} (−251.5), V^{+2} (−226.8), $V(OH)_2$ (−634.3), $V(OH)_3$ (−922.6), V_2O_4 (−1330.5), VO^{+2} (−456.1), VO_2^+ (−596.6), VO_4^{-3} (−907.9), HVO_4^{-2} (−974.9), $H_2VO_4^-$ (−1020.9), HOH (−237.2), H^+ (0.0), and OH^- (−157.3)

Figure 14.1 (Continued)

Equations for the lines:

V^{+2}/V	$E = -1.18 + 0.030 \log[V^{+2}]$
$V(OH)_2/V$	$E = -0.83 - 0.059 \, pH$
V^{+3}/V^{+2}	$E = -0.26$
$V(OH)_3/V^{+2}$	$E = 0.16 - 0.177 \, pH - 0.059 \log[V^{+2}]$
$V(OH)_3/V(OH)_2$	$E = -0.53 - 0.059 \, pH$
VO^{+2}/V^{+3}	$E = 0.34 - 0.118 \, pH$
$VO^{+2}/V(OH)_3$	$E = -0.08 + 0.059 \, pH + 0.059 \log[VO^{+2}]$
$V_2O_4/V(OH)_3$	$E = 0.21 - 0.059 \, pH$
VO_2^+/VO^{+2}	$E = 1.00 - 0.118 \, pH$
VO_2^+/V_2O_4	$E = 0.71 + 0.059 \log[VO_2^+]$
$H_2VO_4^-/V_2O_4$	$E = 1.23 - 0.118 \, pH + 0.059 \log[H_2VO_4^-]$
HVO_4^{-2}/V_2O_4	$E = 1.71 - 0.177 \, pH + 0.059 \log[HVO_4^{-2}]$
$HVO_4^{-2}/V(OH)_3$	$E = 0.96 - 0.118 \, pH + 0.030 \log[HVO_4^{-2}]$
$VO_4^{-3}/V(OH)_3$	$E = 1.31 - 0.148 \, pH + 0.030 \log[VO_4^{-3}]$
$V(OH)_2/V^{+2}2$	$pH = 11.7 - \log[V^{+2}]$
$V(OH)_3/V^{+3}3$	$pH = 7.1 - \log[V^{+3}]$
$V_2O_4/VO^{+2}4$	$pH = 9.8 - 2 \log[VO^{+2}]$
VO_4^{-3}/HVO_4^{-2}	$pH = 11.8$
$HVO_4^{-2}/H_2VO_4^-$	$pH = 8.1$
$H_2VO_4^-/VO_2^+$	$pH = 4.4$

b. Discovery, occurrence, and extraction. V, named after the Scandinavian goddess Vanadis, was initially discovered in Mexico by Andreas Manuel del Rio in 1801, but he erroneously became convinced that it was Cr. He had isolated the element from a Mexican brown lead ore, which was chiefly PbCrO$_4$. In 1831 Nils Gabriel Sefstrom rediscovered the element, and gave it its name. He had isolated V from iron which had been produced from an iron ore in Sweden, and he recognized that it was the same as what del Rio had described. The major sources of V are the minerals patronite VS$_4$, roscoelite K$_2$V$_4$Al$_2$Si$_6$O$_{20}$(OH)$_4$, carnotite K$_2$(UO$_2$)$_2$(VO$_4$)$_2$, and vanadinite Pb$_5$Cl(VO$_4$)$_3$. The ore is roasted with NaCl or Na$_2$CO$_3$, and the resulting

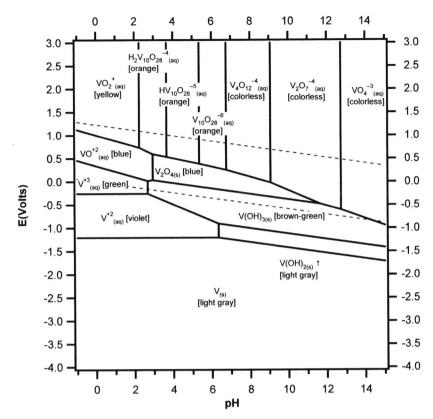

Figure 14.2 E–pH diagram for V species. Soluble species concentrations (except H$^+$) = $10^{-1.0}$ M. Soluble species and most solids are hydrated. No agents producing complexes or insoluble compounds are present other than HOH and OH$^-$.

Species ($\Delta G°$ in kJ/mol): V (0.0), V^{+3} (−251.5), V^{+2} (−226.8), V(OH)$_2$ (−634.3), V(OH)$_3$ (−922.6), V$_2$O$_4$ (−1330.5), VO^{+2} (−456.1), VO$_2^+$ (−596.6), VO$_4^{-3}$ (−907.9), HV$_{10}$O$_{28}^{-5}$ (−7708.2), H$_2$V$_{10}$O$_{28}^{-4}$ (−7729.1), V$_{10}$O$_{28}^{-6}$ (−7677.6), V$_4$O$_{12}^{-4}$ (−3194.5), V$_2$O$_7^{-4}$ (−1728.0), HOH (−237.2), H$^+$ (0.0), and OH$^-$ (−157.3)

Equations for the lines:

V^{+2}/V	E = −1.18 + 0.030 log[V^{+2}]
V(OH)$_2$/V	E = −0.83 − 0.059 pH
V^{+3}/V^{+2}	E = −0.26
V(OH)$_3$/V^{+2}	E = 0.16 − 0.177 pH − 0.059 log[V^{+2}]
V(OH)$_3$/V(OH)$_2$	E = −0.53 − 0.059 pH
VO^{+2}/V^{+3}	E = 0.34 − 0.118 pH

Figure 14.2 (Continued)

$VO^{+2}/V(OH)_3$	$E = -0.08 + 0.059\ pH + 0.059\ log[VO^{+2}]$
$V_2O_4/V(OH)_3$	$E = 0.21 - 0.059\ pH$
VO_2^+/VO^{+2}	$E = 1.00 - 0.118\ pH$
$H_2V_{10}O_{28}^{-4}/VO^{+2}$	$E = 1.14 - 0.201\ pH + 0.006\ log[H_2V_{10}O_{28}^{-4}] - 0.059\ log[VO^{+2}]$
$H_2V_{10}O_{28}^{-4}/V_2O_4$	$E = 0.85 - 0.083\ pH + 0.006\ log[H_2V_{10}O_{28}^{-4}]$
$HV_{10}O_{28}^{-5}/V_2O_4$	$E = 0.87 - 0.089\ pH + 0.006\ log[HV_{10}O_{28}^{-5}]$
$V_{10}O_{28}^{-6}/V_2O_4$	$E = 0.90 - 0.094\ pH + 0.006\ log[V_{10}O_{28}^{-6}]$
$V_4O_{12}^{-4}/V_2O_4$	$E = 1.08 - 0.118\ pH + 0.015\ log[V_4O_{12}^{-4}]$
$V_2O_7^{-4}/V_2O_4$	$E = 1.63 - 0.177\ pH + 0.030\ log[V_2O_7^{-4}]$
$V_2O_7^{-4}/V(OH)_3$	$E = 0.92 - 0.118\ pH + 0.015 log[V_2O_7^{-4}]$
$VO_4^{-3}/V(OH)_3$	$E = 1.31 - 0.148\ pH + 0.030\ log[VO_4^{-3}]$
$VO_4^{-3}/V_2O_7^{-4}$	$2\ pH = 26.2 + 2log[VO_4^{-3}] - log[V_2O_7^{-4}]$
$V_2O_7^{-4}/V_4O_{12}^{-4}$	$4\ pH = 37.4 + 2\ log[V_2O_7^{-4}] - log[V_4O_{12}^{-4}]$
$V_4O_{12}^{-4}/V_{10}O_{28}^{-6}$	$8\ pH = 58.2 + 5\ log[V_4O_{12}^{-4}] - 2\ log[V_{10}O_{28}^{-6}]$
$H_2V_{10}O_{28}^{-4}/VO_2^+$	$14\ pH = 23.6 - 10\ log[VO_2^+] + log[H_2V_{10}O_{28}^{-4}]$
$V(OH)_2/V^{+2}$	$2\ pH = 11.7 - log[V^{+2}]$
$V(OH)_3/V^{+3}$	$3\ pH = 7.1 - log[V^{+3}]$
V_2O_4/VO^{+2}	$4\ pH = 9.8 - 2\ log[VO^{+2}]$
$V_{10}O_{28}^{-6}/HV_{10}O_{28}^{-5}$	$pH = 5.4$
$HV_{10}O_{28}^{-5}/H_2V_{10}O_{28}^{-4}$	$pH = 3.7$

$NaVO_3$ is leached out with water. Then the solution is treated with sulfuric acid to precipitate a red polyvanadate compound, which is fused to V_2O_5. Reduction of the V_2O_5 with Ca or Al gives the pure metal.

c. The element. V is a light gray metal which is insoluble in HOH, bases, and non-oxidizing acids due to its passive coat of oxide. It dissolves in HF to give colorless $VF_6{}^-$, and in hot HNO_3, hot H_2SO_4, or aqua regia to give yellow $VO_2{}^+$. The metal reacts at elevated temperatures with O_2 to yield yellow-red V_2O_5, with F_2 to give colorless VF_5, with Cl_2 to produce red-brown VCl_4, with Br_2 to give gray-brown VBr_3, and with I_2 to yield brown black VI_3.

d. Oxides and hydroxides.

V shows four main oxides: V_2O_5, V_2O_4, V_2O_3, and V_2O_2. V_2O_5 is prepared by heating V in air, or by making a concentrated solution of $VO_2{}^+$ very strongly acidic. Hydrated dark blue V_2O_4 results when $VO_2{}^+$ is gently reduced to VO^{+2}, then treated with OH^-. Similarly, controlled reduction of $VO_2{}^+$ to V^{+3}, or vigorous reduction to V^{+2}, then alkaline precipitation gives brown-green $V(OH)_3$ or light gray $V(OH)_2$. V_2O_5 is amphoteric, dissolving in both acid and base, giving the species as indicated on the E–pH diagram. V_2O_4 is also amphoteric, acid dissolution giving VO^{+2} and basic dissolution producing $V_4O_9{}^{-2}$ which rapidly goes over to V(V) species. Both $V(OH)_3$ and $V(OH)_2$ are basic in that they are soluble in acid, but not in base.

e. Compounds.

Table 14.1 presents a number of V species which are pertinent to the aqueous chemistry of the element. Other than the oxides, the halides are the major simple compounds of V which have aqueous implications. In general, most of the halides may be synthesized by dry methods, but only some of them in solution. And, the method of preparation often leads to different properties, especially solubilities, probably because HOH molecules are in coordination positions in the solution products. The strong tendency of V to attach to O is evidenced by the ready hydrolysis of many of these compounds, particularly those of the higher oxidation states. The extraordinary stabilities of the VO^{+2} and the $VO_2{}^+$ ions are witnesses to this. Those in the lower oxidation states can also be subject to disproportionation and other redox reactions. For example, the reduction potential of V(II) is so low that most of its compounds exist only in the absence of water and air.

The vanadate(V) anions give rise to numerous salts. Among the most common are the orthovanadates ($VO_4{}^{-3}$), metavanadates ($VO_3{}^-$), and pyrovanadates ($V_2O_7{}^{-4}$), these anions being colorless or pale yellow. Variations in the methods of preparation lead to these different species. In general, the alkali metal compounds of these ions are soluble and most others are insoluble. Treatment of a vanadate solution with H_2O_2 results in the $VO_4{}^-$ ion which is red-brown.

f. Redox reactions.

The multivalent characteristics of V in aqueous solution are well illustrated by the E–pH diagram in Figure 14.1. By combining an electron ladder from the V diagram with electron ladders from other elements, reagents which will bring about various transformations can be predicted. In some cases, these predictions will not be manifested experimentally, often because of the passivity of a metal, the slow kinetics of

Table 14.1

Vanadium Species

Name	Formula	State	Color	Solubility	$\Delta G°$ (kJ/mole)
Vanadium	V	s	Light gray		0.0
Vanadium(II)	V^{+2}	aq	Violet		-251.5
Vanadium(III)	V^{+3}	aq	Green		-226.8
Oxovanadium(IV)	VO^{+2}	aq	Blue		-456.1
Dioxovanadium(V)	VO_2^+	aq	Yellow		-596.6
Vanadate(V)	VO_4^{-3}	aq	Colorless		-907.9
Hydrogen vanadate(V)	HVO_4^{-2}	aq	Colorless		-974.9
Dihydrogen vanadate(V)	$H_2VO_4^-$	aq	Colorless		-1020.9
Decavanadate(V)	$V_{10}O_{28}^{-6}$	aq	Orange		-7677.6
Hydrogen decavanadate(V)	$HV_{10}O_{28}^{-5}$	aq	Orange		-7708.2
Dihydrogen decavanadate(V)	$H_2V_{10}O_{28}^{-4}$	aq	Orange		-7729.1
Tetravanadate(V)	$V_4O_{12}^{-4}$	aq	Colorless		-3194.5
Divanadate(V)	$V_2O_7^{-4}$	aq	Colorless		-1728.0
Vanadium(II)					
bromide	VBr_2	s	Orange	Decomp	
chloride	VCl_2	s	Green	Decomp	
fluoride	VF_2	s	Blue	Decomp	
hydroxide	$V(OH)_2$	s	Light gray	Insol	-634.3
iodide	VI_2	s	Red-violet	Decomp	
oxide	V_2O_2	s	Light gray	Insol	-808.4
sulfate	VSO_4	s	Violet	Decomp	
Vanadium(III)					
bromide	VBr_3	s	Gray-brown	Decomp	-418.0
chloride#	VCl_3	s	Red-violet	Decomp	-511.3
chloride	VCl_3	s	Violet	Decomp	
fluoride#	VF_3	s	Yellow-green	Insol	
fluoride	VF_3	s	Green	Decomp	
hydroxide	$V(OH)_3$	s	Brown-green	Insol	-922.6
iodide	VI_3	s	Brown-black	Decomp	-268.0
oxide	V_2O_3	s	Black	Insol	-1139.0
Vanadium(IV)					
chloride	VCl_4	l	Red-brown	Decomp	-503.8
fluoride	VF_4	s	Green	Decomp	
oxide	V_2O_4	s	Blue	Insol	-1330.5
oxychloride	$VOCl_2$	s	Green	Insol	-636.0
oxysulfate	$VOSO_4$	s	Blue	Sol	-1170.0
Vanadium(V)					
fluoride	VF_5	l	Colorless	Decomp	-1373.0
oxide	V_2O_5	s	Yellow-red	Sol	-1419.5
oxychloride	$VOCl_3$	l	Pale yellow	Decomp	
sulfide	V_2S_5	s	Black-green	Insol	

Prepared by a dry method (no HOH involved). See Section 3e on Cr.

a compound, or the competition from other reactions. An example is that VO_2^+ is reduced to V^{+2} by Zn, to V^{+3} by Mg, and to VO^{+2} by SO_3^{-2}. Note that the prediction with regard to Mg is not valid, this probably being due to passivity.

g. Complexes. V evidences complexes in all of its aqueous oxidation states. These are usually prepared by redox reactions on other species in the presence of complexing agents. Illustrative complexes of V(II) are brown-yellow $V(CN)_6^{-4}$ and green $VCl_2(HOH)_4$. The V(II) species tend to be inert. The complexes of V(III) include green VF_6^{-3}, $VX_4(HOH)_2^-$ with X = F, Cl, and Br, red $V(NCS)_6^{-3}$, red-brown $V(NH_3)^{+3}$, green V(acetylacetone)$_3$, and V(edta)$^-$. V(IV) usually complexes as the VO^{+2} ion in such species as $VOX_4(HOH)^{-2}$ with X = F, Cl, Br, SCN, $\frac{1}{2}SO_4$, $\frac{1}{2}C_2O_4$, and blue-green VO(acetylacetone)$_2$. And V(V) enters into complexation ordinarily as VO_2^+, examples being $VO_2(C_2O_4)_2^{-3}$, the red peroxide $VO_2(O_2)_2^{-3}$, $VO_2Cl_4^{-3}$, and $VO_2(edta)^{-3}$. Log β_n values for V^{+2} are: with SCN$^-$ (1.4) and with edta^{-4} (12.7). Log β_n values for V^{+3} are: with SCN$^-$ (2.1), OH$^-$ (11.7), SO_4^{-2} (1.5), edta^{-4} (26.0). The values of log β_n for VO^{+2} are: with SCN$^-$ (2.3, 3.7), OH$^-$ (8.3, 18.3), SO_4^{-2} (2.4), F$^-$ (3.4, 5.7, 7.3, 8.1), $C_2O_4^{-2}$ (6.5, 11.8), edta^{-4} (18.8). Reported values of log β_n for VO_2^+ are: with OH$^-$ (10.7, 20.7, 26.1, 25.8), SO_4^{-2} (1.0), F$^-$ (3.0, 5.6, 6.9, 7.0), $C_2O_4^{-2}$ (5.0, 8.5), edta^{-4} (15.6).

h. Analysis. Preferred methods of analysis for V include ICPAES (limit 1 ppb), ICPMS (limit 10 ppt), and colorimetry using the agent p-amino-N, N–dimethylaniline to treat V(IV) (limit 20 ppm). Pre-separation using ion chromatography, ion exchange, or solvent extraction permits separate species to be measured in the ICPAES and ICPMS techniques.

i. Health aspects. Values reported for the LD50 (oral rat) of V_2O_5 are in the range 10–100 mg/kg, indicating it to be quite toxic. For $NaVO_3$, the average LD50 values (oral rat) are about 200 mg/kg, and those for VCl_3 are about 300 mg/kg.

3. Chromium (Cr) $4s^1 3d^5$

a. The E–pH diagram. Figure 14.3 presents the E–pH diagram for Cr at a total aqueous Cr concentration of $10^{-1.0}$ M. Oxidation states of 0, II, III, and VI appear, with Cr(II) being unstable and Cr(VI) strongly oxidizing. The Cr(II) cation is actually $Cr(HOH)_6^{+2}$ and the Cr(III) cation is $Cr(HOH)_6^{+3}$. The Cr(III) cation substitutes various ligands (L$^-$) for one of the water molecules to give $Cr(HOH)_5L^{+2}$ and the resulting species usually are differently colored than the hexaaqua species. In the region of pH below 6, in addition to $HCrO_4^-$, the dichromate(VI) ion $Cr_2O_7^{-2}$ is also present. At very low pH values, the species $Cr_3O_{10}^{-2}$ and $Cr_4O_{13}^{-2}$ also appear. Note that the general formula $(CrO_3)_n CrO_4^{-2}$ applies to these isopolyanions. H_2CrO_4 is a strong acid as indicated in the upper left corner of the E–pH diagram. The log K_p value for CrO_4^{-2} is 6.4 and that for $HCrO_4^-$ is -0.9. The legend of the figure gives equations for the lines between the species.

b. Discovery, occurrence, and extraction. Nicolas-Louis Vauquelin in 1797 isolated Cr from the mineral crocoite $PbCrO_4$. The mineral was powdered, treated with HCl which precipitated $PbCl_2$, and then the solution was evaporated to yield red CrO_3. Heating of this CrO_3 with charcoal reduced it to give the metal. The new metal was named after the Greek chroma, meaning color and referring to the many colors of its compounds. The major source of Cr is the mineral chromite $FeCr_2O_4$, with crocoite and chrome

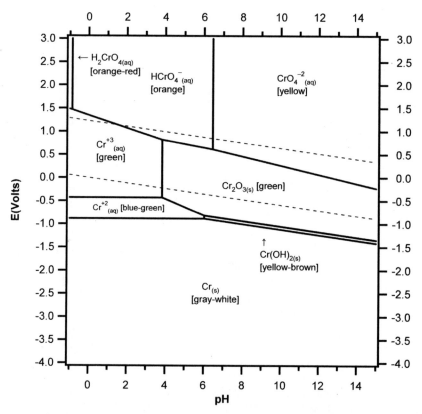

Figure 14.3 E–pH diagram for Cr species. Soluble species concentrations (except H^+) $= 10^{-1.0}$ M. Soluble species and most solids are hydrated. No agents producing complexes or insoluble compounds are present other than HOH and OH^-.

Species ($\Delta G°$ in kJ/mol): Cr (0.0), Cr^{+2} (−164.9), Cr^{+3} (−206.3), Cr_2O_3 (−1004.2), $Cr(OH)_2$ (−576.1), CrO_4^{-2} (−728.0), $HCrO_4^-$ (−764.8), H_2CrO_4 (−759.8), HOH (−237.2), H^+ (0.0), and OH^- (−157.3)

Figure 14.3 (Continued)

Equations for the lines:

Cr^{+2}/Cr	$E = -0.85 + 0.030 \log[Cr^{+2}]$
Cr^{+3}/Cr^{+2}	$E = -0.43$
Cr_2O_3/Cr^{+2}	$E = -0.19 - 0.089 \text{ pH} - 0.059 \log[Cr^{+2}]$
$Cr_2O_3/Cr(OH)_2$	$E = -0.46 - 0.059 \text{ pH}$
H_2CrO_4/Cr^{+3}	$E = 1.37 - 0.118 \text{ pH}$
$HCrO_4^-/Cr^{+3}$	$E = 1.35 - 0.138 \text{ pH}$
$HCrO_4^-/Cr_2O_3$	$E = 1.14 - 0.079 \text{ pH} + 0.020 \log[HCrO_4^-]$
CrO_4^{-2}/Cr_2O_3	$E = 1.27 - 0.098 \text{ pH} + 0.020 \log[CrO_4^{-2}]$
$Cr(OH)_2/Cr$	$E = -0.53 - 0.059 \text{ pH}$
$Cr(OH)_2/Cr^{+2}$	$2 \text{ pH} = 11.1 - \log[Cr^{+2}]$
Cr_2O_3/Cr^{+3}	$6 \text{ pH} = 21.1 - 2 \log[Cr^{+3}]$
$CrO_4^{-2}/HCrO_4^-$	$\text{pH} = 6.4$
$HCrO_4^-/H_2CrO_4$	$\text{pH} = -0.9$

ochre Cr_2O_3 being minor ones. Chromite is put into molten NaOH and air is employed for oxidation to give a residue containing soluble Na_2CrO_4. This is leached, the solution evaporated, and the solid is reduced to Cr_2O_3 with carbon. A further reduction of this Cr_2O_3 with Al gives the metal. The Na_2CrO_4 prepared in this process is the usual basis of further Cr chemistry.

c. The element. Chromium is a silvery-white active metal whose surfaces coat with a resilient layer of Cr_2O_3 which causes it to be inactive toward air and HOH. The metal dissolves in HCl or H_2SO_4 to give the green Cr^{+3} ion. Treatment with oxidizing reagents such as HNO_3 or aqua regia makes the metal passive. When heated in air, Cr is oxidized to green Cr_2O_3.

d. Oxides and hydroxides. Cr forms two oxides of interest to its aqueous chemistry: green Cr_2O_3, and red CrO_3. As indicated above, Cr_2O_3 can be made by heating Cr in air, or alternately, by thermally decomposing $(NH_4)_2Cr_2O_7$. A solution approach is to add OH^- to a solution of Cr^{+3} to obtain green hydrated Cr_2O_3, which can be heated to anhydrous Cr_2O_3. Addition of concentrated H_2SO_4 to a concentrated solution of $Na_2Cr_2O_7$ produces red CrO_3, which is highly poisonous. CrO has not been prepared, but treatment of Cr^{+2} with OH^- yields a yellow-brown precipitate $Cr(OH)_2$, which readily oxidizes to hydrated Cr_2O_3.

e. Compounds. Table 14.2 gives data on numerous compounds of Cr. When compounds of Cr are prepared by dry methods, that is, procedures not involving HOH solutions, the species which result are sometimes quite different from species which come from aqueous reactions. The cause of this is that Cr has a strong complexing power for HOH which results in species prepared in HOH incorporating HOH as a ligand. For example, $CrCl_3$ produced by a dry method (heating Cr in Cl_2) has no water molecules as ligands, but the compound synthesized in HOH could be $[Cr(HOH)_6]Cl_3$ or $[Cr(HOH)_5Cl]Cl_2$ or $[Cr(HOH)_4Cl_2]Cl$. Notice that the coordination number of 6 is carried through these complexes. The $CrCl_3$(dry) and these aqueous-made species will have different properties, including the colors (green versus violet).

The CrO_4^{-2} and $Cr_2O_7^{-2}$ anions give rise to numerous salts. Chromates of the alkali metal cations, NH_4^+, Mg^{+2}, Ca^{+2}, and Mn^{+2} are soluble; most others are insoluble. Almost all dichromates are soluble, that of Ag^+ being insoluble.

f. Redox reactions. The E–pH diagram makes evident a number of redox properties of Cr species. Cr(II) is highly reducing, with complexation stabilizing it somewhat. Cr(III) and Cr(VI) are both stable in HOH solution with Cr(VI) being a strong oxidizing agent, resting just below the upper O_2/HOH line. The narrow confines of the $Cr(OH)_2$ lines indicate its ready oxidation. Cr(III) compounds are oxidized to CrO_4^{-2} in acid solution by ClO_3^-, H_2O_2, MnO_4^-, and PbO_2, and in basic solution by H_2O_2, ClO^-, PbO_2, Ag_2O, HgO, CuO, and MnO_4^-. Chromium(VI) is reduced in acid solution by $C_2O_4^{-2}$, SCN^-, $S_2O_3^{-2}$, and SO_3^{-2}.

g. Complexes. Chromium shows an extensive number of complexes in both the II and III oxidation states, a coordination number of 6 being exhibited most of the time. Among those for Cr(II) are $Cr(NH_3)_6^{+2}$, $Cr(NCS)_6^{-4}$, $Cr(CN)_6^{-4}$, Cr(acetylacetone)$_2$, and $Cr_2(C_2H_3O_2)_4(HOH)_2$, the latter showing a Cr-to-Cr bond. The Cr(III) state is represented by $Cr(NH_3)_6^{+3}$, CrX_6^{-3} with X = F, Cl, NCS, CN, and Cr(acetylacetone)$_3$. There are also many complexes with mixed ligands. Most Cr(III) complexes are inert including the Cr^{+3} cation, which is more fully represented as $Cr(HOH)_6^{+3}$. The rate of reaction of these inert complexes can be increased by the use of catalysts, a common one being Cr^{+2}. It is of interest to note that the $Cr(HOH)_6^{+3}$ ion is violet, but various $CrX(HOH)_5^{+2}$ ions are usually green. Some log β_n values for Cr(II) complexes are: with SCN^- (1.1, 0.8), $C_2O_4^{-2}$ (3.9, 6.8), malonate^{-2} (3.9, 7.1), sulfosalicylate^{-2} (7.1, 12.9), edta^{-4} (13.6). Values of log β_n for complexes of Cr(III) are: with SCN^- (3.1, 3.0), NH_3 (log β_6 = 13.0), OH^- (10.0, 18.3, 24.0, 28.6), SO_4^{-2} (2.6), F^- (5.2, 7.7, 10.2), malonate^{-2} (8.3), sulfosalicylate^{-2} (9.6), edta^{-4} (23.4).

Table 14.2
Chromium Species

Name	Formula	State	Color	Solubility	$\Delta G°$ (kJ/mole)
Chromium	Cr	s	Gray-white		0.0
Chromium(II)	Cr^{+2}	aq	Blue-green		−164.9
Chromium(III)	Cr^{+3}	aq	Green		−206.3
Chromic(VI) acid	H_2CrO_4	aq	Orange-red		−759.8
Hydrogen chromate(VI)	$HCrO_4^-$	aq	Orange		−764.8
Chromate(VI)	CrO_4^{-2}	aq	Yellow		−728.0
Dichromate(VI)	$Cr_2O_7^{-2}$	aq	Red-yellow		−1276.1
Chromium(II)					
acetate	$Cr(C_2H_3O_2)_2$	s	Red	Insol	
bromide	$CrBr_2$	s	White	Sol	−289.0
chloride#	$CrCl_2$	s	White	Sol	−356.1
chloride	$CrCl_2$	s	Blue	Sol	
fluoride	CrF_2	s	Green	Insol	−736.0
hydroxide	$Cr(OH)_2$	s	Yellow-brown	Decomp	
iodide	CrI_2	s	Gray	Sol	−163.0
oxalate	CrC_2O_4	s	Yellow	Insol	
oxide	CrO	s	Black	Insol	
sulfate	$CrSO_4$	s	Blue	Sol	
sulfide	CrS	s	Black	Insol	
Chromium(III)					
acetate	$Cr(C_2H_3O_2)_3$	s	Gray-green	Sol	
bromide#	$CrBr_3$	s	Green	Insol	
bromide	$CrBr_3$	s	Violet	Sol	
chloride#	$CrCl_3$	s	Green	Insol	−486.2
chloride	$CrCl_3$	s	Violet	Sol	
fluoride#	CrF_3	s	Green	Insol	−1088.0
fluoride	CrF_3	s	Violet	Insol	
nitrate	$Cr(NO_3)_3$	s	Purple	Sol	
oxalate	$Cr_2(C_2O_4)_3$	s	Red	Sol	
oxide#	Cr_2O_3	s	Green	Insol	−1058.1
oxide	Cr_2O_3	s	Green	Insol	−1004.2
phosphate	$CrPO_4$	s	Violet	Insol	
perchlorate	$Cr(ClO_4)_3$	s	Blue-green	Sol	
sulfate#	$Cr_2(SO_4)_3$	s	Peach	Insol	
sulfate	$Cr_2(SO_4)_3$	s	Violet	Sol	
sulfide	Cr_2S_3	s	Black	Decomp	
sulfite	$Cr_2(SO_3)_3$	s	Green	Sol	
Chromium(VI)					
oxide	CrO_3	s	Red	Decomp	−510.0
oxychloride	CrO_2Cl_2	l	Red	Decomp	−510.9

Prepared by a dry method (no HOH involved). See Section 3e.

h. Analysis. ICPAES for Cr has a limit of 0.5 ppb, ICPMS 10 ppt, and colorimetric determination of CrO_4^{-2} using diphenycarbazide 10 ppm. Pre-separation using ion chromatography, ion exchange, or solvent extraction permits separate species to be measured in the ICPAES and ICPMS methods.

i. Health aspects. The LD50 (oral, rat) for CrO_3 is 80 mg/kg, that for Na_2CrO_4 is 130 mg/kg, and that for $Na_2Cr_2O_7$ is 50 mg/kg. The LD50 (oral, rat) for $CrCl_3$ is about 1800 mg/kg and that for $Cr(NO_3)_3$ is about 3000 mg/kg. These figures point up the high toxicity of Cr(VI) as compared to the markedly decreased toxicity of Cr(III).

4. Manganese (Mn) $4s^2 3d^5$

a. The E–pH diagram. Figure 14.4 presents the E–pH diagram for Mn at a total aqueous concentration of $10^{-1.0}$ M and with no complexing agent other than HOH. Oxidation numbers of 0, II, III, IV, VI, and VII are represented in the diagram. In addition the mixed oxidation state Mn_3O_4 (II and III) is shown. Mn is displayed as an active metal, the faint pink Mn^{+2} ion is the predominant aqueous-stable ionic species, and the main high oxidation state species is purple MnO_4^-. The aqueous Mn(II) cation is more accurately represented as $Mn(HOH)_6^{+2}$. Insoluble compounds in the basic region show the progression from Mn(II) to Mn(III) to Mn(IV) as oxidation occurs. In the lower right corner of the MnO_4^- domain, the green anionic Mn(VI) species MnO_4^{-2} appears. The strong acid character of $HMnO_4$ is also indicated. The legend of the figure gives equations for the lines between the species.

b. Discovery, occurrence, and extraction. The mineral pyrolusite MnO_2 has been known since antiquity, but it was thought to be an iron ore. In 1740, Johann Heinrich Pott proved that it did not contain iron, and that from it, salts different from iron salts could be prepared. Torbern Olaf Bergman in 1770 recognized pyrolusite to be a compound of a new metal, which Carl Wilhelm Scheele confirmed after a 3-year investigation during 1771–1774. Scheele gave the new element its name after a district in Greece called Magnesia where Mn minerals are common. Bergman's assistant, Johann Gottlieb Gahn, isolated the metal in 1774 by heating the mineral pyrolusite in charcoal and oil.

The most important sources of Mn are pyrolusite MnO_2, hausmannite Mn_3O_4, and rhodochrosite $MnCO_3$. The metal is produced by dissolution of the minerals in acid, then electrolysis of the Mn(II) solution. An alternate method is to reduce MnO_2 with Al.

c. The element. Mn is a red-gray active metal which reacts slowly with HOH and rapidly with dilute acids to give Mn^{+2}, and burns in air to black Mn_3O_4 (can be considered as $MnO + Mn_2O_3$).

d. Oxides and hydroxides. Five oxides of Mn are known—MnO, Mn_3O_4, Mn_2O_3, MnO_2, and Mn_2O_7. MnO can be prepared by igniting $Mn(OH)_2$, $MnCO_3$, or MnC_2O_4 in the absence of O_2. Green MnO is soluble in acids to give faint pink Mn^{+2}, and treatment of the latter with OH^- yields pink $Mn(OH)_2$. When MnO is exposed to air it oxidizes to black Mn_3O_4, and when $Mn(OH)_2$ is exposed to air, it goes to black-brown $MnO(OH)$. Heating the latter produces Mn_2O_3. MnO_2 can be prepared by the reduction of MnO_4^- or the oxidation of Mn^{+2} (for example, with ClO_3^-). When concentrated H_2SO_4 is added to a very cold, concentrated solution of MnO_4^-, a red, oily liquid Mn_2O_7 separates. It explodes even on the slightest warming. When any oxide of Mn with an oxidation state greater than II is treated with HCl,

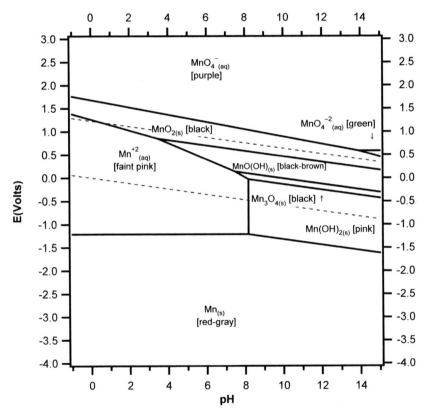

Figure 14.4 E–pH diagram for Mn species. Soluble species concentrations (except H^+) $= 10^{-1.0}$ M. Soluble species and most solids are hydrated. No agents producing complexes or insoluble compounds are present other than HOH and OH^-. Species include Mn, Mn^{+2}, $Mn(OH)_2$, Mn_3O_4, $MnO(OH)$, MnO_2, MnO_4^-, and MnO_4^{-2}.

Species ($\Delta G°$ in kJ/mol): Mn (0.0), MnO_4^- (−447.3), MnO_4^{-2} (−503.7), MnO_2 (−465.3), $MnO(OH)$ (−567.1), Mn_3O_4 (−1283.0), $Mn(OH)_2$ (−615.0), Mn^{+2} (−228.4), HOH (−237.2), H^+ (0.0), and OH^- (−157.3)

Figure 14.4 (Continued)

Equations for the lines:

MnO_4^-/MnO_4^{-2} $E = 0.58$

MnO_4^-/MnO_2 $E = 1.70 - 0.079\ pH + 0.020\ log[MnO_4^-]$

MnO_4^{-2}/MnO_2 $E = 2.26 - 0.118\ pH + 0.030\ log[MnO_4^{-2}]$

MnO_2/Mn^{+2} $E = 1.23 - 0.118\ pH - 0.030\ log[Mn^{+2}]$

$MnO_2/MnO(OH)$ $E = 1.06 - 0.059\ pH$

$MnO(OH)/Mn^{+2}$ $E = 1.41 - 0.177\ pH - 0.059\ log[Mn^{+2}]$

$MnO(OH)/Mn_3O_4$ $E = 0.58 - 0.059\ pH$

Mn_3O_4/Mn^{+2} $E = 1.82 - 0.236\ pH - 0.089\ log[Mn^{+2}]$

$Mn_3O_4/Mn(OH)_2$ $E = 0.45 - 0.059\ pH$

Mn^{+2}/Mn $E = -1.18 + 0.030\ log[Mn^{+2}]$

$Mn(OH)_2/Mn$ $E = -0.73 - 0.059\ pH$

$Mn(OH)_2/Mn^{+2}$ $2\ pH = 15.4 - log[Mn^{+2}]$

Cl_2 is evolved and Mn^{+2} results. And when any such oxide is treated with concentrated H_2SO_4, O_2 is given off and Mn^{+2} appears.

e. Compounds. Table 14.3 is a compilation of some of the major compounds of Mn along with several of their properties. To be noted are the numerous salts of Mn^{+2}. The permanganate ion also exhibits a number of salts, almost all of which are soluble, dissolving to give purple solutions. When insoluble Mn compounds are precipitated from aqueous solutions, the initial products may be considerably hydrated, some of the water molecules being waters of constitution such as $Mn(OH)_2$, some of the water molecules being ligands, and some of them being simple hydration. Aging often leads to the loss of HOH, as does isolation and heating.

f. Redox reactions. MnO_4^- is a powerful oxidizing agent which is used in many processes. Aqueous solutions of it are kinetically stabilized; however with time, it slowly decomposes to MnO_2. The Mn E–pH diagrams in Chapter 2 (Figures 2.11 through 2.14) should be reviewed to observe the more complex behavior of Mn when some further species are taken into account.

g. Complexes. The complexes of Mn(II) are fairly weak. They include MnF_3^-, $MnCl_4^{-2}$, $Mn(CN)_6^{-4}$, $Mn(en)_3^{+2}$, $Mn(ox)_2^{-2}$, $MnBr_4^{-2}$, MnI_4^{-2}, $Mn(NCS)_6^{-4}$, and $Mn(edta)^{-2}$. Complexation in many cases stabilizes Mn(III). Examples are MnF_5^{-2}, $MnCl_5^{-2}$, $Mn(CN)_6^{-3}$, $Mn(ox)_3^{-3}$,

Table 14.3
Manganese Species

Name	Formula	State	Color	Solubility	$\Delta G°$ (kJ/mole)
Manganese	Mn	s	Red-gray		0.0
Manganese(II)	Mn^{+2}	aq	Faint pink		−228.4
Manganese(III)	Mn^{+3}	aq	Red-brown		−85.0
Manganate(VI)	MnO_4^{-2}	aq	Green		−503.7
Permanganate(VII)	MnO_4^{-}	aq	Purple		−447.3
Manganese(II)					
acetate	$Mn(C_2H_3O_2)_2$	s	Brown	Sol	
bromide	$MnBr_2$	s	Rose	Sol	−371.0
carbonate	$MnCO_3$	s	White	Insol	−816.7
chloride	$MnCl_2$	s	Rose	Sol	−440.5
fluoride	MnF_2	s	Red	Insol	−751.0
hydroxide	$Mn(OH)_2$	s	Pink	Insol	−615.0
iodide	MnI_2	s	Rose	Sol	−271.0
nitrate	$Mn(NO_3)_2$	s	Rose	Sol	−385.9
oxalate	MnC_2O_4	s	Red-white	Insol	
oxide	MnO	s	Green	Insol	−362.8
phosphate	$Mn_3(PO_4)_2$	s	Pink	Insol	
sulfate	$MnSO_4$	s	Pink	Sol	−957.4
sulfide	MnS	s	Green	Insol	−218.4
Manganese(III)					
fluoride	MnF_3	s	Red-purple		
oxide	Mn_2O_3	s	Black	Insol	−878.9
oxyhydroxide	$MnO(OH)$	s	Black-brown	Insol	−567.1
Manganese(IV)					
dioxide	MnO_2	s	Black	Insol	−465.3
Manganese(VII)					
oxide	Mn_2O_7	l	Red	Sol	

$Mn(edta)^{-}$, and $Mn(acac)_3$. Complexation also stabilizes Mn(IV) as evidenced by MnF_6^{-2}, $MnCl_6^{-2}$, $Mn(CN)_6^{-2}$, and $Mn(IO_3)_6^{-2}$. Insolubility (as in MnO_2) also stabilizes Mn(IV). Log β_n values with Mn^{+2} are as follows: $SCN^{-}(1.2)$, $NH_3(1.0, 1.5, 1.7, 1.3)$, $OH^{-}(3.4, 5.8, 7.2, 7.7)$, $SO_4^{-2}(2.3)$, $S_2O_3^{-2}(2.0)$, $F^{-}(0.7)$, ethylenediamine (2.8, 4.9, 5.8), oxinate^{-}(6.2), 1,10-phenanthroline (4.0, 7.3, 10.3), $C_2O_4^{-2}(3.2, 4.4)$, $C_2H_3O_2^{-}(1.4)$, tart^{-2}(2.5), cit^{-3}(4.2), salicylate^{-2}(5.9, 9.8), glycinate^{-}(3.2, 4.7, 5.7), nta^{-3}(8.6, 11.0), and edta^{-4}(13.9). With Mn^{+3} log β_n takes these values: OH^{-} (14.4), $SO_4^{-2}(1.2)$, F^{-}(5.7), Cl^{-}(0.9), $C_2O_4^{-2}$(10.0, 16.6, 18.4), nta^{-3}(20.3), and edta^{-4}(25.3).

h. Analysis. For Mn, ICPAES has a detection limit of 50 ppt and ICPMS shows a limit of 10 ppt. Pre-separation using ion chromatography, ion exchange, or solvent extraction permits separate species to be measured.

Colorimetric analysis using $NaIO_4$ for oxidation to MnO_4^- allows a determination down to 20 ppm.

i. Health aspects.   The LD50 (oral rat) for $MnCl_2$ is 950 mg/kg, that for MnO_2 3.5 g/kg, and that for $KMnO_4$ is 1.1 g/kg. These data indicate that Mn compounds are only mildly toxic.

15

The Fe–Co–Ni Group

1. Introduction

The three elements to be treated in this chapter (Fe, Co, Ni) are the sixth, seventh, and eighth members of the first transition series. The first five members (Sc, Ti, V, Cr, Mn) have been treated in previous chapters (Chapters 12, 13, and 14).The ten elements of this first transition series (Sc through Zn) are characterized by electron activity in the 4s–3d levels. All elements in the 3d transition series are metals, and many of their compounds tend to be colored as a result of unpaired electrons. Most of the elements have a strong tendency to form complex ions due to participation of the d electrons in bonding. Unlike the previous three elements (V, Cr, Mn), these three do not show a variety of oxidation states. The higher oxidation states are almost absent in compounds, Fe showing principally the II and III, Co the II and III, and Ni only the II. The III states are less stable than the II states unless they are stabilized by complex formation. The resemblance of these three elements is notable, they being more like each other than they are to the elements below them.

2. Iron (Fe) $4s^2 3d^6$

a. E–pH diagram. The E–pH diagram in Figure 15.1 shows Fe in oxidation states of 0, II, and III. This diagram, which involves iron at

351

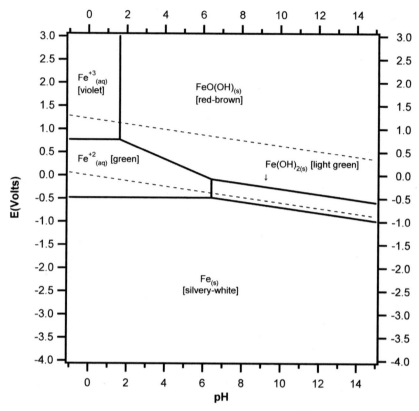

Figure 15.1 E–pH diagram for Fe species. Soluble species concentrations (except H^+) = $10^{-1.0}$ M. Soluble species and most solids are hydrated. No agents producing complexes or insoluble compounds are present other than HOH and OH^-.

Species ($\Delta G°$ in kJ/mol): Fe (0.0), Fe^{+2} (−86.6), Fe^{+3} (−12.1), $Fe(OH)_2$ (−493.3), FeO(OH) (−464.4), HOH (−237.2), H^+ (0.0), and OH^- (−157.3)

Equations for the lines:

$FeO(OH)/Fe^{+2}$	$E = 1.00 - 0.177\, pH - 0.059 \log [Fe^{+2}]$
$FeO(OH)/Fe(OH)_2$	$E = 0.30 - 0.059\, pH$
$Fe(OH)_2/Fe$	$E = -0.10 - 0.059\, pH$
Fe^{+3}/Fe^{+2}	$E = 0.77$
Fe^{+2}/Fe	$E = -0.45 + 0.030 \log [Fe^{+2}]$
$Fe(OH)_2/Fe^{+2}$	$2\, pH = 11.9 - \log [Fe^{+2}]$
$FeO(OH)/Fe^{+3}$	$3\, pH = 3.9 - \log [Fe^{+3}]$

$10^{-1.0}$ M is oversimplified in several ways. The Fe(II) and Fe(III) cations are more properly designated as $Fe(HOH)_6^{+2}$ and $Fe(HOH)_6^{+3}$, reflecting the coordination number of 6. The region just to the left of the $Fe^{+2}/Fe(OH)_2$ line involves $Fe(OH)^+$, and the region just to the left of the $Fe^{+3}/FeO(OH)$ line involves numerous hydroxo complexes such as $Fe(HOH)_5OH^{+2}$ and $Fe(HOH)_4(OH)_2^+$. At very high pH values, there is a tendency for $Fe(OH)_2$ to be converted to $HFeO_2^-$, and at very high E values, the powerful oxidant, red-purple FeO_4^{-2} can be produced. When these latter five species are introduced, Figure 15.2 results. This, too, is probably simplified since it is known that other species are present, such as $Fe_2(OH)_4^{+4}$ and $Fe_3(OH)_4^{+5}$. The species FeO_4^{-2} can only be prepared in a very strong base. It is apparently kinetically stabilized under these conditions, but it rapidly decomposes to $FeO(OH)$ with even a slight drop in pH. The legends of the figures show equations for the lines separating the species.

b. Discovery, occurrence, and extraction.

Iron has been known from pre-historic times, dating to at least 7000 BC. The name iron probably stems from the word isarn which occurs in many older Germanic and Celtic languages. Latin designates the metal as ferrum, from which the symbol is derived. The major ores are red-brown hematite Fe_2O_3, brown limonite $2Fe_2O_3 \cdot 2HOH$, and black magnetite Fe_3O_4. The pure metal is prepared by the reduction of oxides with H_2 or CO. Steel (impure Fe) is prepared by reducing ores with coke C and air, then reducing the C content from about 4% to lower values by controlled oxidation.

c. The element.

Pure Fe is a soft, silvery-white active metal. It oxidizes in moist air and in air-saturated HOH to give red-brown Fe_2O_3 and red-brown $FeO(OH)$. Fe dissolves in non-oxidizing acids to yield green Fe^{+2} and H_2, and in dilute oxidizing acids like HNO_3 to give yellow-brown Fe(III), but concentrated oxidizing acids render it passive. The yellow-brown Fe(III) is due to partly hydrolyzed Fe^{+3} as shown in Figure 15.2. Fe does not dissolve in oxygen-free HOH or dilute OH^- because an impervious oxide coat is formed. However, oxygen-containing HOH and OH^- attack Fe to produce $FeO(OH)$.

d. Oxides and hydroxides.

Fe exhibits three main oxides FeO, Fe_2O_3, and Fe_3O_4, the latter exhibiting Fe(II) and Fe(III) in a 1:2 ratio. FeO is generally only a product of dry methods, since it decomposes in HOH. The Fe(II) hydroxide $Fe(OH)_2$ is produced by the treatment of Fe^{+2} with OH^-. The light green compound is oxidized by air to red-brown $FeO(OH)$. This latter compound may also be prepared by the treatment of Fe^{+3} salts with OH^-. Fe_3O_4 is ordinarily prepared by high-temperature solid-state reactions, but it can be prepared in solution only under carefully controlled conditions using specific reagents, for example, by oxidation of Fe(II) in a solution containing KOH, KNO_3, and KH_2PO_3. Impure Fe_3O_4 may be prepared by basic precipitation of stoichiometric amounts of Fe^{+2} and Fe^{+3}.

e. Compounds.   Table 15.1 presents a number of Fe species which are pertinent to the aqueous chemistry of the element. The solution species listed are hydrated and most of the solid compounds listed are the hydrated variety. When dry methods are used to prepare a number of Fe compounds, these species will have different structures than their counterparts prepared by aqueous methods. This reflects the fact that HOH is often a constituent of the aqueous-derived compounds. Sometimes when the anhydrous dry-prepared compounds are put into HOH, they hydrate, and in other cases they do not.

f. Redox reactions.   Figure 15.2 reflects the various redox relationships of Fe species when no complexing agent except OH^- is present. In some cases,

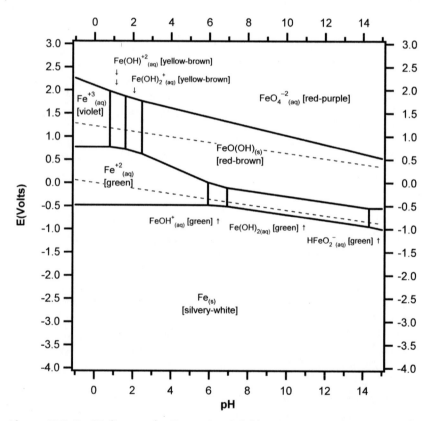

Figure 15.2 E–pH diagram for Fe species. Soluble species concentrations (except H^+) $= 10^{-1.0}$ M. Soluble species and most solids are hydrated. No agents producing complexes or insoluble compounds are present other than HOH and OH^-.

Species ($\Delta G°$ in kJ/mol): Fe (0.0), Fe^{+2} (-86.6), Fe^{+3} (-12.1), $Fe(OH)_2$ (-493.3), $FeO(OH)$ (-464.4), $FeOH^{+2}$ (-244.8), $Fe(OH)_2{}^+$ (-472.8), $HFeO_2{}^-$ (-405.8), $FeO_4{}^{-2}$ (-352.3), $FeOH^+$ (-290.0), HOH (-237.2), H^+ (0.0), and OH^- (-157.3)

Figure 15.2 (Continued)

Equations for the lines:

FeO_4^{-2}/Fe^{+3}	$E = 2.10 - 0.157 \, pH$
$FeO_4^{-2}/FeOH^{+2}$	$E = 2.09 - 0.138 \, pH$
$FeO_4^{-2}/Fe(OH)_2^{+}$	$E = 2.06 - 0.118 \, pH$
$FeO_4^{-2}/FeO(OH)$	$E = 2.03 - 0.098 \, pH + 0.020 \log [FeO_4^{-2}]$
$FeO(OH)/Fe^{+2}$	$E = 1.00 - 0.177 \, pH - 0.059 \log [Fe^{+2}]$
$FeO(OH)/FeOH^{+}$	$E = 0.65 - 0.118 \, pH - 0.059 \log [FeOH^{+}]$
$FeO(OH)/Fe(OH)_2$	$E = 0.30 - 0.059 \, pH$
$FeO(OH)/HFeO_2^{-}$	$E = -0.61 - 0.059 \log [HFeO_2^{-}]$
$Fe(OH)_2/Fe$	$E = -0.10 - 0.059 \, pH$
$HFeO_2^{-}/Fe$	$E = 0.36 - 0.089 \, pH + 0.030 \log [HFeO_2^{-}]$
Fe^{+3}/Fe^{+2}	$E = 0.77$
$FeOH^{+2}/Fe^{+2}$	$E = 0.82 - 0.059 \, pH$
$Fe(OH)_2^{+}/Fe^{+2}$	$E = 0.91 - 0.118 \, pH$
Fe^{+2}/Fe	$E = -0.45 + 0.030 \log [Fe^{+2}]$
$FeOH^{+}/Fe$	$E = -0.27 - 0.030 \, pH + 0.030 \log [FeOH^{+}]$
$HFeO_2^{-}/Fe(OH)_2$	$pH = 15.4 + \log [HFeO_2^{-}]$
$Fe(OH)_2/FeOH^{+}$	$pH = 5.9 - \log [FeOH^{+}]$
$FeOH^{+}/Fe^{+2}$	$pH = 5.9$
$FeO(OH)/Fe(OH)_2^{+}$	$pH = 1.5 - \log [Fe(OH)_2^{+}]$
$Fe(OH)_2^{+}/FeOH^{+2}$	$pH = 1.6$
$FeOH^{+2}/Fe^{+3}$	$pH = 0.8$

Fe does not exhibit its theoretical $E°$ value because of a thin refractory oxide coat, which causes it to show an effective $E°$ value somewhat higher. Fe^{+2} is oxidized to Fe^{+3} by numerous agents including O_2, Cl_2, Br_2, HNO_3, aqua regia, MnO_4^{-}, $Cr_2O_7^{-2}$, Ag^{+}, H_2O_2, MnO_2, HNO_2, and $HClO$. Fe^{+3} is reduced to Fe^{+2} by Zn or Al or Mg in acid, H_2S, SO_3^{-2}, I^{-}, and Cu^{+}. The strongly oxidizing FeO_4^{-2} with an $E°$ $(FeO_4^{-2}/FeO(OH)) = 2.03$ v requires drastic oxidation to produce, as described above. And, if there were not some kinetic stabilization, it would not persist.

Table 15.1
Iron Species

Name	Formula	State	Color	Solubility	ΔG° (kJ/mole)
Iron	Fe	s	Silvery-white		0.0
Iron(II) ion	Fe^{+2}	aq	Green		−86.6
Hydroxoiron(II)	$Fe(OH)^+$	aq	Green		−290.0
Iron(III) ion	Fe^{+3}	aq	Violet		−12.1
Mono(X^-)iron(III) ion	FeX^{+2}	aq	Yellow-brown		
Hydroxoiron(III)	$FeOH^{+2}$	aq	Yellow-brown		−244.8
Dihydroxoiron(III)	$Fe(OH)_2{}^+$	aq	Yellow-brown		−472.8
Hydrogen dioxoironate(II)	$HFeO_2{}^-$	aq	Green		−405.8
Tetraoxoironate(VI)	$FeO_4{}^{-2}$	aq	Red-purple		−352.3
Iron(II)					
acetate	$Fe(C_2H_3O_2)_2$	s	Green	Sol	
bromide	$FeBr_2$	s	Green-yellow	Sol	−237.4
carbonate	$FeCO_3$	s	Gray	Insol	−666.7
chloride	$FeCl_2$	s	Blue-green	Sol	−302.8
cyanide	$Fe(CN)_2$	s	Green-yellow	Sol	
fluoride	FeF_2	s	White	Insol	−663.6
hydroxide	$Fe(OH)_2$	s	Light green	Insol	−493.3
iodide	FeI_2	s	Gray	Sol	−119.9
nitrate	$Fe(NO_3)_2$	s	Green	Sol	
oxalate	FeC_2O_4	s	Yellow	Insol	
oxide	FeO	s	Black	Insol	
perchlorate	$Fe(ClO_4)_2$	s	Green	Sol	
phosphate	$Fe_3(PO_4)_2$	s	White	Insol	
sulfate	$FeSO_4$	s	Green	Sol	−821.0
sulfide	FeS	s	Black-brown	Insol	
sulfite	$FeSO_3$	s	Light green	Insol	
thiocyanate	$Fe(SCN)_2$	s	Green	Sol	
thiosulfate	FeS_2O_3	s	Green	Sol	
Iron(II-III)					
oxide	Fe_3O_4	s	Black	Insol	−1015.5
Iron(III)					
bromide	$FeBr_3$	s	Green	Sol	−242.9
chloride	$FeCl_3$	s	Brown-yellow	Sol	−334.0
fluoride	FeF_3	s	Red-purple	Insol	
hydroxyacetate	$FeOH(C_2H_3O_2)_2$	s	Brown-red	Insol	
iodate	$Fe(IO_3)_3$	s	Green-yellow	Insol	
nitrate	$Fe(NO_3)_3$	s	Violet	Sol	
oxalate	$Fe_2(C_2O_4)_3$	s	Yellow	Sol	
oxide	Fe_2O_3	s	Red-brown	Insol	−742.2
oxyhydroxide	FeO(OH)	s	Red-brown	Insol	−464.4
perchlorate	$Fe(ClO_4)_3$	s	Pink	Sol	
phosphate	$FePO_4$	s	Pink	Insol	−1183.3
sulfate	$Fe_2(SO_4)_3$	s	Yellow	Sol	−2263.0
sulfide	Fe_2S_3	s	Black	Insol	
thiocyanate	$Fe(SCN)_3$	s	Red	Sol	

g. Complexes. In both the II and the III oxidation states, Fe forms many complexes, a coordination number of 6 being predominant, although a few entities show a coordination number of 4, for example, $FeCl_4^-$. $E°$ values of the Fe(III)/Fe(II) couple can be drastically altered by complexation as the following systems indicate:

$$Fe(C_2O_4)_3^{-3} + e^- \rightarrow Fe(C_2O_4)_3^{-4} \quad E° = 0.01 \text{ v}$$

$$Fe(CN)_6^{-3} + e^- \rightarrow Fe(CN)_6^{-4} \quad E° = 0.36 \text{ v}$$

$$Fe(HOH)_6^{+3} + e^- \rightarrow Fe(HOH)_6^{+2} \quad E° = 0.77 \text{ v}$$

$$Fe(dipy)_3^{+3} + e^- \rightarrow Fe(dipy)_3^{+2} \quad E° = 1.11 \text{ v}$$

$$Fe(phen)_3^{+3} + e^- \rightarrow Fe(phen)_3^{+2} \quad E° = 1.12 \text{ v}$$

2,2′-Dipyridyl is symbolized as dipy and 1,10-phenanthroline as phen. These provide examples of the stabilization of oxidation states by complexation. Fe(II) complexes with a coordination number of 4 include : FeX_4^{-2} with $X = $ Cl, Br, SCN, and some with 6 as the coordination number are $Fe(HOH)_6^{+2}$, $Fe(CN)_6^{-4}$, $Fe(NH_3)_6^{+2}$, $Fe(en)_3^{+2}$, $Fe(acac)_2$, and $Fe(ox)_3^{-4}$. Typical Fe(III) complexes are violet $Fe(HOH)_6^{+3}$, yellow-brown $Fe(HOH)_5(OH)^{+2}$, $Fe(edta)^-$, $Fe(PO_4)_2^{-3}$, $Fe(ox)_3^{-3}$, $Fe(acac)_3$, $FeCl_4^-$, $Fe(HOH)_5(SCN)^{+2}$, $Fe(CN)_6^{-3}$, $Fe(HOH)F_5^{-2}$, $Fe(cit)_2^{-3}$, $Fe(tart)_3^{-3}$, and $Fe(salicylate)_3$.

Log β_n values with Fe^{+2} are: CN^- (log $\beta_6 = 35.4$) SCN^- (1.3), NH_3(1.4, 2.2, 2.9, 3.7), OH^- (4.5, 7.4, 10.0, 9.6), SO_4^{-2}(2.2), $S_2O_3^{-2}$(2.0), F^-(0.8), ethylenediamine(4.3, 7.7, 9.7), pyridine(0.6, 0.9), dipyridyl (4.4, 7.9, 17.2), 1,10-phenanthroline(5.9, 11.2, 21.0), $C_2O_4^{-2}$(3.1, 5.1), $C_2H_3O_2^-$(1.4), succinate^{-2}(1.4), tart^{-2}(2.2, 2.5), cit^{-3}(4.4) salicylate^{-2}(6.6, 11.2), glycinate$^-$(4.3, 7.7), nta^{-3}(8.3, 12.8), edta$^-$(14.3), acac$^-$(5.1, 8.7). Values with Fe^{+3} are: $B(OH)_4^-$(8.5, 15.6, 20.6, 20.3), CN^-(log $\beta_6 = 43.6$), SCN^-(3.0, 4.6, 5.0, 6.3, 6.2, 6.1), NO_3^-(1.0), OH^-(11.8, 22.3, 30.0, 34.4), SO_4^{-2}(4.0, 5.4), $S_2O_3^{-2}$(2.0), F^-(6.0, 9.1, 11.9), Cl^-(1.5, 2.1, 1.1), Br^-(0.6) I^-(2.9, 1.6), oxinate$^-$(14.5, 26.3, 36.9), 1,10-phenanthroline(6.5, 11.4, 13.8), formate$^-$(3.1), $C_2O_4^{-2}$(7.5, 13.6, 18.5), $C_2H_3O_2^-$(3.4, 6.5, 8.3), malonate^{-2}(7.5), succinate^{-2}(6.9), tart^{-2}(6.5), cit^{-3}(11.5), salicylate^{-2}(17.4, 27.5, 35.3), glycinate$^-$(10.0), nta^{-3}(15.9, 24.3), edta^{-4}(25.1), acac$^-$(9.8, 18.8, 26.2), salicylaldehyde$^-$(8.8, 15.6).

h. Analysis. Preferred methods of analysis for Fe include ICPAES (limit 0.1ppb), ICPMS (limit 10 ppt), and colorimetry using the agent KSCN to treat Fe(III) (limit 10 ppm). Pre-separation using ion chromatography, ion exchange, or solvent extraction permits separate species to be measured in the ICPAES and ICPMS techniques.

i. Health aspects. Humans require Fe, the average amount in the adult body being about 4 g, 3 g of which is in hemoglobin. About 1–2 mg per day is sufficient to maintain this level. The LD50 (oral rat) for $FeCl_3 \cdot 6HOH$ is 450 mg/kg and that for $FeCl_2 \cdot 4HOH$ is also 450 mg/kg. The insolubility of Fe_2O_3 is reflected in its LD50 (oral rat) value of 10g/kg.

3. Cobalt (Co) $4s^2 3d^7$

a. The E–pH diagram. Figure 15.3 presents the E–pH diagram for Co at a soluble compound concentration of $10^{-1.0}$ M. No complexing agents other than OH^- or HOH and no species which would produce insoluble compounds are present. The resemblance to Fe is obvious, except that the M^{+2} species is more stable and the M^{+3} species is less stable. The Co(II) and Co(III)

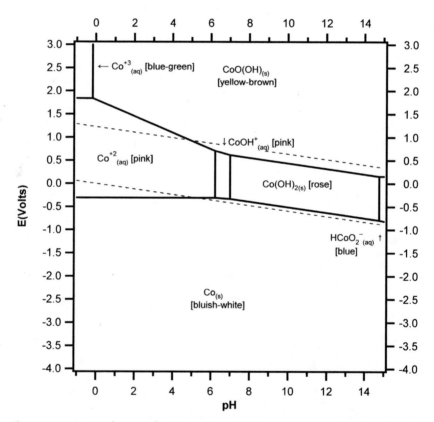

Figure 15.3 E–pH diagram for Co species. Soluble species concentrations (except H^+) $= 10^{-1.0}$ M. Soluble species and most solids are hydrated. No agents producing complexes or insoluble compounds are present other than HOH and OH^-.

Species ($\Delta G°$ in kJ/mol): Co (0.0), Co^{+2} (−53.6), Co^{+3} (+123.8), $Co(OH)_2$ (−458.1), $CoO(OH)$ (−359.0), $CoOH^+$ (−255.2), $HCoO_2^-$ (−368.2), HOH (−237.2), H^+ (0.0), and OH^- (−157.3)

Figure 15.3 (Continued)

Equations for the lines:

$CoO(OH)/Co^{+2}$ $E = 1.75 - 0.177\,pH - 0.059\log[Co^{+2}]$

$CoO(OH)/CoOH^+$ $E = 1.38 - 0.118\,pH - 0.059\log[CoOH^+]$

$CoO(OH)/Co(OH)_2$ $E = 1.03 - 0.059\,pH$

$CoO(OH)/HCoO_2^-$ $E = 0.10 - 0.059\log[HCoO_2^-]$

Co^{+3}/Co^{+2} $E = 1.84$

Co^{+2}/Co $E = -0.28 + 0.030\log[Co^{+2}]$

$CoOH^+/Co$ $E = -0.09 - 0.030\,pH + 0.030\log[CoOH^+]$

$Co(OH)_2/Co$ $E = 0.08 - 0.059\,pH$

$HCoO_2^-/Co$ $E = 0.55 - 0.089\,pH + 0.030\log[HCoO_2^-]$

$CoO(OH)/Co^{+3}$ $3\,pH = -1.5 - \log[Co^{+3}]$

$HCoO_2^-/Co(OH)_2$ $pH = 15.8 + \log[HCoO_2^-]$

$Co(OH)_2/CoOH^+$ $pH = 6.0 - \log[CoOH^+]$

$CoOH^+/Co^{+2}$ $pH = 6.2$

cations are more properly represented as the hexaaqua species $Co(HOH)_6^{+2}$ and $Co(HOH)_6^{+3}$. $Co(OH)^+$ can be better written as $Co(OH)(HOH)_5^+$, the solids $Co(OH)_2$ and $CoO(OH)$ are hydrated, and the anion $HCoO_2^-$ is aquated. The legend of the figure shows equations for the lines separating the species.

b. Discovery, occurrence, and extraction.   Cobalt compounds have been known since about 2600 BC, being found as the blue glaze on pottery in Egyptian tombs. In 1735, Georg Brandt, attributed this blue color to a half-metal which he called cobalt regulus. Fire assay of an ore called cobalt gave him the metal. The name of the element is derived from the German word Kobold which means goblin, a name possibly assigned because of the stench in roasting its arsenic-containing ores. The most important of its many ores are smaltite $CoAs_2$, cobaltite $CoAsS$, and linneaite Co_3S_4. These are generally mixed with compounds of Ni and Cu.

The ores are blended with Na_2CO_3 and KNO_3, then roasted. Part of the S and As are removed as volatile compounds, leaving Co, Ni, and Cu oxides, with some sulfates, and arsenates. The sulfates and arsenates are leached with HOH, then the oxides are dissolved in hot H_2SO_4. The solution is treated with the oxidant NaClO, and the hydroxides are selectively precipitated by careful adjustment of the pH with $Ca(OH)_2$. The Co precipitates as Co_3O_4, and it is reduced to the metal by heating with C.

c. The element. Co is a hard lustrous bluish-white metal which acquires a somewhat inert coat of Co_3O_4 in air or aerated HOH. It is not further reactive with HOH, air, or bases, but dissolves slowly in non-oxidizing acids and rapidly in dilute oxidizing acids to give pink Co^{+2}.When finely powdered, Co heated in air gives black Co_3O_4.

d. Oxides and hydroxides. The addition of OH^- to solutions of Co^{+2} yields the rose-colored hydrated $Co(OH)_2$. Thermal decomposition of this hydroxide gives green-brown CoO. As Figure 15.3 indicates $Co(OH)_2$ is amphoteric, dissolving in both acids and bases, the latter reagent yielding blue $HCoO_2^-$. In alkaline solution, $Co(OH)_2$ is readily oxidized by the O_2 in air or other oxidizing agents to yellow-brown CoO(OH). This Co(III) oxyhydroxide can also be prepared by the treatment of Co(III) complexes with OH^-. Treatment of CoO(OH) with H_2SO_4 decomposes it into Co^{+2} and O_2, whereas the use of HCl yields Co^{+2} and Cl_2.

e. Compounds. A number of species which are pertinent to the aqueous chemistry of Co are presented in Table 15.2. The solution species listed are hydrated and most of the solid compounds listed are the hydrated variety. When dry methods are used to prepare a number of Co compounds, these species will often have different structures and properties than their counterparts prepared by aqueous methods.

f. Redox reactions. The E–pH diagram in Figure 15.3 indicates the redox characteristics of the Co system when no complexing agent other than OH^- is present. Co^{+3} is an exceedingly strong oxidizing agent which is signified by its occupancy of a very high and limited region of the diagram. It and CoO(OH) will attack HOH in the acid region. The latter compound attains stability in the basic region as shown by the dropping of the $CoO(OH)/Co(OH)_2$ line below the O_2/HOH line. Co metal is somewhat refractory toward dissolution in non-oxidizing acids which means that the Co^{+2}/Co line may effectively rest slightly higher. Agents which will reduce Co^{+2} to Co include Zn, Cd, Mg, and Al. Co^{+2} can be oxidized to Co(III) in base by Cl_2, Br_2, H_2O_2, and in acetic acid by HClO. Co(III) is reduced to Co^{+2} by $H_2C_2O_4$, H_2SO_3, HCl, and HBr.

g. Complexes. Of the Co aqueous ions, the most stable form is Co^{+2}, Co^{+3} oxidizing HOH rapidly. However, when Co^{+3} is strongly complexed, it is so greatly stabilized that Co(III) becomes the most stable form. This phenomenon is illustrated by the following couples:

$$Co(HOH)_6{}^{+3} + e^- \rightarrow Co(HOH)_6{}^{+2} \qquad E^\circ = 1.84 \text{ v}$$

$$Co(C_2O_4)_3{}^{-3} + e^- \rightarrow Co(C_2O_4)_3{}^{-4} \qquad E^\circ = 0.57 \text{ v}$$

Table 15.2
Cobalt Species

Name	Formula	State	Color	Solubility	$\Delta G°$ (kJ/mole)
Cobalt	Co	s	Bluish-white		0.0
Cobalt(II) ion	Co^{+2}	aq	Pink		-53.6
Hydroxocobalt(II) ion	$Co(OH)^{+}$	aq	Pink		-255.2
Hydrogen dioxocobaltate(II)	$HCoO_2^{-}$	aq	Blue		-368.2
Cobalt(III) ion	Co^{+3}	aq	Blue-green		123.8
Cobalt(II)					
acetate	$Co(C_2H_3O_2)_2$	s	Red-violet	Sol	
bromate	$Co(BrO_3)_2$	s	Red	Sol	
bromide	$CoBr_2$	s	Red-violet	Sol	
carbonate	$CoCO_3$	s	Red	Insol	
chlorate	$Co(ClO_3)_2$	s	Red	Sol	
chloride	$CoCl_2$	s	Black	Sol	
chromate	$CoCrO_4$	s	Gray-black	Insol	
cyanide	$Co(CN)_2$	s	Blue-violet	Insol	
fluoride	CoF_2	s	Pink	Sol	
hydroxide	$Co(OH)_2$	s	Rose-red	Insol	-458.1
iodate	$Co(IO_3)_2$	s	Red	Sol	
iodide	CoI_2	s	Green	Sol	
nitrate	$Co(NO_3)_2$	s	Red	Sol	
oxalate	CoC_2O_4	s	Pink	Insol	
oxide	CoO	s	Green-brown	Insol	
perchlorate	$Co(ClO_4)_2$	s	Red	Sol	
phosphate	$Co_3(PO_4)_2$	s	Reddish-white	Insol	
sulfate	$CoSO_4$	s	Red-pink	Sol	
sulfide	CoS	s	Reddish-white	Insol	
sulfite	$CoSO_3$	s	Red	Insol	
thiocyanate	$Co(SCN)_2$	s	Violet	Sol	
thiosulfate	CoS_2O_3	s	Green	Sol	
Cobalt(II-III)					
oxide	Co_3O_4	s	Black	Insol	
Cobalt(III)					
acetate	$Co(C_2H_3O_2)_3$	s	Green	Decomp	
chloride	$CoCl_3$	s	Red	Decomp	
fluoride	CoF_3	s	Green	Decomp	
oxyhydroxide	$CoO(OH)$	s	Yellow-brown	Insol	-359.0
sulfate	$Co_2(SO_4)_3$	s	Blue-green	Decomp	
sulfide	Co_2S_3	s	Black	Insol	

$$Co(edta)^- + e^- \rightarrow Co(edta)^{-2} \qquad E° = 0.37 \text{ v}$$

$$Co(dipy)_3{}^{+3} + e^- \rightarrow Co(dipy)_3{}^{+2} \qquad E° = 0.31 \text{ v}$$

$$Co(en)_3{}^{+3} + e^- \rightarrow Co(en)_3{}^{+2} \qquad E° = 0.18 \text{ v}$$

$$Co(NH_3)_6{}^{+3} + e^- \rightarrow Co(NH_3)_6{}^{+2} \qquad E° = 0.11 \text{ v}$$

$$Co(CN)_6{}^{-3} + e^- \rightarrow Co(CN)_5(HOH)^{-3} \qquad E° = -0.80 \text{ v}$$

In these equations, edta = ethylenediaminetetracetate^{-4}, dipy = dipyridyl, and en = ethylenediamine. Co(II) complexes are of two types: those with a coordination number of 4 and those with 6. The former tend to be blue, while the latter are often pink to violet. Co(II) complexes are usually unstable, since they tend to be oxidized by air to Co(III) complexes.

$CoX_4{}^{-2}$ complexes include those in which X = F, Cl, Br, I, SCN, CN, and NO_2. Some of them with a coordination number of 6 are those shown above in the Co(III)/Co(II) couples. Co(III) has an exceptionally strong tendency to form complexes, $CoX_6{}^{-3}$ with X = F, Cl, Br, I, CN, NO_2, $\frac{1}{2}CO_3$, $\frac{1}{2}$ox, $\frac{1}{2}SO_3$ (X's can be mixed), $Co(NH_3)_6{}^{+3}$ (orange brown), $Co(acac)_3$, $Co(nta)_2{}^{-3}$, and $Co(edta)^-$. Co(III) complexes are usually inert, that is, in many cases are exceptionally slow to react. Log β_n values with Co^{+2} complexes are: $B(OH)_4{}^-$ (log $\beta_4 = 10.0$), SCN^-(1.7, 1.3), NH_3(2.0, 3.5, 4.4, 5.1, 5.1, 4.4), $NO_3{}^-$ (0.2), $P_2O_7{}^{-4}$(6.1), OH^-(4.3, 8.4, 9.7, 10.2), $SO_4{}^{-2}$(2.4), $S_2O_3{}^{-2}$(2.1), F^-(0.4), ethylenediamine(5.6, 10.5, 13.8), oxinate$^-$(8.7), dipyridyl(5.8, 11.2, 15.9), 1,10-phenanthroline(7.1, 13.7, 19.8), formate$^-$(0.7, 1.2), $C_2O_4{}^{-2}$(3.3, 5.6), $C_2H_3O_2{}^-$(1.5), malonate^{-2}(3.0, 4.4), cit^{-3}(5.0), salicylate^{-2}(6.7, 11.4, 13.4, 17.4), glycinate$^-$(5.1, 9.0, 11.6), picolinate$^-$(5.7, 10.4, 14.1), nta^{-3}(10.4, 14.4), edta^{-4}(16.3). Values with Co^{+3} are: NH_3(7.3, 14.0, 21.0, 25.7, 30.8, 35.2), OH^-(13.5), ethylenediamine (log $\beta_3 = 48.7$), edta^{-4}(41.4).

h. Analysis. Co may be determined down to about 1.0 ppb by ICPAES and 10 ppt by ICPMS. Ion chromatography has a detection limit of about 5.0 ppb.

i. Health aspects. A low level of Co, chiefly in the vitamin B12 complex, is necessary for human health. The LD50 (oral rat) for $CoCl_2 \cdot 6HOH$ is about 750 mg/kg, values for other soluble Co(II) salts are similar, and that for insoluble Co_3O_4 is about 1.7 g/kg.

4. Nickel (Ni) $4s^2 3d^8$

a. E–pH diagrams. Figure 15.4 depicts an E–pH diagram for Ni with soluble species at $10^{-1.0}$ M, with no complexing agent other than HOH and OH^- present, and with no species that could produce insoluble compounds other than OH^-. The cation Ni^{+2} is an abbreviated form of $Ni(HOH)_6{}^{+2}$, and the two compounds $Ni(OH)_2$ and $NiO(OH)$ are hydrated. If the Ni

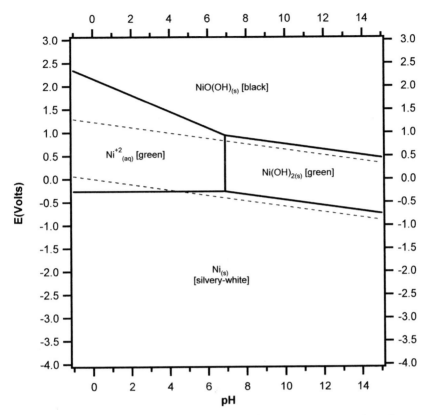

Figure 15.4 E–pH diagram for Ni species. Soluble species concentrations (except H^+) $= 10^{-1.0}$ M. Soluble species and most solids are hydrated. No agents producing complexes or insoluble compounds are present other than HOH and OH^-.

Species ($\Delta G°$ in kJ/mol): Ni (0.0), Ni^{+2} (−45.7), $Ni(OH)_2$ (−446.9), NiO(OH) (−316.9), $NiOH^+$ (−227.3), $HNiO_2^-$ (−350.0), HOH (−237.2), H^+ (0.0), and OH^- (−157.3)

Equations for the lines:

$NiO(OH)/Ni^{+2}$ $E = 2.11 - 0.177\,pH - 0.059\,\log[Ni^{+2}]$

$NiO(OH)/Ni(OH)_2$ $E = 1.35 - 0.059\,pH$

Ni^{+2}/Ni $E = -0.24 + 0.030\,\log[Ni^{+2}]$

$Ni(OH)_2/Ni$ $E = 0.14 - 0.059\,pH$

$Ni(OH)_2/Ni^{+2}$ $2\,pH = 12.8 - \log[Ni^{+2}]$

concentration is reduced to $10^{-3.0}$ M, the diagram in Figure 15.5 results, with $HNiO_2^-$, an aquated species, appearing. The diagram shows that the major aqueous oxidation states of Ni are 0 and II, with strong oxidizing agents necessary to give the III state. The legends of the figures show equations for the lines separating the species.

b. Discovery, occurrence, and extraction.
In 1751 Axel Fredrick Cronstedt, a Swedish mineralogist, produced an impure metal from what was probably the mineral gersdorffite (NiAsS). Later, he was sent a mineral from Germany which had been called kupfernickel. Miners had noted its resemblance to Cu ores, but were unable to isolate Cu from it. Hence the name, which means devil's copper, or false copper. Cronstedt isolated a metal from it, and found it to be the same as the metal he had obtained from gersdorffite. He adopted the word nickel for the new element. The major commercial sources of Ni are pentlandite $NiFe_2S_2$ and garnierite $Ni_3Mg_3Si_4O_{10}(OH)_8$.

Pentlandite usually occurs as grains in rock ores along with grains of some other sulfides, chiefly Fe and Cu. These grains are isolated by crushing, flotation, and magnetic separation. The concentrate is treated by controlled roasting and smelting to produce Fe_3O_4, CuS, and NiS. Heating of this mixture with SiO_2 converts the Fe_3O_4 to slag, and the separated sulfides are dissolved in acid. Ni may be obtained from the resulting solution by selective electrodeposition or reduction with H_2.

c. The element.
Ni is a lustrous, silvery-white metal that is unreactive in air, HOH, and OH^-, dissolves only slowly in HCl and H_2SO_4, and dissolves readily in dilute HNO_3 to give green Ni^{+2}. Concentrated HNO_3 renders it passive. If Ni is very finely powdered and heated in air, it goes over to green-black NiO.

d. Oxides and hydroxides.
NiO is prepared as indicated above or by heating $Ni(OH)_2$, $NiCO_3$, or $Ni(NO_3)_2$. Green $Ni(OH)_2$ can be prepared by the addition of OH^- to a solution of Ni^{+2}. Both NiO and $Ni(OH)_2$ dissolve in acid to yield Ni^{+2}. Black NiO(OH) results when $Ni(OH)_2$ is treated with Cl_2. Prolonged treatment of $Ni(OH)_2$ with $S_2O_8^{-2}$ gives black, hydrated NiO_2. Treatment of both of these higher oxidation state oxides with acid results in decomposition.

e. Compounds.
Table 15.3 is a compilation of the more common Ni compounds which are related to the element's aqueous chemistry. The salts can usually be prepared by the treatment of $Ni(OH)_2$, NiO, or $NiCO_3$ with appropriate acids. Most of the species in the table are hydrated, and it needs to be recognized that hydrated and anhydrous species quite often behave differently. For example, anhydrous $NiCl_2$ is yellow, and dissolves in HOH very slowly, whereas $NiCl_2 \cdot 6HOH$ is green, contains the complex ion $[NiCl_2(HOH)_4]$, and dissolves in HOH readily.

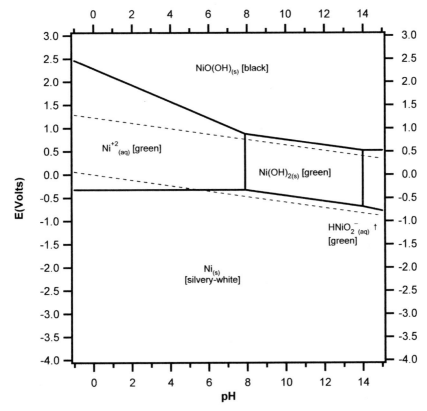

Figure 15.5 E–pH diagram for Ni species. Soluble species concentrations (except H^+) = $10^{-3.0}$ M. Soluble species and most solids are hydrated. No agents producing complexes or insoluble compounds are present other than HOH and OH^-.

Species ($\Delta G°$ in kJ/mol): Ni (0.0), Ni^{+2} (−45.7), $Ni(OH)_2$ (−446.9), NiO(OH) (−316.9), $NiOH^+$ (−227.3), $HNiO_2^-$ (−350.0), HOH (−237.2), H^+ (0.0), and OH^- (−157.3)

Equations for the lines:

$NiO(OH)/Ni^{+2}$	$E = 2.11 - 0.177\,pH - 0.059\log[Ni^{+2}]$
$NiO(OH)/Ni(OH)_2$	$E = 1.35 - 0.059\,pH$
$NiO(OH)/HNiO_2^-$	$E = 0.34 - 0.059\log[HNiO_2^-]$
Ni^{+2}/Ni	$E = -0.24 + 0.030\log[Ni^{+2}]$
$HNiO_2^-/Ni$	$E = 0.65 - 0.089\,pH + 0.030\log[HNiO_2^-]$
$Ni(OH)_2/Ni$	$E = 0.14 - 0.059\,pH$
$HNiO_2^-/Ni(OH)_2$	$pH = 17.0 + \log[HNiO_2^-]$
$Ni(OH)_2/Ni^{+2}$	$2\,pH = 12.8 - \log[Ni^{+2}]$

Table 15.3
Nickel Species

Name	Formula	State	Color	Solubility	$\Delta G°$ (kJ/mole)
Nickel	Ni	s	Silvery-white	Insol	0.0
NIckel(II) ion	Ni^{+2}	aq	Green		−46.4
Hydroxonickel(II) ion	$Ni(OH)^+$	aq	Green		−227.3
Hydrogen dioxonickelate(II)	$HNiO_2^-$	aq	Green		−350.0
Nickel(II)					
acetate	$Ni(C_2H_3O_2)_2$	s	Green	Sol	
bromate	$Ni(BrO_3)_2$	s	Green	Sol	
bromide	$NiBr_2$	s	Yellow-green	Sol	
carbonate	$NiCO_3$	s	Pale green	Insol	
chlorate	$Ni(ClO_3)_2$	s	Green	Sol	
chloride	$NiCl_2$	s	Green	Sol	−294.8
chromate	$NiCrO_4$	s	Brown	Insol	
cyanide	$Ni(CN)_2$	s	Gray-blue	Insol	
fluoride	NiF_2	s	Green	Sol	−640.2
formate	$Ni(HCOO)_2$	s	Green	Sol	
hydroxide	$Ni(OH)_2$	s	Green	Insol	−446.9
iodate	$Ni(IO_3)_2$	s	Yellow	Sol	
iodide	NiI_2	s	Blue-green	Sol	
nitrate	$Ni(NO_3)_2$	s	Green	Sol	
oxalate	NiC_2O_4	s	Green	Insol	
oxide	NiO	s	Green-black	Insol	
perchlorate	$Ni(ClO_4)_2$	s	Green	Sol	
phosphate	$Ni_3(PO_4)_2$	s	Green	Insol	
sulfate	$NiSO_4$	s	Green	Sol	−798.8
sulfide	NiS	s	Black	Insol	
sulfite	$NiSO_3$	s	Green	Insol	
thiocyanate	$Ni(SCN)_2$	s	Yellow	Sol	
thiosulfate	NiS_2O_3	s	Green	Sol	
Nickel(III)					
oxyhydroxide	$NiO(OH)$	s	Black	Insol	−316.9
Nickel(IV)					
dioxide	NiO_2	s	Black	Insol	

f. Redox reactions. Reduction of Ni(II) to Ni is easily attained by treatment with Zn, Cd, Mg, Al, or Sn. Oxidation of Ni(II) to Ni(III) and NI(IV) requires very strong oxidants such as Cl_2, ClO^-, and $S_2O_8^{-2}$.

g. Complexes. Ni(II) forms many complexes, coordination numbers of 4, 5, and 6 being exhibited, the latter the most prevalent. The complexes with coordination number 4 may exhibit two dispositions of the ligands, tetrahedral and square planar. Tetrahedral coordination number 4 complexes tend to be blue or green, and those with coordination number 6 are usually yellow, red, or brown. Ni(II) complexes with a coordination number of 6 include : $Ni(NH_3)_6^{+2}$, $Ni(dipy)_3^{+2}$, $Ni(en)_3^{+2}$, $Ni(SCN)_6^{-4}$, and $Ni(NO_2)_6^{-4}$. Those with a coordination number of 4 are exemplified by NiX_4^{-2} with X = F, Cl, Br, I, CN, SCN, and $Ni(C_2O_4)_2^{-2}$. Complexes with Ni(III) and Ni(IV) are NiF_6^{-3}, NiF_6^{-2}, both instantly liberating O_2 from HOH.

Log β_n values for a number of Ni(II) complexes are as follows: $B(OH)_4^-$ (log $\beta_3 = 8.4$), SCN^-(1.8, 1.6, 1.5), NH_3(2.7, 4.9, 6.6, 7.7, 8.3, 8.3), NO_3^-(0.4), OH^-(4.1, 8.9, 11.0, 12.0), SO_4^{-2}(2.3), $S_2O_3^{-2}$(2.1), F^-(0.5), ethylenediamine (7.3, 13.5, 17.6), pyridine(1.9, 3.1, 3.7), oxinate$^-$(9.3), dipyridyl(7.0, 13.9, 20.2), 1,10-phenanthroline(8.6, 16.7, 24.3), formate$^-$(0.5), $C_2O_4^{-2}$(5.2), acetate$^-$(1.4), malonate^{-2}(3.2, 4.9), lactate$^-$(1.6, 2.8, 3.1), maleate^{-2}(2.0), succinate^{-2}(2.3), cit^{-3}(5.4), benzoate$^-$(0.9), salicylate^{-2}(7.0, 11.7), glycinate$^-$(6.2, 11.1, 14.2), nta^{-3}(11.5, 16.4), edta^{-4}(18.6).

h. Analysis.   Ni may be determined down to about 1.0 ppb by ICPAES and 10 ppt by ICPMS. Ion chromatography has a detection limit of about 50 ppb, and use of a colorimetric reagent permits analysis down to 10 ppb.

i. Health aspects.   The LD50 (oral rat) for $NiSO_4 \cdot 6HOH$ is reported as 260 mg/kg, that for $NiCl_2 \cdot 6HOH$ is 170 mg/kg, and that for the insoluble NiO is well over 1g/kg.

16

The Cu Group

1. Introduction

The elements of this group, copper Cu, silver Ag, and gold Au, often called the coinage metals, resemble each other in some ways, particularly their tendency to nobility, but it cannot be said that the properties of Ag are intermediate between those of Cu and Au. Even though the d shell is full, the d electrons are active, particularly in Cu and Au. The most stable oxidation states are II for Cu, I for Ag, and III for Au. For Cu, Cu(I) as the simple ion Cu^+ disproportionates in HOH, and Cu(III) is so powerfully oxidizing that it is reduced by HOH. Stability may be brought to Cu(I) and Cu(III) only by complexation or insolubility. For Ag, Ag(II) and Ag(III) are reduced by HOH, stability resulting only by forming complex species or insoluble compounds. For Au, the simple Au^+ cation disproportionates in HOH, and Au(II) is not known.

2. Copper (Cu) $4s^1 3d^{10}$

a. E–pH diagram. Figure 16.1 sets out the E–pH diagram for Cu at a soluble species concentration of $10^{-1.0}$ M. It is assumed that there is no complexing agent or any insoluble compound producing agent other than OH^- or HOH. Further, almost all species are being considered in their hydrated forms, that is, the forms that they take in the presence of HOH.

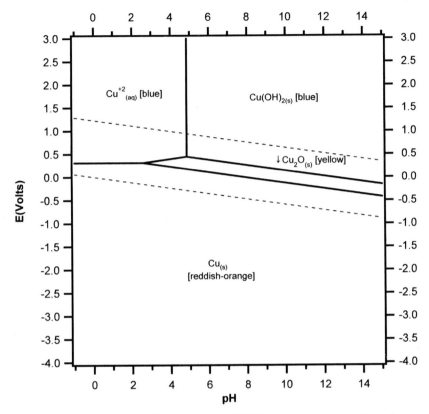

Figure 16.1 E–pH diagram for Cu species. Soluble species concentrations (except H$^+$) = 10$^{-1.0}$ M. Soluble species and most solids are hydrated. No agents producing complexes or insoluble compounds are present other than HOH and OH$^-$.

Species ($\Delta G°$ in kJ/mol): Cu (0.0), Cu(OH)$_2$ (−359.0), Cu$_2$O (−147.7), Cu^{+2} (65.7), Cu$^+$ (49.8), HCuO$_2^-$ (−264.4), HOH (−237.2), H$^+$ (0.0), and OH$^-$ (−157.3)

Equations for the lines:

$$Cu^{+2}/Cu \qquad E = 0.34 + 0.030 \log[Cu^{+2}]$$

$$Cu^{+2}/Cu_2O \qquad E = 0.22 + 0.059\ pH + 0.059 \log[Cu^{+2}]$$

$$Cu(OH)_2/Cu_2O \qquad E = 0.71 - 0.059\ pH$$

$$Cu_2O/Cu \qquad E = 0.46 - 0.059\ pH$$

$$Cu(OH)_2/Cu^{+2} \qquad 2\ pH = 8.7 - \log[Cu^{+2}]$$

Oxidation states of 0, I, and II are present. The reddish-orange Cu is a fairly noble metal, and the sole Cu(I) compound is shown as yellow Cu_2O since CuOH is unstable. The Cu^+ ion does not appear, even though it has been entered in the construction of the diagram. This reflects its strong tendency to disproportionate into Cu(II) and Cu, as predicted by $\Delta G°$ values. The compound which results when OH^- is added to a blue Cu^{+2} solution is blue $Cu(OH)_2$, not black CuO. Cu^{+2} is more properly written as $Cu(HOH)_6^{+2}$, and just off to the left of the $Cu^{+2}/Cu(OH)_2$ line, hydrolyzed species like $Cu_2(OH)_2^{+2}$ occur. The legend of the figure shows equations for the lines separating the species.

b. Discovery, occurrence, and extraction. Cu was probably the first metal to be utilized by man. Items made of the metal can be dated back to 3500 BC. The ancient Egyptians made tools, bowls, nails, and pipes out of Cu. Investigations indicate that they were obtaining the metal by charcoal reduction of Cu ores. The name and the symbol derive from Cyprus where there were Cu mines, from which a substance called chalkos cuprios was extracted. The major ores of Cu are chalcopyrite $CuFeS_2$, chalcocite Cu_2S, malachite $Cu_2(OH)_2CO_3$, cuprite Cu_2O, and free Cu.

Most Cu is isolated from sulfide ores containing Fe. They are crushed, concentrated by froth flotation, then SiO_2 is added. The mixture is melted to convert the Fe to the oxide which separates from the melt as a slag. The impure residue is mixed with more SiO_2, melted, and air is blasted through the melt to produce liquid Cu. The resulting Cu can be purified electrolytically.

c. The element. Cu is a reddish-orange metal which coats with black CuO and/or green $Cu_2(OH)_2SO_4$ in moist urban air. Otherwise, it is unreactive with HOH and air, and is insoluble in bases and non-oxidizing acids. It dissolves in oxidizing acids like HNO_3, and will go very slowly into solution in non-oxidizing acids in the presence of air. Powdered Cu burns to black CuO in air.

d. Oxides and hydroxides. When Cu(II) solutions are treated with OH^-, $Cu(OH)_2$ usually separates as a blue gel. It dissolves readily in acids, and in concentrated OH^- as blue $HCuO_2^-$. The black CuO is an insoluble compound produced by heating Cu hydroxide, carbonate, or nitrate. The yellow Cu_2O can be prepared by reducing Cu(II) in basic solution with sugar or hydrazine. Heating of the yellow compound leads to a red variety of Cu_2O.

e. Compounds. Table 16.1 presents many species of Cu. It should be noted that all Cu(I) compounds listed are insoluble. This is a reflection that the Cu^+ ion is unstable in HOH solution except at very low concentrations. As indicated in the E–pH diagram, all Cu(I) salts are oxidized by air. Cu(I) salts are generally prepared by reduction of Cu(II) species in the presence of an

Table 16.1
Copper Species

Name	Formula	State	Color	Solubility	$\Delta G°$ (kJ/mole)
Copper	Cu	s	Reddish-orange	Insol	0.0
Copper(I) ion	Cu^+	aq	Colorless		49.8
Copper(II) ion	Cu^{+2}	aq	Blue		65.7
Hydroxocopper(II) ion	$Cu(OH)^+$	aq	Blue		−126.4
Hydrogen dioxocopperate(II)	$HCuO_2^-$	aq	Blue		−264.4
Copper(I)					
bromide	CuBr	s	White	Insol	−101.0
chloride	CuCl	s	White	Insol	−119.5
cyanide	CuCN	s	White	Insol	108.5
iodide	CuI	s	White	Insol	−69.6
oxide	Cu_2O	s	Yellow	Insol	−147.7
sulfate	Cu_2SO_4	s	Gray	Decomp	
sulfide	Cu_2S	s	Black	Insol	−53.2
sulfite	Cu_2SO_3	s	White	Insol	
thiocyanate	CuSCN	s	White	Insol	63.0
Copper(II)					
acetate	$Cu(C_2H_3O_2)_2$	s	Green	Sol	
bromate	$Cu(BrO_3)_2$	s	Blue-green	Sol	
bromide	$CuBr_2$	s	Green	Sol	−123.0
carbonate	$CuCO_3$	s	Yellow	Insol	−516.9
chlorate	$Cu(ClO_3)_2$	s	Green	Sol	
chloride	$CuCl_2$	s	Blue-green	Sol	−186.8
chromate	$CuCrO_4$	s	Yellow-brown	Insol	
cyanide	$Cu(CN)_2$	s	Yellow-green	Decomp	
fluoride	CuF_2	s	Blue	Sol	−519.5
formate	$Cu(HCOO)_2$	s	Blue	Sol	
hydroxide	$Cu(OH)_2$	s	Blue	Insol	−359.0
iodate	$Cu(IO_3)_2$	s	Blue	Insol	−231.8
nitrate	$Cu(NO_3)_2$	s	Blue	Sol	−109.0
oxalate	CuC_2O_4	s	Blue-white	Insol	−651.1
oxide	CuO	s	Black	Insol	
perchlorate	$Cu(ClO_4)_2$	s	Blue	Sol	
phosphate	$Cu_3(PO_4)_2$	s	Blue	Insol	
sulfate	$CuSO_4$	s	Blue	Sol	−694.0
sulfide	CuS	s	Black	Insol	−53.8
thiocyanate	$Cu(SCN)_2$	s	Black	Decomp	
thiosulfate	CuS_2O_3	s	Green	Sol	

insolubilizing agent. Some Cu(III) compounds have been prepared by fusion procedures, but they decompose in HOH.

f. Redox reactions. Cu^+ is unstable in HOH solution at the $10^{-1.0}$ M level as Figure 16.1 indicates. However, numerous insoluble Cu(I) compounds are stable in the presence of HOH since they are in equilibrium with very small concentrations of Cu^+ in solution. These small concentrations are stable in HOH solution as can be seen from Figure 16.2 which shows the Cu system at the $10^{-7.0}$ M level. The use of other E–pH diagrams along with that of Cu predicts in many instances redox reactions of Cu species. For example, neither CuI_2 nor $Cu(CN)_2$ can be prepared because the anions have the capability to reduce the Cu(II) to yield CuI and CuCN. Further, Cu^{+2} is reduced by Zn, Cd, Al, Zn, Sn, Pb, Fe, Co, Ni, and Mg, and metallic Cu is oxidized by oxidizing acids and solutions of Hg^{+2}, Ag^+, Pt(IV), and Au(III). Cu is dissolved in cyanide solutions in the presence of air according to $4Cu + 2HOH + 8CN^- + O_2 \rightarrow 4Cu(CN)_2^- + 4OH^-$, this reaction occurring because of the high stability of the complex.

g. Complexes. Not only does insolubility stabilize Cu(I) species, complexation can also bring this about. Again, the stabilization is due to the decrease of the Cu^+ ion in the solution, most of the Cu(I) being tied up in the complex. Among the more stable of the Cu(I) complexes are $Cu(NH_3)_2^+$, $Cu(thiourea)_3^+$, $Cu(CN)_3^{-2}$, $CuCl_2^-$, $CuBr_2^-$, $Cu(py)_4^+$, and $Cu(dipy)_2^+$. Some log β_n values for Cu(I) complexes are : CN^- (16.3, 21.6, 23.1, 23.5), SCN^- (−, 11.0, 10.9, 10.4), NH_3(5.9, 10.6), SO_3^{-2}(7.9, 8.7, 9.4), $S_2O_3^{-2}$(10.4, 12.3, 13.7), Cl^-(2.7, 5.5, 5.7), Br^-(−, 5.9), I^-(−, 8.9, 9.4, 9.7), ethylenediamine(−, 11.2), pyridine(4.8, 7.6, 8.2, 8.5), 1,10-phenanthroline(−, 15.8), glycinate$^-$(10.1).

Cu(II) also forms numerous complexes, some of the more important ones being $Cu(NH_3)_4^{+2}$, $Cu(en)_2^{+2}$, $Cu(dipy)_2^{+2}$, $Cu(glycinate)_2$, $Cu(acac)_2$, $Cu(py)_4^{+2}$, $Cu(edta)^{-2}$, CuX_4^{-2} with X = Cl, Br, NO_2, ½tart, and ¹⁄₂₀ox. Log β_n values for Cu^{+2} are: $B(OH)_4^-$(7.1, 12.4, 15.2), CO_3^{-2}(6.8, 9.9), CN^-(log β_4 = 25.0), SCN^-(2.3, 3.7), NH_3(4.0, 7.5, 10.3, 11.8, 12.4), NO_3^-(0.5), OH^-(6.3, 10.7, 14.2, 16.4), SO_4^{-2}(2.4), $S_2O_3^{-2}$(−, 12.3), F^-(1.2), Cl^-(0.4), Br^-(0.0), ethylenediamine(10.5, 19.6), pyridine(2.5, 4.3, 5.2, 6.0), oxinate$^-$(12.6), formate$^-$(1.4, 2.3, 2.2, 1.9), $C_2O_4^{-2}$(4.8, 9.2), acetate$^-$(2.2, 3.6, 3.1, 2.9), malonate^{-2}(5.1, 7.8), citrate^{-3}(5.9), salicylate^{-2}(10.6, 18.5), phthalate^{-2}(4.0, 5.3), glycinate$^-$(8.6, 15.6), nta^{-3}(13.0, 17.4), edta^{-4}(18.8), acetylaetonate$^-$(8.3, 15.0).

Two of the rare complexes of Cu(III) are CuO_2^- and CuF_6^{-3}. Both attack HOH to liberate O_2.

h. Analysis. Cu may be analyzed down to the 1.0 ppb level by ICPAES and down to the 10 ppt level by ICPMS. Ion chromatography may be

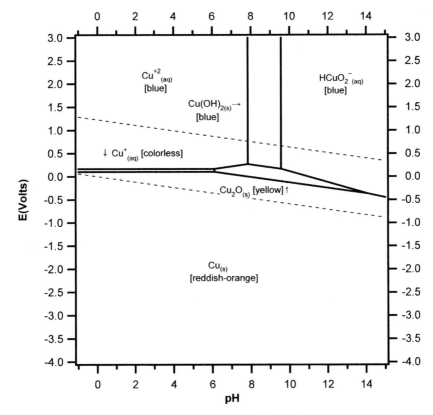

Figure 16.2 E–pH diagram for Cu species. Soluble species concentrations (except H^+) $= 10^{-7.0}$ M. Soluble species and most solids are hydrated. No agents producing complexes or insoluble compounds are present other than HOH and OH^-.

Species ($\Delta G°$ in kJ/mol): Cu (0.0), Cu(OH)$_2$ (−359.0), Cu$_2$O (−147.7), Cu^{+2} (65.7), Cu$^+$ (49.8), HCuO$_2^-$ (−264.4), HOH (−237.2), H$^+$ (0.0), and OH$^-$ (−157.3)

Equations for the lines:

Cu^{+2}/Cu^+	$E = 0.16$
Cu^+/Cu	$E = 0.52 + 0.059 \log[Cu^+]$
Cu^{+2}/Cu_2O	$E = 0.22 + 0.059\ pH + 0.059 \log[Cu^{+2}]$
$Cu(OH)_2/Cu_2O$	$E = 0.71 - 0.059\ pH$
Cu_2O/Cu	$E = 0.46 - 0.059\ pH$
$HCuO_2^-/Cu_2O$	$E = 1.71 - 0.118\ pH + 0.059 \log[HCuO_2^-]$
$HCuO_2^-/Cu$	$E = 1.09 - 0.089\ pH + 0.030 \log[HCuO_2^-]$
Cu_2O/Cu^+	$2\ pH = -1.77 - 2 \log[Cu^+]$
$HCuO_2^-/Cu(OH)_2$	$pH = 16.6 + \log[HCuO_2^-]$
$Cu(OH)_2/Cu^{+2}$	$2\ pH = 8.7 - \log[Cu^{+2}]$

employed down to 10 ppb, with colorimetry using neocuproine going down to 100 ppb.

i. Health aspects. The LD50 (oral rat) for $CuSO_4$ is 300 mg/kg and that of CuO is 470 mg/kg.

3. Silver (Ag) $5s^1 4d^{10}$

a. The E–pH diagrams. Figures 16.3 and 16.4 present E–pH diagrams for Ag, one at $10^{-1.0}$ M, and one at $10^{-3.0}$ M. Oxidation states of 0, I, and II are shown, with 0 and I being predominant, the II state appearing only at high potential. The cations $Ag(HOH)_4^+$ and $Ag(HOH)_6^{+2}$ are represented simply as Ag^+ and Ag^{+2}, and $Ag(OH)_2^-$ is sometimes written for AgO^-. It is believed that insoluble AgOH does not exist, a more accurate formula for the brown precipitate being Ag_2O. This oxide is amphoteric, being soluble in acid and in very concentrated base. The black species Ag_2O_2 is a mixed oxidation-state compound, there being one Ag(I) and one Ag(III) involved. An Ag(III) species, namely $Ag(OH)_4^-$ can be prepared in 10.0 M NaOH by electrolytic oxidation, but it has a very short life. The legends of the figures show equations for the lines separating the species.

b. Discovery, occurrence, and extraction. Silver dates back to prehistoric times due to the facts that it is sometimes found free in nature, and that its compounds are easily reduced. Its earliest mention is in the book of Genesis (13:2), and slag wastes in Asia Minor give evidence that it was separated from lead about 3000 BC. The major minerals of Ag are argentite Ag_2S, horn silver AgCl, pyrargyrite Ag_3SbS_3, and free Ag. These minerals, among others, often appear as minor constituents in conjunction with ores of Cu, Pb, and Zn. As a general procedure, the residues from the processing of these metals are treated with hot H_2SO_4 which solubilizes Cu, Pb, and Zn, then the resulting solid is heated with lime and/or silica to cause the remaining Cu, Pb, and Zn to become slag. Following removal of the slag and HNO_3 dissolution, electrolysis is used to plate out the Ag. The name silver derives from the Old Saxon silubar or Old German silbar, and the symbol reflects the Latin term argentum.

c. The element. Ag is a silver-white metal, unreactive with air, HOH, bases, and non-oxidizing acids, but dissolves readily in oxidizing acids such as HNO_3 and in non-oxidizing acids plus an oxidizing agent. Slow dissolution comes about if the added oxidizing agent is air. The product of these dissolutions is colorless Ag^+. Ag will also dissolve in a solution of CN^- in the presence of air to yield $Ag(CN)_2^-$. Heating of finely-divided Ag in air gives Ag_2O. Small amounts of S in various forms in the

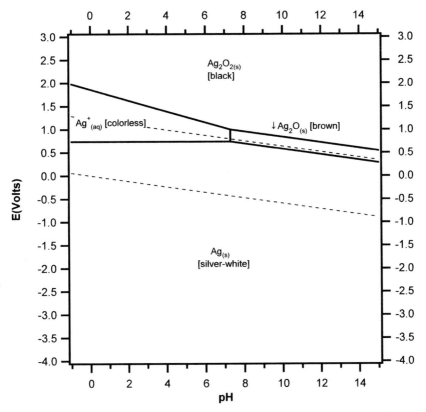

Figure 16.3 E–pH diagram for Ag species. Soluble species concentrations (except H^+) $= 10^{-1.0}$ M. Soluble species and most solids are hydrated. No agents producing complexes or insoluble compounds are present other than HOH and OH^-.

Species ($\Delta G°$ in kJ/mol): Ag (0.0), Ag_2O_2 (27.6), Ag_2O (−11.2), Ag^{+2} (268.6), Ag^+ (77.1), AgO^- (−22.6), HOH (−237.2), H^+ (0.0), and OH^- (−157.3)

Equations for the lines:

$$Ag_2O_2/Ag^+ \qquad E = 1.80 - 0.118\ pH - 0.059\ log[Ag^+]$$

$$Ag_2O_2/Ag_2O \qquad E = 1.43 - 0.059\ pH$$

$$Ag^+/Ag \qquad E = 0.80 + 0.059\ log[Ag^+]$$

$$Ag_2O/Ag \qquad E = 1.17 - 0.059\ pH$$

$$Ag_2O/Ag^+ \qquad 2\ pH = 12.6 - 2\ log[Ag^+]$$

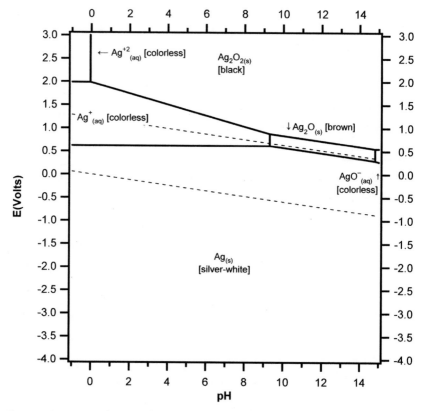

Figure 16.4 E–pH diagram for Ag species. Soluble species concentrations (except H^+) = $10^{-3.0}$ M. Soluble species and most solids are hydrated. No agents producing complexes or insoluble compounds are present other than HOH and OH^-.

Species ($\Delta G°$ in kJ/mol): Ag (0.0), Ag_2O_2 (27.6), Ag_2O (−11.2), Ag^{+2} (268.6), Ag^+ (77.1), AgO^- (−22.6), HOH (−237.2), H^+ (0.0), and OH^- (−157.3)

Equations for the lines:

Ag^{+2}/Ag^+	$E = 1.98$
Ag_2O_2/Ag^+	$E = 1.80 - 0.118\ pH - 0.059\ log[Ag^+]$
Ag_2O_2/Ag_2O	$E = 1.43 - 0.059\ pH$
Ag_2O_2/AgO^-	$E = 0.38 - 0.059\ log[AgO^-]$
AgO^-/Ag	$E = 2.22 - 0.118\ pH + 0.059\ log[AgO^-]$
Ag^+/Ag	$E = 0.80 + 0.059\ log[Ag^+]$
Ag_2O/Ag	$E = 1.17 - 0.059\ pH$
AgO^-/Ag_2O	$2\ pH = 35.6 + 2\ log[AgO^-]$
Ag_2O_2/Ag^{+2}	$4\ pH = -6.2 - 2\ log[Ag^{+2}]$
Ag_2O/Ag^+	$2\ pH = 12.6 - 2\ log[Ag^+]$

atmosphere will slowly coat Ag with a black film of Ag_2S, commonly called tarnish.

d. Oxides and hydroxides. The main oxides of aqueous interest which Ag forms are Ag_2O and Ag_2O_2. Brown Ag_2O results when Ag^+ is treated with OH^-. When this oxide or other simple $Ag(I)$ compounds are treated with a vigorous oxidizing agent, such as $S_2O_8^{-2}$ or O_3, black insoluble Ag_2O_2 is produced. Three other oxides of Ag have been prepared (Ag_3O, Ag_2O_3, Ag_3O_4) under strenuous conditions, but they all decompose in HOH.

e. Compounds. Table 16.2 list numerous Ag compounds along with states, colors, solubilities, and standard free energies. Many of the solid compounds can be prepared as anhydrous species from HOH solution. K_{sp} values for most of the insoluble compounds in Table 16.2 can be calculated from the free energy values. Values of K_{sp} are generally given for an ionic strength of 0.0, which essentially means that all ions are at infinite dilution. Log K_{sp} values of AgCl at different ionic strengths are: $-9.8[0.0]$, $-9.6[0.5]$, $-9.7[1.0]$, $-10.1[3.0]$, and $-10.4[4.0]$.

f. Redox reactions. The E–pH diagrams emphasize the nobility of Ag, which implies that it is quite inactive, and that its compounds may be readily reduced. Among the appropriate reducing agents are Pb, Sn, Sn^{+2}, Bi, Cu, Cu^+, Cd, Fe, Al, Mn, Zn, Mg, H_3PO_2, H_2SO_3, H_2O_2, and glucose. An important redox reaction represents the solubility of Ag metal in CN^- in the presence of air: $4Ag + 2HOH + 8CN^- + O_2 \rightarrow 4Ag(CN)_2^- + 4OH^-$.

g. Complexes. Ag^+ is a soft cation which means that it has small affinity for oxygen, but complexes readily with softer ligands such as N and S. In these complexes, it usually displays a coordination number of 2 or 4. Examples are $Ag(NH_3)_2^+$, $Ag(thiourea)_2^+$, $Ag(CN)_2^-$, $AgCl_2^-$, $AgBr_2^-$, $Ag(py)_2^+$, $Ag(OH)_2^-$, and $Ag(S_2O_3)_2^{-3}$. Log β_n values for Ag^+ complexes are: $B(OH)_4^-(0.5)$, $CN^-(-, 20.5, 21.4, 20.8)$, $SCN^-(4.8, 8.2, 9.5, 9.7)$, $NH_3(3.3, 7.2)$, $OH^-(2.0, 4.0)$, $SO_3^{-2}(5.6, 8.7, 9.0)$, $SO_4^{-2}(1.3)$, $S_2O_3^{-2}(8.8, 13.7, 14.2)$, $F^-(0.4)$, $Cl^-(3.3, 5.3, 6.0)$, $Br^-(4.7, 7.7, 8.7, 9.0)$, $I^-(6.6, 11.7, 13.1, 14.2)$, ethylenediamine(4.7, 7.7), pyridine(2.1, 4.1), oxinate$^-$(5.2, 9.6), acetate$^-$(0.7, 0.6), glycinate$^-$(3.5, 6.9), edta^{-4}(7.3). $Ag(II)$ complexes include $Ag(py)_4^{+2}$ and $Ag(dipy)_2^{+2}$, and $Ag(III)$ complexes are $H_5Ag(IO_6)_2^{-2}$ and $H_7Ag(TeO_6)_2^{-2}$. These upper oxidation state complexes are produced by vigorous oxidation of $Ag(I)$ in the presence of the appropriate complexing agents.

h. Analysis. Ag may be analyzed down to the 1.0 ppb level by ICPAES and down to the 1 ppt level by ICPMS. Ion chromatography may be employed down to 100 ppb, and colorimetry using p-diethyl-aminobenzylidenerhodamine will go down to 50 ppb.

Table 16.2
Silver Species

Name	Formula	State	Color	Solubility	$\Delta G°$ (kJ/mole)
Silver	Ag	s	Silver-white	Insol	0.0
Silver(I) ion	Ag^+	aq	Colorless		77.1
Silver(II) ion	Ag^{+2}	aq	Colorless		268.6
Oxosilverate(I)	AgO^-	aq	Colorless		−22.6
Silver(I)					
acetate	$AgC_2H_3O_2$	s	White	Insol	−296.2
bromate	$AgBrO_3$	s	White	Sol	54.4
bromite	$AgBrO_2$	s	White	Sol	
bromide	AgBr	s	Yellow	Insol	−97.0
carbonate	Ag_2CO_3	s	Yellow	Insol	−437.0
chlorate	$AgClO_3$	s	White	Sol	71.2
chloride	AgCl	s	White	Insol	−109.9
chlorite	$AgClO_2$	s	White	Sol	75.8
chromate	Ag_2CrO_4	s	Red	Insol	−640.5
cyanate	AgCNO	s	White	Insol	−58.2
cyanide	AgCN	s	White	Insol	157.0
fluoride	AgF	s	Yellow	Sol	−188.0
iodate	$AgIO_3$	s	White	Insol	−93.8
iodide	AgI	s	Yellow	Insol	−66.2
nitrate	$AgNO_3$	s	White	Sol	−33.5
nitrite	$AgNO_2$	s	White	Sol	19.1
oxalate	$Ag_2C_2O_4$	s	White	Insol	−584.4
oxide	Ag_2O	s	Brown	Insol	−11.3
perchlorate	$AgClO_4$	s	White	Sol	88.3
periodate	$AgIO_4$	s	Orange-yellow	Decomp	
permanganate	$AgMnO_4$	s	Purple	Sol	
phosphate	Ag_3PO_4	s	Yellow	Insol	−879.1
sulfate	Ag_2SO_4	s	White	Insol	−618.8
sulfide	Ag_2S	s	Black	Insol	−40.7
thiocyanate	AgSCN	s	White		101.4
thiosulfate	$Ag_2S_2O_3$	s	White	Insol	
Silver(I, III)					
oxide	$Ag'Ag'''O_2$	s	Black	Insol	27.6
Silver(II)					
fluoride	AgF_2	s	Brown	Decomp	−302.1

i. Health aspects. The LD50 (oral rat) for $AgNO_3$ is about 1170 mg/kg which has such a high value because it goes over to the insoluble AgCl in the gastrointestinal tract. This is probably the case since the intraperitoneal value is 83 mg/kg. This is somewhat supported by the LD50 (oral rat) value for AgI of about 2800 mg/kg, AgI being more insoluble than AgCl.

4. Gold (Au) $6s^1 5d^{10}$

a. E–pH diagram.   Figure 16.5 depicts the E–pH diagram of $10^{-3.0}$ M Au in the presence of $10^{0.0}$ M Cl. This diagram has been presented because most Au chemistry begins with the dissolution of Au metal in aqua regia (HNO_3 + 3HCl) to give $AuCl_4^-$. The element shows oxidation numbers of 0, I, and III, but only 0 and III appear on the diagram because of the strong tendency of Au(I) to disproportionate to Au and Au(III). The simple aquated ions $Au(HOH)_n^+$ and $Au(HOH)_n^{+3}$ are not believed to exist, but it is possible that hydrolyzed species of Au(III) exist in the very high acid and very high potential regions of the E–pH diagram. Practically all of the aqueous chemistry of Au involves complex ions. The legend of the figure shows equations for the lines separating the species.

b. Discovery, occurrence, and extraction.   Numerous archaeological finds indicate that Au was used as precious jewelry and coinage in Egypt as of 3400 BC. This Au was probably found in the native state in river sands which were the product of weathering and was separated by density-dependent methods such as panning. Au usually occurs in the native form, but also in calaverite $AuTe_2$. The native-occurring metal is generally dispersed in rock (about 10 ppm), which is powdered, then treated with CN^- and air to extract the Au as the soluble complex $Au(CN)_2^-$. To this is added Zn powder which precipitates the Au. Further refining can be carried out electrolytically. The word gold was employed in older Germanic languages, and it is related to the Sanskrit gelwa, which means shining or yellow. The Latin word aurum, which also means yellow, is the origin of the element's symbol.

c. The element.   Au is a lustrous, yellow metal which is unaffected by air, HOH, bases, and acids. To put it into solution it is necessary to use a mixture of an acid whose anion forms a strong complex and a powerful oxidizing agent, such as HCl plus HNO_3, Cl_2, ClO^-, ClO_2^-, ClO_3^-, or H_2O_2 to produce $AuCl_4^-$. Dissolution in CN^- and air gives $Au(CN)_2^-$.

d. Oxides and hydroxides.   Neither Au_2O nor AuOH exists; when OH^- is added to Au(I) solutions (such as $AuCl_2^-$), the result is a mixture of Au and brown Au_2O_3. The addition of OH^- to a solution of $AuCl_4^-$ gives a brown precipitate of hydrated Au_2O_3; the hydroxide apparently only exists as a low concentration neutral species in solution. Au_2O_3 is soluble in strong acids to give complex species such as $AuCl_4^-$, $AuBr_4^-$, and $Au(NO_3)_4^-$, and soluble in concentrated base to give $Au(OH)_4^-$.

e. Compounds.   Table 16.3 lists a number of Au compounds. It is to be noted that the simple Au(I) halides (which are produced by dry methods) are unstable in HOH, decomposing by disproportionation into Au and Au(III).

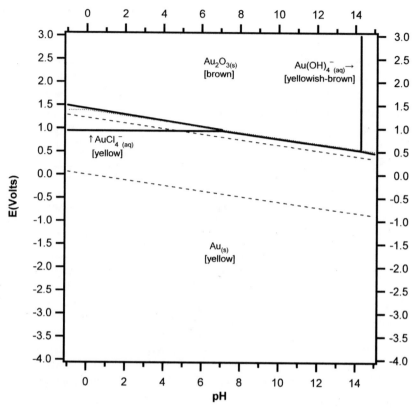

Figure 16.5 E–pH diagram for Au/Cl species. Au soluble species concentrations (except H^+) = $10^{-3.0}$ M. Cl soluble species concentrations (except H^+) = $10^{0.0}$ M. Soluble species and most solids are hydrated. No agents producing complexes or insoluble compounds are present other than HOH, OH^- and Cl^-. The finely-dashed line (which runs along the bottom of the $Au(OH)_4^-$ and Au_2O_3 lines) represents the oxidation of Cl^- to Cl_2 and/or ClO_4^- and hence the disappearance of all species containing Cl^-.

Species ($\Delta G°$ in kJ/mol): Au (0.0), Au_2O_3 (77.8), $Au(OH)_4^-$ (−455.2), $AuCl_4^-$ (−234.7), Cl_2 (7.1), ClO_4^- (−8.4), Cl^- (−131.4), HOH (−237.2), H^+ (0.0), and OH^- (−157.3)

Equations for the lines:

$Au_2O_3/AuCl_4^-$ $E = 1.42 - 0.065 \text{ pH} + 0.007 \log[ClO_4^-] - 0.002$

                                   $\log[AuCl_4^-]$

$AuCl_4^-/Au$ $E = 1.01 - 0.079 \log[Cl^-] + 0.020 \log[AuCl_4^-]$

Au_2O_3/Au $E = 1.36 - 0.059 \text{ pH}$

$Au(OH)_4^-/Au$ $E = 1.71 - 0.079 \text{ pH} + 0.020 \log[Au(OH)_4^-]$

$Au(OH)_4^-/Au_2O_3$ $2 \text{ pH} = 34.7 + 2 \log[Au(OH)_4^-]$

Table 16.3
Gold Species

Name	Formula	State	Color	Solubility	$\Delta G°$ (kJ/mole)
Gold	Au	s	Yellow	Insol	0.0
Dibromogoldate(I)	$AuBr_2^-$	aq			−115.0
Dichlorogoldate(I)	$AuCl_2^-$	aq			−151.0
Dicyanogoldate(I)	$Au(CN)_2^-$	aq			286.0
Diiodogoldate(I)	AuI_2^-	aq			−47.6
Tetrabromogoldate(III)	$AuBr_4^-$	aq	Purple		−167.0
Tetrachlorogoldate(III)	$AuCl_4^-$	aq	Yellow		−234.7
Tetrahydroxogoldate(III)	$Au(OH)_4^-$	aq	Yellowish-brown		−455.2
Tetraiodogoldate(III)	AuI_4^-	aq			−45.0
Gold(I)					
bromide	AuBr	s	Yellow-gray	Insol, decomp	
chloride	AuCl	s	Yellow	Insol, decomp	−14.6
cyanide	AuCN	s	Yellow	Insol	
iodide	AuI	s	Yellow	Insol, decomp	
sulfide	Au_2S	s	Brown-black	Insol	
Gold(III)					
bromide	$AuBr_3$	s	Red-brown	Sol, hydrol	
chloride	$AuCl_3$	s	Red	Sol, hydrol	−53.6
cyanide	$Au(CN)_3$	s	White	Sol, hydrol	
fluoride	AuF_3	s	Orange	Sol, hydrol	
iodide	AuI_3	s	Green	Insol, decomp	
nitrate	$Au(NO_3)_3$	s	Yellow	Sol, hydrol	
oxide	Au_2O_3	s	Brown	Insol	77.8
sulfide	Au_2S_3	s	Brown-black	Insol	

Au salts of oxyacids are rare, and when they do exist, they readily decompose. AuCN is stable and insoluble in HOH, but $Au(CN)_3$ is unknown even though the complex $Au(CN)_4^-$ can be prepared.

f. Redox reactions. Au is obviously highly resistant to oxidation, which also means that Au compounds are highly subject to reduction. Au(III) is readily reduced to the element by most metals, $H_2C_2O_4$, aldehydes, hydrazine, formic acid, HNO_2, H_2SO_3, Fe^{+2}, Sn^{+2}, I^-, and alkaline H_2O_2. Reduction of Au(III) using Hg takes the Au down only to the Au(I) state. Many Au compounds are decomposed by light, and all by heat. Most simple salts of Au(I), with the exception of AuCN, are unstable in the presence of HOH, although decomposition may be slow in some cases.

g. Complexes. The aqueous chemistry of Au(I) is almost completely that of its complex compounds. In these compounds, the Au usually shows a coordination number of 2, although 4 is found in a few instances. Some exemplary complexes are $Au(NH_3)_2^+$, $Au(thiourea)_2^+$, $Au(CN)_2^-$, $AuCl_2^-$, $Au(SCN)_2^-$, and $Au(S_2O_3)_2^{-3}$. Some log β_n values for Au(I) complexes are CN^- (β_2 38.3), SCN^- (15.3, 17.0), NH_3(β_4 30.0).

For complexes of Au(III), the coordination number is most often 4, with higher coordination numbers being seen in some cases. Complexes include $Au(NH_3)_4^{+3}$, $Au(en)_2^{+3}$, AuX_4^- with X = Cl, Br, I, CN, SCN, NO_3. Included among known log β_n values are the following: CN^- (β_4 56.0), SCN^- (β_4 42.0), NH_3(β_2 27.0), OH^- (15.5, 29.0, 42.0, 44.2, 44.9, 42.9).

h. Analysis. Au can be measured down to 1.0 ppb using ICPAES and down to 1 ppt employing ICPMS. Colorimetry using o-tolidine can go down to 50 ppb, and IC down to 100 ppb.

i. Health aspects. The toxicity of Au compounds is possibly indicated by reports that the LD50 (oral mouse) is > 1.4 g/kg for Au nanoparticles.

17

The Zn Group

1. Introduction

The elements of this group (zinc Zn, cadmium Cd, mercury Hg) all exhibit a II oxidation state in aqueous systems, and Hg also shows a I oxidation state as indicated by the unusual cation Hg_2^{+2}. None of the elements shows oxidation states greater than II, which indicates that the d electrons are not involved. Within the group Zn and Cd resemble each other more closely than Cd and Hg. This is especially evident in the nobility of Hg ($E°$ positive for Zn and Cd, negative for Hg), the lack of an Hg hydroxide, the thermal instability of HgO, and the greater stabilities of many Hg complexes as compared to those of Zn and Cd.

2. Zinc (Zn) $4s^2 3d^{10}$

a. E–pH diagram.   Figure 17.1 shows the E–pH diagram for Zn at a $10^{-1.0}$ M concentration for soluble species except H^+ (and OH^-). No complexing agent other than HOH and OH^- is assumed to be present. The Zn^{+2} ion is more properly expressed as $Zn(HOH)_6^{+2}$, and the hydroxo complexes probably have enough HOH attached to realize a coordination number of 6. In aqueous solution, Zn acts only in the oxidation states of 0 and II. The legend of the figure shows equations for the lines separating the species.

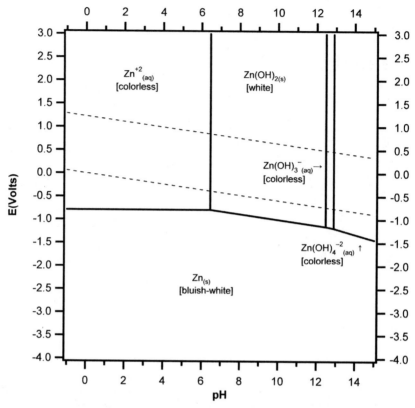

Figure 17.1 E–pH diagram for Zn species. Soluble species concentrations (except H^+) $= 10^{-1.0}$ M. Soluble species and most solids are hydrated. No agents producing complexes or insoluble compounds are present other than HOH and OH^-.

Species ($\Delta G°$ in kJ/mol): Zn (0.0), $Zn(OH)_2$ (−554.4), Zn^{+2} (−146.9), $ZnOH^+$ (−341.8), $Zn(OH)_3^-$ (−713.8), $Zn(OH)_4^{-2}$ (−877.6), HOH (−237.2), H^+ (0.0), and OH^- (−157.3)

Equations for the lines:

Zn^{+2}/Zn	$E = -0.76 + 0.030 \log[Zn^{+2}]$
$Zn(OH)_2/Zn$	$E = -0.41 - 0.059 \, pH$
$Zn(OH)_3^-/Zn$	$E = -0.01 - 0.089 \, pH + 0.030 \log[Zn(OH)_3^-]$
$Zn(OH)_4^{-2}/Zn$	$E = 0.37 - 0.118 \, pH + 0.030 \log[Zn(OH)_4^{-2}]$
$Zn(OH)_4^{-2}/Zn(OH)_3^-$	$pH = 12.9$
$Zn(OH)_3^-/Zn(OH)_2$	$pH = 13.6 + \log[Zn(OH)_3^-]$
$Zn(OH)_2/Zn^{+2}$	$2 \, pH = 11.7 - \log[Zn^{+2}]$

b. Discovery, occurrence, and extraction. Brass, an alloy of Cu and Zn, dates back to pre-historic times. Indications are that it was produced by heating calamine ($ZnCO_3$) with Cu and C. Pure Zn was being produced in India and China during the thirteenth and fourteenth centuries, and by about 1600, was being imported by Europe. Even before this, some reports indicate that zinc was recognized in Europe, but it was generally believed that it was a mixture of metals. In 1746, Andreas Marggraf published a book describing the production of Zn from its mineral calamine $ZnCO_3$ (now also known as smithsonite), and soon it was recognized by Lavoisier as an element. The name zinc derives from the German zink, which means sharp point, a name designating its appearance when it deposits in a smelter.

The major ores of Zn are zinc blende ZnS (also known as sphalerite) and calamine $ZnCO_3$. Much more Zn is now produced from ZnS which is concentrated from its rock matrix by flotation or sedimentation. The concentrate is then roasted to generate ZnO, which is smelted with coke, or which is dissolved in H_2SO_4 and the solution is electrolyzed.

c. The element. Zn is an active, bluish-white metal which does not dissolve in pure HOH, but dissolves in non-oxidizing acids to give colorless Zn^{+2}. It is also soluble in OH^- to give $Zn(OH)_2$ or $Zn(OH)_3^-$, and $Zn(OH)_4^{-2}$ depending upon the base concentration. In air, it takes on a thin coat of oxide which is somewhat refractory, and it burns in air to yield ZnO. Zn dissolves in very dilute HNO_3 to give chiefly NH_4^+, in cold, moderately dilute HNO_3 primarily to evolve N_2O, and in more concentrated HNO_3 to give off NO as the main product. Hot, concentrated H_2SO_4 dissolves Zn with the production of SO_2.

d. Oxides and hydroxides. When solutions of Zn^{+2} are treated with a stoichiometric amount of OH^-, white $Zn(OH)_2$ is precipitated. This hydroxide is amphoteric, dissolving in H^+ to give Zn^{+2} and dissolving in excess OH^- to yield $Zn(OH)_3^-$ and/or $Zn(OH)_4^{-2}$. When these latter species are treated with H_2O_2, the white peroxide ZnO_2 results. The white oxide ZnO may be made by ignition of $Zn(OH)_2$, $ZnCO_3$, $Zn(NO_3)_2$, $ZnSO_4$, or ZnC_2O_4, or by heating Zn in air.

e. Compounds. Treatment of Zn, ZnO, or $ZnCO_3$ with appropriate acids leads to numerous compounds of Zn. The most common of these are presented in Table 17.1. ZnS is interesting in that it is an insoluble transition-metal sulfide which is white, others being deeply colored.

f. Redox reactions. Zn is a good reducing agent, precipitating many metal ions as the metal. Metal ions with reduction potentials greater (more positive) than the Zn^{+2} reduction potential will usually deposit out, there being a few passivity exceptions. Zn^{+2} can be transformed to the metal by reducing agents with E° values below the Zn^{+2} reduction potential, such as Mg.

Table 17.1
Zinc Species

Name	Formula	State	Color	Solubility	$\Delta G°$ (kJ/mole)
Zinc	Zn	s	Bluish-white	Insol	0.0
Zinc(II) ion	Zn^{+2}	aq	Colorless		−146.9
Monohydroxozinc(II)	$Zn(OH)^+$	aq	Colorless		−341.8
Trihydroxozincate(II)	$Zn(OH)_3{}^-$	aq	Colorless		−713.8
Tetrahydroxozincate(II)	$Zn(OH)_4{}^{-2}$	aq	Colorless		−877.6
Zinc(II)					
acetate	$Zn(C_2H_3O_2)_2$	s	White	Sol	−885.8
bromate	$Zn(BrO_3)_2$	s	White	Sol	
bromide	$ZnBr_2$	s	White	Sol	−312.1
carbonate	$ZnCO_3$	s	White	Insol	−731.6
chlorate	$Zn(ClO_3)_2$	s	White	Sol	
chloride	$ZnCl_2$	s	White	Sol	−369.4
chromate	$ZnCrO_4$	s	Yellow	Insol	
cyanide	$Zn(CN)_2$	s	White	Insol	121.0
fluoride	ZnF_2	s	White	Sol	−713.4
hydroxide	$Zn(OH)_2$	s	White	Insol	−554.4
iodate	$Zn(IO_3)_2$	s	White	Sol	−438.8
iodide	ZnI_2	s	White	Sol	−209.0
nitrate	$Zn(NO_3)_2$	s	White	Sol	−298.8
oxalate	ZnC_2O_4	s	White	Insol	−820.9
oxide	ZnO	s	White	Insol	−318.3
perchlorate	$Zn(ClO_4)_2$	s	White	Sol	−164.3
permanganate	$Zn(MnO_4)_2$	s	Purple-brown		
phosphate	$Zn_3(PO_4)_2$	s	White	Insol	
sulfate	$ZnSO_4$	s	White	Sol	−871.5
sulfide	ZnS	s	White	Insol	−181.0
sulfite	$ZnSO_3$	s	White	Sol	
thiocyanate	$Zn(SCN)_2$	s	White	Sol	

g. Complexes. Complexes of Zn usually show a coordination number of 4, but a few attain the value of 6. Illustrative examples are $Zn(NH_3)_4{}^{+2}$, $Zn(en)_2{}^{+2}$, $Zn(py)_4{}^{+2}$, $Zn(glycinate)_3{}^-$, $Zn(acac)_2$, $Zn(edta)^{-2}$, $Zn(cit)^-$, $Zn(nta)^-$, and $ZnX_4{}^{-2}$ with X = Cl, Br, I, NO_2, CN, SCN, ½tartrate, ½oxalate. Log β_n values for complexation follow: $B(OH)_4{}^-$ (β_4 11.8), CN^- (5.3, 11.1, 16.1, 19.6), SCN^- (1.3, 1.9, 2.0, 1.6), NH_3 (2.2, 4.5, 6.9, 8.9), $NO_3{}^-$ (0.4), $P_2O_7{}^{-4}$ (8.7, 11.0), OH^- (5.0, 11.1, 13.6, 14.8), $SO_4{}^{-2}$ (0.9, 1.9, 1.7, 1.7), S_2O_3 (2.4, 1.9), F^- (1.2), Cl^- (0.4), ethylenediamine(5.7, 10.6, 13.9), pyridine(1.0, 1.6, 1.9), oxinate$^-$ (8.6, 15.8), dipyridyl(5.1, 9.5, 13.2), 1,10-phenanthroline(6.2, 12.1, 17.3), formate$^-$ (0.7, 1.1, 1.2), $C_2O_4{}^{-2}$ (3.9, 6.4),

$C_2H_3O_2^-$(1.6, 1.9, 1.6), tart^{-2}(3.8, 5.0), cit^{-3}(4.3, 5.9), salicylate^{-2}(6.9), glycinate$^-$(5.4, 9.8, 12.3), nta^{-3}(10.7, 14.3), edta$^-$(16.5).

h. Analysis. Zn may be analyzed down to the 100 ppt level by ICPAES and down to the 10 ppt level by ICPMS. Ion chromatography may be employed down to 10 ppb, and colorimetry using dithizone down to 100 ppb.

i. Health aspects. The LD50 (oral rat) for $ZnCl_2$ is 350 mg/kg, and that for ZnO is >5 g/kg.

3. Cadmium (Cd) $5s^2 4d^{10}$

a. E–pH diagram. Figure 17.2 displays the E–pH diagram for $10^{-1.0}$ M soluble Cd species with no complexing agent present except HOH and OH$^-$. Most of the soluble species are assumed to be hydrated. The Cd^{+2} cation is more accurately represented as $Cd(HOH)_6^{+2}$. Notice that the species $Cd(OH)^+$, $Cd(OH)_3^-$, and $Cd(OH)_4^{-2}$ have been entered into the calculations, but they do not show up as the predominant species in the diagram. This does not mean that they are not in the solution; it means only that they are not predominant. If the system is run at $10^{-6.0}$ M, these three additional species appear as Figure 17.3 illustrates. The E–pH diagram clearly indicates that the oxidation states stable in the presence of HOH are 0 and II. The legends of the figures show equations for the lines separating the species.

b. Discovery, occurrence, and extraction. In the years 1817–18, several German investigators appear to have discovered cadmium independently. The reports are somewhat interlocked and it is difficult to come to a decision regarding priority, but Friedrich Stromeyer is generally recognized as the discoverer. He subjected some impure ZnO to several separation steps and was able to isolate a brown oxide. This brown material was mixed with lampblack and ignited to produce a metal which he named Kadmium. Stromeyer took this name from the Greek kadmeia, an ancient name for calamine ($ZnCO_3$), the Zn mineral from which the Cd had been derived.

The most important source of Cd is the small amount (about 0.3%) found in many Zn ores. The Zn ore is converted to ZnO which is dissolved in dilute H_2SO_4. Zn powder is added to precipitate the Cd. Further purification can be achieved by dissolution of the Cd followed by electrolysis.

c. The element. Cd is an active, silvery-white metal which forms a thin coat of CdO in moist air. It is insoluble in HOH and bases, but dissolves in non-oxidizing acids to give Cd^{+2} and H_2. With oxidizing acids, the reaction is more complicated, for example, with HNO_3 there is a strong resemblance to Zn in that oxides of N and even NH_4^+ are produced. The metal burns in air to yield CdO.

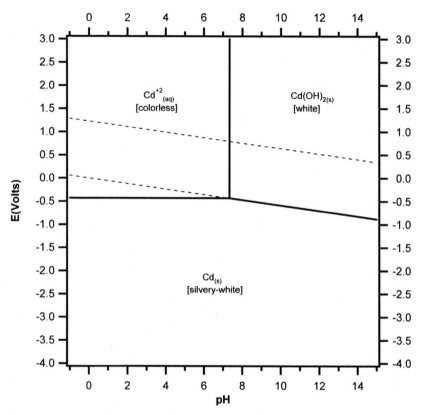

Figure 17.2 E–pH diagram for Cd species. Soluble species concentrations (except H^+) = $10^{-1.0}$ M. Soluble species and most solids are hydrated. No agents producing complexes or insoluble compounds are present other than HOH and OH^-.

Species ($\Delta G°$ in kJ/mol): Cd (0.0), $Cd(OH)_2$ (−473.6), Cd^{+2} (−77.6), $CdOH^+$ (−261.1), $Cd(OH)_3^-$ (−600.7), $Cd(OH)_4^{-2}$ (−758.4), HOH (−237.2), H^+ (0.0), and OH^- (−157.3)

Equations for the lines:

$$Cd^{+2}/Cd \qquad E = -0.40 + 0.030 \log[Cd^{+2}]$$

$$Cd(OH)_2/Cd \qquad E = 0.00 - 0.059\,pH$$

$$Cd(OH)_2/Cd^{+2} \qquad 2\,pH = 13.7 - \log[Cd^{+2}]$$

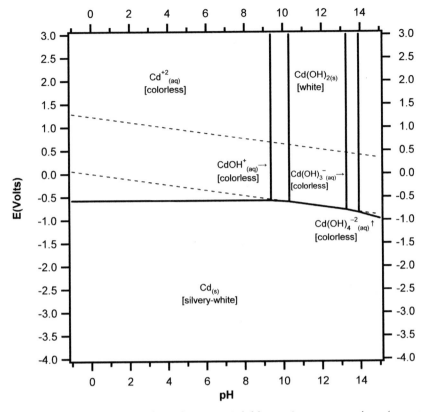

Figure 17.3 E–pH diagram for Cd species. Soluble species concentrations (except H^+) = $10^{-6.0}$ M. Soluble species and most solids are hydrated. No agents producing complexes or insoluble compounds are present other than HOH and OH^-.

Species ($\Delta G°$ in kJ/mol): Cd (0.0), $Cd(OH)_2$ (−473.6), Cd^{+2} (−77.6), $CdOH^+$ (−261.1), $Cd(OH)_3^-$ (−600.7), $Cd(OH)_4^{-2}$ (−758.4), HOH (−237.2), H^+ (0.0), and OH^- (−157.3)

Equations for the lines:

Cd^{+2}/Cd	$E = -0.40 + 0.030 \log[Cd^{+2}]$
$CdOH^+/Cd$	$E = -0.12 - 0.030 \, pH + 0.030 \log[CdOH^+]$
$Cd(OH)_2/Cd$	$E = 0.00 - 0.059 \, pH$
$Cd(OH)_3^-/Cd$	$E = 0.57 - 0.089 \, pH + 0.030 \log[Cd(OH)_3^-]$
$Cd(OH)_4^{-2}/Cd$	$E = 0.99 - 0.118 \, pH + 0.030 \log[Cd(OH)_4^{-2}]$
$Cd(OH)_4^{-2}/Cd(OH)_3^-$	$pH = 13.9$
$Cd(OH)_3^-/Cd(OH)_2$	$pH = 19.3 + \log[Cd(OH)_3^-]$
$Cd(OH)_2/CdOH^+$	$pH = 4.3 - \log[CdOH^+]$
$CdOH^+/Cd^{+2}$	$pH = 9.4$

d. Oxides and hydroxides. $Cd(OH)_2$ can be prepared by the addition of OH^- to a solution of Cd^{+2}. Ignition of this hydroxide or the burning of Cd in air yields CdO. CdO may also be prepared by the ignition of $CdCO_3$, $Cd(NO_3)_2$, or $CdSO_4$. Treatment of $Cd(OH)_2$ with highly concentrated base gives a solution from which $Na_2Cd(OH)_4$ can be isolated. If H_2O_2 is added to $Cd(OH)_2$, hydrated peroxides are formed.

e. Compounds. Treatment of Cd, CdO, $Cd(OH)_2$, or $CdCO_3$ with appropriate acids yields a wide variety of Cd compounds, a number of which are listed in Table 17.2. Of interest are CdO and CdS which, unlike most other Cd compounds, are colored.

f. Redox reactions. Comparison of Figures 17.1 and 17.2 shows that Cd is not as good a reducing agent as Zn, which means that metals such as Zn, Mg, and Al can displace it provided there are no kinetic limitations. Conversely, Cd precipitates many metals form their solutions as comparisons with other E–pH diagrams indicate (Ag, Hg, Bi, Cu, Pb, Sn, Co).

g. Complexes. The coordination number usually found in Cd complexes is 4, although higher numbers are known. Exemplary Cd(II) complexes are CdX_4^{-2} with X = Cl, Br, I, CN, $\frac{1}{2}$tart, SCN, NO_2, $\frac{1}{2}$ox, $Cd(NH_3)_4^{+2}$, $Cd(en)_2^{+2}$, $Cd(oxine)_2$, $Cd(edta)^{-2}$, $Cd(acac)_2$, and $Cd(cit)^-$. Log β_n values for Cd^{+2} follow: $B(OH)_4^-$(log β_4 10.6), CO_3^{-2}(log β_3 6.2), CN^-(6.0, 11.1, 15.7, 17.9), SCN^-(1.9, 2.8, 2.8, 2.3), NH_3(2.6, 4.6, 5.9, 6.7, 6.9, 5.4), NO_3^-(0.5, 0.2), $P_2O_7^{-4}$(8.7), OH^-(3.9, 7.7, 8.7, 8.7), SO_3^{-2}(4.2), SO_4^{-2}(1.0, 2.0, 2.7, 2.3), $S_2O_3^{-2}$(3.9, 6.3, 6.7, 7.1), F^-(0.4, 0.5), Cl^-(2.0, 2.6, 2.4, 1.7), Br^-(2.1, 3.0, 3.0, 2.9), I^-(2.3, 3.9, 5.0, 6.0), ethylenediamine(5.4, 9.9, 11.7), pyridine(1.3, 2.0, 2.3), oxinate$^-$(7.8), dipyridyl(4.2, 7.7, 10.3), 1,10-phenanthroline(5.8, 10.6, 14.6), formate$^-$(1.0, 1.4, 1.8), oxalate^{-2}(2.8), acetate$^-$(1.9, 3.2, 2.2, 2.0), maleate^{-2}(2.2, 3.6, 3.8), succinate^{-2}(1.7, 2.8), cit^{-3}(3.2, 4.5), salicylate^{-2}(5.6), phthalate^{-2}(2.5), glycinate$^-$(4.7, 8.4, 10.7), aspartate^{-2}(4.4, 7.6), picolinate$^-$(4.8, 8.3, 10.8), nta^{-3}(9.8, 14.6), dipicolinate^{-2}(6.8, 11.2), edta^{-4}(16.5), thiourea(1.5, 2.2, 2.6, 3.1), acac$^-$(3.8, 6.7).

h. Analysis. Cd may be analyzed down to the 100 ppt level by ICPAES and down to the 10 ppt level by ICPMS. Ion chromatography may be employed down to 10 ppb, with colorimetry down to 20 ppb using such agents as 4-nitronaphthalene-diazoamine-azo-benzene.

i. Health aspects. The LD50 (oral rat) for $CdCl_2$ is 88 mg/kg, and that for CdO is 72 mg/kg. This emphasizes the extreme toxicity of Cd, its action leading to dysfunction of the kidneys.

Table 17.2
Cadmium Species

Name	Formula	State	Color	Solubility	$\Delta G°$ (kJ/mole)
Cadmium	Cd	s	Silvery-white	Insol	0.0
Cadmium(II) ion	Cd^{+2}	aq	Colorless		−77.6
Monohydroxocadmium(II)	$Cd(OH)^+$	aq	Colorless		−261.1
Trihydroxocadmate(II)	$Cd(OH)_3^-$	aq	Colorless		−600.7
Tetrahydroxocadmate(II)	$Cd(OH)_4^{-2}$	aq	Colorless		−758.4
Cadmium(II)					
acetate	$Cd(C_2H_3O_2)_2$	s	White	Sol	
borate	$Cd(BO_3)_2$	s	White	Sol	
bromate	$Cd(BrO_3)_2$	s	White	Sol	
bromide	$CdBr_2$	s	White	Sol	−296.3
carbonate	$CdCO_3$	s	White	Insol	−669.4
chlorate	$Cd(ClO_3)_2$	s	White	Sol	
chloride	$CdCl_2$	s	White	Sol	−343.9
cyanide	$Cd(CN)_2$	s	White	Sol	207.9
fluoride	CdF_2	s	White	Sol	−647.7
hydroxide	$Cd(OH)_2$	s	White	Insol	−473.6
iodate	$Cd(IO_3)_2$	s	White	Sol	
iodide	CdI_2	s	Yellow	Sol	−201.4
molybdate	$CdMoO_4$	s	Yellow	Insol	
nitrate	$Cd(NO_3)_2$	s	White	Sol	−255.0
oxalate	CdC_2O_4	s	White	Insol	
oxide	CdO	s	Brown	Insol	−228.4
perchlorate	$Cd(ClO_4)_2$	s	White	Sol	
permanganate	$Cd(MnO_4)_2$	s	Purple-brown	Sol	
phosphate	$Cd_3(PO_4)_2$	s	White	Insol	
sulfate	$CdSO_4$	s	White	Sol	−822.7
sulfide	CdS	s	Yellow-orange	Insol	−156.5
sulfite	$CdSO_3$	s	White	Sol	
thiocyanate	$Cd(SCN)_2$	s	White	Sol	
tungstate	$CdWO_4$	s	Yellow	Insol	

4. Mercury (Hg) $6s^2 5d^{10}$

a. E–pH diagram. Figure 17.4 sets out the E–pH diagram for $10^{-1.0}$ M soluble Hg species in the presence of no complexing agent except OH^- or HOH. The Hg_2^{+2} ion is actually $Hg_2(HOH)_2^{+2}$ and the Hg^{+2} ion is $Hg(HOH)_6^{+2}$. Since the E–pH diagram expresses the Hg concentration in terms of single Hg atoms, this implies that the concentration of Hg_2^{+2} is $10^{-1.3}$ in the diagram. For Figure 17.4, several species have been entered

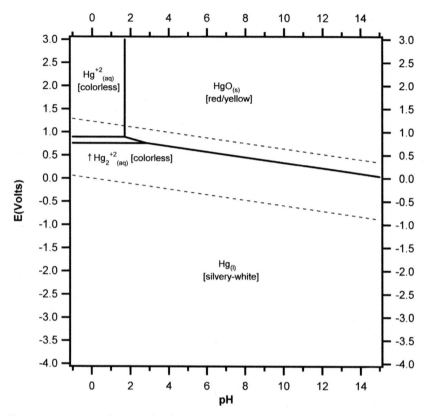

Figure 17.4 E–pH diagram for Hg species. Soluble species concentrations (except H$^+$) = $10^{-1.0}$ M. Soluble species and most solids are hydrated. No agents producing complexes or insoluble compounds are present other than HOH and OH$^-$.

Species ($\Delta G°$ in kJ/mol): Hg (0.0), HgO (−58.6), Hg^{+2} (164.8), Hg$_2^{+2}$ (153.6), HgOH$^+$ (−51.9), Hg(OH)$_3^-$ (−427.2), HOH (−237.2), H$^+$ (0.0), and OH$^-$ (−157.3).

Equations for the lines:

Hg^{+2}/Hg$_2^{+2}$ $\quad$ $E = 0.91 + 0.059 \log[\text{Hg}^{+2}] - 0.030 \log[\text{Hg}_2^{+2}]$

Hg$_2^{+2}$/Hg $\quad$ $E = 0.79 + 0.030 \log[\text{Hg}_2^{+2}]$

HgO/Hg $\quad$ $E = 0.92 - 0.059 \,\text{pH}$

HgO/Hg$_2^{+2}$ $\quad$ $E = 1.05 - 0.118 \,\text{pH} - 0.030 \log[\text{Hg}_2^{+2}]$

HgO/Hg^{+2} $\quad$ $2 \,\text{pH} = 2.4 - \log[\text{Hg}^{+2}]$

into the calculation, but they do not appear as predominant species in the diagram. Figure 17.5, which uses a soluble Hg species concentration of $10^{-5.0}$ M, shows $Hg(OH)^+$ and $Hg(OH)_3^-$ in addition to the four species in Figure 17.4. The legends of the figures show equations for the lines separating the species.

b. Discovery, occurrence, and extraction.   Hg was among the metals known in the ancient world, being found in Egyptian graves dating to 1500 BC, and also early in China and India. Cinnabar, HgS, a widely used pigment known as vermilion, was the mineral from which it was prepared. The name derives from the Roman god, and the symbol reflects the Latin word hydrargyrum, which translates as liquid silver. Cinnabar is the only important source of Hg. The ore is heated in a stream of air and the resulting Hg vapor is condensed, or reduction of the ore can be carried out by heating with Ca or Fe.

c. The element.   Hg, in sharp contrast to Zn and Cd, is a noble metal. The heavy, shiny, liquid is unreactive with air, HOH, bases, and non-ionizing acids. Dissolution is brought about with oxidizing acids to produce Hg_2^{+2} if Hg is in excess, or Hg^{+2} if acid is in excess. When the metal is gently heated in air, the red/yellow HgO results.

d. Oxides and hydroxides.   No hydroxide solids are known for Hg. Hg_2O can be made by adding base to Hg_2^{+2}, but the oxide is subject to disproportionation to Hg and Hg^{+2}. Red HgO is obtained by dissolving Hg in HNO_3, isolating $Hg(NO_3)_2$ by evaporation, and then by thermal decomposition of the nitrate. The yellow form of HgO is produced when OH^- is added to a solution of Hg^{+2}. HgO is insoluble in HOH and bases, but dissolves in acids to give Hg^{+2}.

e. Compounds.   Table 17.3 lists a number of the more important Hg compounds. The major characteristic of Hg(I) salts is their insolubility, only the nitrate, chlorate, and perchlorate being soluble. Hg(I) compounds can be prepared by reduction of Hg(II) compounds, often by Hg. The equilibrium constant for the disproportionation reaction $Hg_2^{+2} \rightarrow Hg^{+2} + Hg$ is about $10^{-2.0}$. Any agent that reduces the concentration of Hg^{+2} more than the concentration of Hg_2^{+2} tends to displace the equilibrium to the right. The mechanisms of such a concentration change include complexation, insolubility, and a weakly dissociated soluble species (such as $HgCl_2$ which is covalent in solution).

Acids react with HgO to produce corresponding Hg(II) compounds. Two classes of Hg(II) compounds may be defined: covalent and ionic. The covalent compounds $HgCl_2$, $HgBr_2$, HgI_2, and $Hg(CN)_2$ go into HOH solution chiefly as undissociated molecules, which undergo little hydrolysis. The ionic compounds which include HgF_2, $Hg(NO_3)_2$, $HgSO_4$, and $Hg(ClO_4)_2$ go into

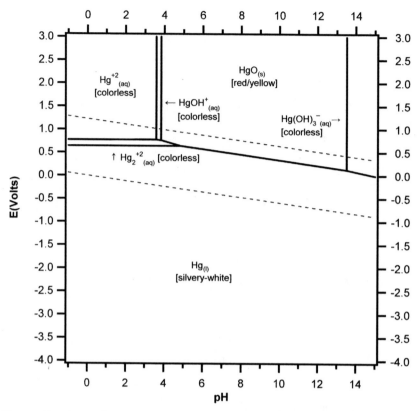

Figure 17.5 E–pH diagram for Hg species. Soluble species concentrations (except H^+) = $10^{-5.0}$ M. Soluble species and most solids are hydrated. No agents producing complexes or insoluble compounds are present other than HOH and OH^-.

Species ($\Delta G°$ in kJ/mol): Hg (0.0), HgO (−58.6), Hg^{+2} (164.8), Hg_2^{+2} (153.6), $HgOH^+$ (−51.9), $Hg(OH)_3^-$ (−427.2), HOH (−237.2), H^+ (0.0), and OH^- (−157.3)

Equations for the lines:

Hg^{+2}/Hg_2^{+2}	$E = 0.91 + 0.059 \log[Hg^{+2}] - 0.030 \log[Hg_2^{+2}]$
Hg_2^{+2}/Hg	$E = 0.79 + 0.030 \log[Hg_2^{+2}]$
HgO/Hg	$E = 0.92 - 0.059 \, pH$
$Hg(OH)_3^-/Hg$	$E = 1.47 - 0.089 \, pH + 0.030 \log[Hg(OH)_3^-]$
HgO/Hg_2^{+2}	$E = 1.05 - 0.118 \, pH - 0.030 \log[Hg_2^{+2}]$
$HgOH^+/Hg_2^{+2}$	$E = 1.12 - 0.059 \, pH + 0.059 \log[HgOH^+] - 0.030 \log[Hg_2^{+2}]$
$Hg(OH)_3^-/HgO$	$pH = 18.6 + \log[Hg(OH)_3^-]$
$HgO/HgOH^+$	$pH = -1.2 - \log[HgOH^+]$
$HgOH^+/Hg^{+2}$	$pH = 3.6$

Table 17.3
Mercury Species

Name	Formula	State	Color	Solubility	ΔG° (kJ/mole)
Mercury	Hg	l	Silvery-white	Insol	0.0
Dimercury(I) ion	Hg_2^{+2}	aq	Colorless		153.6
Mercury(II) ion	Hg^{+2}	aq	Colorless		164.8
Hydroxomercury(II) ion	$Hg(OH)^+$	aq	Colorless		−51.9
Trihydroxomercurate(II)	$Hg(OH)_3^-$	aq	Colorless		−427.2
Dimercury(I)					
acetate	$Hg_2(C_2H_3O_2)_2$	s	White	Sol	−640.1
bromate	$Hg_2(BrO_3)_2$	s	White	Decomp	
bromide	Hg_2Br_2	s	White	Insol	−181.1
carbonate	Hg_2CO_3	s	White	Insol	−468.2
chlorate	$Hg_2(ClO_4)_2$	s	White	Sol	
chloride	Hg_2Cl_2	s	White	Insol	−210.4
chromate	Hg_2CrO_4	s	Red	Insol	−623.8
fluoride	Hg_2F_2	s	White	Decomp	−431.0
iodate	$Hg_2(IO_3)_2$	s	Yellow	Insol	−179.9
iodide	Hg_2I_2	s	Yellow	Insol	−111.0
nitrate	$Hg_2(NO_3)_2$	s	White	Sol	
oxalate	$Hg_2C_2O_4$	s	White	Insol	
perchlorate	$Hg_2(ClO_4)_2$	s	White	Sol	
sulfate	Hg_2SO_4	s	White	Insol	−626.3
thiocyanate	$Hg_2(SCN)_2$	s	White	Insol	226.4
tungstate	Hg_2WO_4	s	Yellow	Insol	
Mercury(II)					
acetate	$Hg(C_2H_3O_2)_2$	s	White	Sol	
bromate	$Hg(BrO_3)_2$	s	White	Sol	
bromide	$HgBr_2$	s	White	Sol	
carbonate	$HgCO_3$	s	Brown-red	Insol	
chlorate	$Hg(ClO_3)_2$	s	White	Sol	
chloride	$HgCl_2$	s	White	Sol	−180.3
chromate	$HgCrO_4$	s	Red	Insol	
cyanate	$Hg(CNO)_2$	s	White	Insol	
cyanide	$Hg(CN)_2$	s	White	Insol	
fluoride	HgF_2	s	White	Decomp	
iodate	$Hg(IO_3)_2$	s	White	Insol	−167.2
iodide	HgI_2	s	Yellow	Insol	−101.7
nitrate	$Hg(NO_3)_2$	s	Yellow	Sol	
oxalate	HgC_2O_4	s	White	Insol	
oxide	HgO	s	Red/yellow	Insol	−58.6
perchlorate	$Hg(ClO_4)_2$	s	White	Sol	
sulfate	$HgSO_4$	s	White	Decomp	
sulfide	HgS	s	Black	Insol	−44.4
thiocyanate	$Hg(SCN)_2$	s	White	Insol	
tungstate	$HgWO_4$	s	Yellow	Insol	

HOH solution as ions, and undergo considerable hydrolysis. The exceptionally low K_{sp} for HgS ($10^{-51.8}$) indicates that treatment of most compounds of Hg with H_2S will precipitate the black HgS.

f. Redox reactions.

Comparison of E–pH diagrams shows that Hg will precipitate Ag, Au, and Pt from their ions in solution, and will reduce Hg^{+2} to Hg_2^{+2}. Metallic Hg and Hg_2^{+2} are oxidized to Hg^{+2} by such agents as Cl_2, Br_2, I_2, HNO_3, and hot H_2SO_4. Agents such as Pb, Sn, Sn^{+2}, Bi, Cu, Cd, Zn, Mg, and H_2SO_3 readily reduce Hg_2^{+2} and Hg^{+2} to the metal.

g. Complexes.

The number of complexes of Hg(I) that have been prepared is few, mainly because the corresponding complex of Hg(II) is much more stable. This situation brings about the disproportionation of the Hg(I) species. Log β_n values for a few complexes are known: SO_4^{-2} (1.3, 3.5), $C_2O_4^{-2}$ (7.0), $P_2O_7^{-4}$ (9.3, 12.3), succinate^{-2} (–, 7.3), and phthalate^{-2} (4.9).

Hg(II) forms complexes chiefly with coordination numbers 2 and 4. Some of them are HgS_2^{-2}, $Hg(NH_3)_2^{+2}$, $Hg(NH_3)_4^{+2}$, HgX_4^{-2} with X = Cl, Br, I, CN, SCN, NO_2, $\frac{1}{2}$ox, $\frac{1}{2}$tart. Log β_n values are compounds: CN^-(17.0, 32.8, 36.3, 39.0), SCN^-(9.1, 17.3, 20.0, 21.8), NH_3(8.8, 17.4, 18.4, 19.3), NO_3^-(0.2, 0.0), OH^-(10.6, 21.8, 20.9), SO_3^{-2}(24.1), SO_4^{-2}(1.3, 2.4), $S_2O_3^{-2}$ (–, 29.2, 30.6), Cl^-(6.7, 13.2, 14.1, 15.1), Br^-(9.0, 17.1, 19.4, 21.0), I^-(12.9, 23.8, 27.6, 29.8), ethylenediamine(14.3, 23.2), pyridine(5.1, 10.0, 10.3, 10.6), dipyridyl(9.6, 16.7, 19.5), 1,10-phenanthroline(19.7, 23.3), formate$^-$(5.4), oxalate^{-2}(9.7), acetate$^-$(5.6, 9.3, 13.3, 17.1), cit^{-3}(10.9), glycinate$^-$(10.3, 19.2), picolinate$^-$(7.7, 15.6), nta^{-3}(14.6), dipicolinate^{-2}(20.3), edta^{-4}(21.7), thiourea(–, 21.3, 24.2, 25.8), acac$^-$(21.5).

h. Analysis.

Hg may be analyzed down to the 10 ppb level by ICPAES and down to the 10 ppt level by ICPMS. Colorimetry using di-β-naphthylthiocarbazone can be employed down to 30 ppb.

i. Health aspects.

Hg metal and Hg compounds are very toxic, especially the $HgCH_3^+$ ion formed by microorganisms which bring about biological methylation. This latter compound can produce irreversible damage to the central nervous system. The LD50 (oral rat) for the soluble compound $HgCl_2$ is 1 mg/kg. However, toxicity is detectable at much lower levels than this.

18

The Actinoid Metals

1. Introduction

The elements making up the Actinoid Metals are those with atomic numbers from 89 through 103: Ac, Th, Pa, U, Np, Pu, Am, Cm, Bk, Cf, Es, Fm, Md, No, and Lr. The name is meant to parallel the lanthanoids. They are generally abbreviated as An. Their valence electron structures are $7s^2 6d^{0-2} 5f^{0-14}$. These elements resemble the lanthanoids somewhat, but they have a much wider variation in oxidation states. Nor do they resemble each other to the extent that the lanthanoids do, this being a result of the oxidation state variations. Ac resembles La greatly, but Th, Pa, and U resemble their vertical congeners (Hf, Ta, W) more than they resemble Ce, Pr, and Nd. From Np onwards, the resemblance to the lanthanoids increases such that by Am, the actinoid elements are behaving very similarly, showing a predominant oxidation state of III. All of this occurs because the 7s, 6d, and 5f levels are much closer in energy than the 6s, 5d, and 4f levels. Table 18.1 lists the actinoids with several of their pertinent characteristics.

2. Availabilities of the Elements

No stable isotopes of any of these elements exist, the last element in the Periodic Table with a stable isotope being Bi (Bi-209). However, some of the An elements have isotopes with very long half lives, which means that they are found in nature in relative abundance, most notably as Th-232 ($10^{10.1}$ years),

397

Table 18.1
Actinoid Metals

Element	Most used isotope	Log Half life(year)	Source or production	Avail amt (g)	Electronic structure	Oxidation states	Origin of name
Actinium	Ac-227	1.34	Ra-226 + n	mg	$5f^0 6d^1 7s^2$	III	Greek aktinos
Thorium	Th-232	10.14	natural	kg	$5f^0 6d^2 7s^2$	IV	God Thor
Protactinium	Pa-231	4.52	natural	cg	$5f^2 6d^1 7s^2$	V	Before Ac
Uranium	U-238	9.65	natural	kg	$5f^3 6d^1 7s^2$	IV, VI	Planet Uranus
Neptunium	Np-237	6.33	U-235 + n's	kg	$5f^5 6d^0 7s^2$	V	Planet Neptune
Plutonium	Pu-239	4.33	U-238 + n	kg	$5f^6 6d^0 7s^2$	IV	Planet Pluto
Americium	Am-243	3.87	Pu-239 + 4n	cg	$5f^7 6d^0 7s^2$	III	America (like Eu)
Curium	Cm-244	1.26	Pu-239 + 5n	cg	$5f^7 6d^1 7s^2$	III	P. & Marie Curie
Berkelium	Bk-249	0.06	Cm-244 + n's	mg	$5f^8 6d^1 7s^2$	III	Berkeley, CA
Californium	Cf-252	0.42	Cm-244 + n's	mg	$5f^{10} 6d^0 7s^2$	III	California
Einsteinium	Es-253	-1.25	Cm-244 + n's	mg	$5f^{11} 6d^0 7s^2$	III	Albert Einstein
Fermium	Fm-257	-0.56	Cm-244 + n's	µg	$5f^{12} 6d^0 7s^2$	III	Enrico Fermi
Mendelevium	Md-258	-0.82	Es-253 + He-4	tr	$5f^{13} 6d^0 7s^2$	III	Dimitri Mendeleev
Nobelium	No-259	-3.94	Cm-248 + O-18	tr	$5f^{14} 6d^0 7s^2$	II	Alfred Nobel
Lawrencium	Lr-260	-5.24	Bk-249 + O-18	tr	$5f^{14} 6d^1 7s^2$	III	Ernest Lawrence

U-235 ($10^{8.8}$ years), and U-238 ($10^{9.7}$ years). Others are products of the decay of the above isotopes, so even though they are shorter lived, they persist in nature since they are continually being produced. The most important nuclides of this type are Ac-227 (21.8 years) and Pa-231 ($10^{4.5}$ years), both coming from U-235 decay. In U ores, very small amounts of Np-237 ($10^{6.3}$ years), Np-239 (2.4 days), and Pu-239($10^{4.3}$ years) arise from the interaction of neutrons with U isotopes.

Isotopes of the elements beyond U are produced artificially, Np and Pu by neutron capture by U, Am and Cm by multiple neutron capture by Pu, and elements beyond Cm by further neutron captures or bombardment of lower atomic number actinoids with ions of He, B, C, N, or O. As the atomic number increases, the elements become more unstable and thus tend to have shorter half lives. Np-237 and Pu-239 are available in multikilogram amounts; Am-241 (430 years), Am-243 (7650 years), and Cm-244 (18.1 years) in 100-g amounts; Bk-249 (320 days), Cf-252 (2.6 years), and Es-253 (20 days), in milligram amounts; Fm-257 in microgram quantities; and Md-258 (55 days), No-259 (1.0 h), and Lr (3.0 min) in trace amounts.

As the half lives of isotopes become shorter, they get more difficult to study. This is because short-lived isotopes are highly radioactive, which causes the emitted radiations to produce redox reactions, even in dilute solutions. Special apparatus is needed to study them, because of their intense radioactivity and the need to work rapidly. Further, due to limited quantities of the heavier elements, they can be studied only in very dilute solutions which sometimes behave strangely.

3. Discoveries and Extractions

Ac, actinium, was initially identified in 1899 by André-Louis Debierne, a French chemist, who separated it from pitchblende. He dissolved the mineral in acid, then added NH_4OH, and found that a radioactive species was carried down with the rare earth hydroxides. He named the element actinium after the Greek aktinos which means ray. Because of its low abundance in U, the element is usually not obtained by isolation from U. It can be obtained in mlligram amounts by irradiation of Ra-226 in a nuclear reactor. The preparation of Ac metal involves reduction of AcF_3 by Li at high temperature.

Th, thorium, was discovered in 1829 by Jöns Jakob Berzelius, who isolated a new oxide from a recently discovered mineral which Jens Esmark had sent to him. He called the oxide thoria and the mineral thorite ($ThSiO_4$) after the Scandinavian god Thor. Berzelius subsequently made the metal by the reduction of ThF_4 with Na. Th now is extracted from monazite, a phosphate of rare earths and Th. The mineral is heated in concentrated NaOH to give hydrous oxides, which are filtered out. HCl is then added to dissolve the solids and when the pH is adjusted to 3.5, ThO_2 precipitates and the rare earths remain in solution. The ThO_2 is solubilized and purified by solvent extraction.

The element is obtained by reduction of ThO_2 with Ca or reduction of $ThCl_4$ with Mg or Ca.

Pa, protactinium, was first identified in 1913 in the decay products of U-238 as the Pa-234 isotope (6.7 h) by Kasimir Fajans and Otto H. Göhring. In 1916, two groups, Otto Hahn and Lisa Meitner, and Frederick Soddy and John A. Cranston, found Pa-231 ($10^{4.5}$ years) as a decay product of U-235. This isotope is the parent of Ac-227 in the U-235 decay series, hence it was named protactinium (before actinium). Isolation from U extraction sludges yielded over 100 g in 1960.

U, uranium, was discovered in 1789 by Martin Heinrich Klaproth in the mineral pitchblende. He dissolved pitchblende in HNO_3, then neutralized the solution with KOH. A yellow precipitate resulted which was taken into solution by excess KOH. The yellow precipitate was heated with C, and a black powder was obtained. Klaproth believed this to be elemental U, but it actually was an oxide. Klaproth named the element after the recently discovered planet Uranus. It was not until 1841 that metallic U was made by Eugène-Melchior Peligot who reduced the chloride with K. U is extracted from such ores as pitchblende U_3O_8 and carnotite $K_2(UO_2)_2(VO_4)_2$ by crushing, concentration, roasting, then dissolution in H_2SO_4 plus an oxidizing agent such as MnO_2 or $NaClO_3$. The U ends up as an anionic sulfate complex with UO_2^{+2} which is purified by anion exchange and then solvent extraction. The element is made by thermal reduction of UF_4 with Mg.

Np through Lr are all prepared artificially by bombardment with neutrons and/or light element ions (He-4, B-10, B-11, C-12, O-16, O-18, Ca-48, Fe-56). Some routes are presented in Table 18.1. The elements have been separated from the targets and other product species by redox reactions, ion exchange, and solvent extraction. In a typical separation, a sulfonic acid ion exchange resin is placed in a column, the tripositive ions of Am through Lr are poured into the column where they are taken up, then the column is eluted with a solution of ammonium α-hydroxybutyrate. As elution proceeds, the An^{+3} ions come off in this order Lr–Md–Fm–Es–Cf–Bk–Cm–Am. They are detected by the distinctive energies of their radioactive emissions.

4. Elements and Compounds

Table 18.1 presents the electronic structures of the actinoids along with their major aqueous oxidation states. The structures of the M^{+2}, M^{+3}, and M^{+4} ions can be derived by double, triple, and quadruple removal of electrons from the 7s and 6d levels, thus leaving only $5f^{0-14}$ electrons. In general, the metals of all the elements may be prepared by the reduction of fluorides, chlorides, or oxides with Li, Mg, Ca, Ba, or Zn. The metals are ordinarily silvery-white, very active ($E°$ values below -1.48 v, except for $No^{+2}/No = -1.26$ v), and form an oxide coat in air, which protects Th, but is less effective for the others. Th, Pa, and U tend to react slowly with acids, or to be inert, with later members showing more lanthanoid behavior. The reactions with HOH parallel those with acids. In general, all members of the series are resistant to alkali attack.

The parallelism with the lanthanoids and the transition elements, Hf, Ta, W is important to recognize:

La—Ce—Pr—Nd—Pm—Sm—Gd—Tb—Dy—Ho—Er—Tm—Yb—Lu

Hf—Ta—W

Ac*–Th—Pa—U—Np—Pu—Am*–Cm*–Bk*–Cf*–Es*–Fm*–No*–Lr

As mentioned previously, Th resembles Hf, Pa resembles Ta, and U resembles W, Ac and those from Am through Lr (marked with *) resemble the rare earths, while Np and Pu function in an intermediate manner.

Table 18.2 presents the predominant aqueous species and solid compounds of the actinoids along with their colors. Some of the $\Delta G°$ values are only estimates, and therefore the E–pH diagrams based upon them are subject to amendment.

Table 18.2
Actinoid Species
Soluble species and most solids are hydrated.

Name	Formula	State	Color	Solubility	$\Delta G°$ (kJ/mole)
Actinium	Ac	s	Silvery-white	Decomp	0.0
Actinium(III) ion	Ac^{+3}	aq	Colorless		−639.3
Actinum(III)					
bromide	$AcBr_3$	s	White	Sol	
chloride	$AcCl_3$	s	White	Sol	−1015.7
fluoride	AcF_3	s	White	Insol	−1683.0
hydroxide	$Ac(OH)_3$	s	White	Insol	−1231.4
iodide	AcI_3	s	White	Sol	
oxalate	$Ac_2(C_2O_4)_3$	s	White	Insol	
oxide	Ac_2O_3	s	White	Insol	−1755.8
Thorium	Th	s	Silvery-white	Insol	0.0
Thorium(IV) ion	Th^{+4}	aq	Colorless		−705.4
Trihydroxothorium(IV)	$Th(OH)_3{}^+$	aq	Colorless		−918.5
Thorium(IV)					
acetate	$Th(C_2H_2O_2)_4$	s	White	Sol	
bromide	$ThBr_4$	s	White	Sol	−925.5
carbonate	$Th(CO_3)_2$	s	White	Insol	
chloride	$ThCl_4$	s	White	Sol	−1094.1
fluoride	ThF_4	s	White	Insol	−2003.3
hydroxide	$Th(OH)_4$	s	White	Insol	−1599.5
iodate	$Th(IO_3)_4$	s	White	Insol	

Continued

Table 18.2
(Continued)

Name	Formula	State	Color	Solubility	$\Delta G°$ (kJ/mole)
Thorium(IV)—*cont'd*					
iodide	ThI_4	s	Yellow	Sol	-661.5
nitrate	$Th(NO_3)_4$	s	White	Sol	-1139.5
oxalate	$Th(C_2O_4)_2$	s	White	Insol	
oxide	ThO_2	s	White	Insol	-1168.8
perchlorate	$Th(ClO_4)_4$	s	White	Sol	
phosphate	$Th_3(PO_4)_4$	s	White	Insol	
sulfate	$Th(SO_4)_2$	s	White	Sol	-2306.0
sulfide	ThS_2	s	Brown	Insol	
Protactinium	Pa	s	Silvery-white		0.0
Protactinium(III) ion	Pa^{+3}	aq	Colorless	Decomp	-430.5
Protactinium(IV) ion	Pa^{+4}	aq	Colorless	Decomp	-564.0
Oxohydroxoprotactinium(V)	$PaOOH^{+2}$	aq	Colorless	Decomp	-1049.8
Protactinium(III)					
iodide	PaI_3	s	Black	Decomp	
Protactinium(IV)					
bromide	$PaBr_4$	s	Orange-red	Decomp	-794.0
chloride	$PaCl_4$	s	Yellow-green	Decomp	-954.0
fluoride	PaF_4	s	Red-brown	Decomp	-1853.0
iodide	PaI_4	s	Black	Decomp	-516.0
oxide	PaO_2	s	Black	Decomp	-1054.4
Protactinium(V)					
bromide	$PaBr_5$	s	Orange-brown	Decomp	-820.0
chloride	$PaCl_5$	s	Yellow	Decomp	-1032.0
fluoride	PaF_5	s	White	Decomp	
iodide	PaI_5	s	Black	Decomp	
oxide	Pa_2O_5	s	White	Insol	-2372.2
Uranium	U	s	Silvery-white	Insol	0.0
Uranium(III) ion	U^{+3}	aq	Purple	Decomp	-480.7
Uranium(IV) ion	U^{+4}	aq	Green		-535.6
Hydoxouranium(IV) ion	$U(OH)^{+3}$	aq	Green		-765.4
Dioxouranium(V) ion	UO_2^+	aq	Colorless	Decomp	-968.6
Dioxouranium(VI) ion	UO_2^{+2}	aq	Yellow		-952.7
Dioxohydroxouranium(VI) ion	UO_2OH^+	aq	Yellow		-1158.2
Uranium(III)					
bromide	UBr_3	s	Red	Sol	-673.2
chloride	UCl_3	s	Green	Sol	-798.7
fluoride	UF_3	s	Black	Insol	-1433.4
iodide	UI_3	s	Black	Sol	-466.5
Uranium(IV)					
bromide	UBr_4	s	Brown	Sol	-767.5
chloride	UCl_4	s	Green	Sol	-929.7
fluoride	UF_4	s	Green	Insol	-1823.4

Table 18.2
(Continued)

Name	Formula	State	Color	Solubility	$\Delta G°$ (kJ/mole)
Uranium(IV)—*cont'd*					
iodide	UI_4	s	Black	Sol	−513.2
oxalate	$U(C_2O_4)_2$	s	Green	Insol	
oxide	UO_2	s	Brown	Insol	−1031.8
oxodibromide	$UOBr_2$	s	Yellow-green	Sol	−929.7
oxodichloride	$UOCl_2$	s	Green	Sol	−996.2
oxodiflouride	UOF_2	s			−1434.3
sulfate	$U(SO_4)_2$	s	Green	Sol	−2085.0
Uranium(V)					
bromide	UBr_5	s	Brown	Decomp	−769.4
chloride	UCl_5	s	Red-brown	Decomp	−950.0
fluoride	UF_5	s	White-yellow	Decomp	−1968.7
oxotribromide	$UOBr_3$	s	Black	Decomp	−900.0
oxotrichloride	$UOCl_3$	s	Brown	Decomp	−1071.0
Uranium(IV+2VI)					
oxide	U_3O_8	s	Green	Insol	−3369.6
Uranium(VI)					
chloride	UCl_6	s	Green	Decomp	−962.0
fluoride	UF_6	s	White	Decomp	−2068.6
oxide	UO_3	s	Orange-yellow	Insol	−1146.0
Uranyl(VI)					
acetate	$UO_2(C_2H_3O_2)_2$	s	Yellow	Sol	
bromide	UO_2Br_2	s	Green-yellow	Sol	−1066.5
carbonate	UO_2CO_3	s	Yellow	Insol	
chloride	UO_2Cl_2	s	Yellow	Sol	−1146.0
fluoride	UO_2F_2	s	Yellow	Sol	−1557.3
hydroxide	$UO_2(OH)_2$	s	Yellow	Insol	−1394.0
iodate	$UO_2(IO_3)_2$	s		Insol	
iodide	UO_2I_2	s	Red	Decomp	
nitrate	$UO_2(NO_3)_2$	s	Yellow	Sol	−1106.4
oxalate	$UO_2C_2O_4$	s	Yellow	Insol	
sulfate	UO_2SO_4	s	Yellow-green	Sol	−1685.7
Neptunium	Np	s	Silvery-white		0.0
Neptunium(III) ion	Np^{+3}	aq	Purple		−553.1
Neptunium(IV) ion	Np^{+4}	aq	Green		−540.6
Neptunyl(V) ion	NpO_2^+	aq	Green		−915.0
Neptunyl(VI) ion	NpO_2^{+2}	aq	Pink-red		−795.4
Pentoxoneptunate(VII)	NpO_5^{-3}	aq	Green		
Neptunium(III)					
bromide	$NpBr_3$	s	Green	Decomp	−706.0
chloride	$NpCl_3$	s	Green	Decomp	−831.7

Continued

Table 18.2
(Continued)

Name	Formula	State	Color	Solubility	$\Delta G°$ (kJ/mole)
Neptunium(III)—*cont'd*					
fluoride	NpF_3	s	Purple	Insol	-1460.3
iodide	NpI_3	s	Brown	Decomp	-513.4
hydroxide	$Np(OH)_3$	s	Green	Insol	-1447.0
Neptunium(IV)					
bromide	$NpBr_4$	s	Red	Hydrol	-737.6
chloride	$NpCl_4$	s	Red-brown	Hydrol	-896.2
fluoride	NpF_4	s	Green	Insol	-1784.0
iodate	$Np(IO_3)_4$	s	Brown	Insol	
nitrate	$Np(NO_3)_4$	s	Green	Hydrol	
oxalate	$Np(C_2O_4)_2$	s	Green	Insol	-1975.0
oxide	NpO_2	s	Green	Insol	-1021.7
sulfate	$Np(SO_4)_2$	s	Green	Hydrol	
Neptunium(VI)					
fluoride	NpF_6	s	Orange		
oxide	NpO_3	s	Brown	Insol	-1018.0
Neptunyl(VI)					
hydroxide	$NpO_2(OH)_2$	s		Insol	
nitrate	$NpO_2(NO_3)_2$	s		Hydrol	-1006.0
Plutonium	Pu	s	Silvery-white		0.0
Plutonium(III) ion	Pu^{+3}	aq	Purple		-579.1
Plutonium(IV) ion	Pu^{+4}	aq	Tan		-518.8
Plutonyl(V) ion	PuO_2^+	aq	Rose		-849.8
Plutonyl(VI) ion	PuO_2^{+2}	aq	Orange		-757.3
Pentoxoplutonate(VII)	PuO_5^{-3}	aq	Green		
Plutonium(III)					
bromide	$PuBr_3$	s	Green	Sol	-768.0
chloride	$PuCl_3$	s	Green	Sol	-892.1
fluoride	PuF_3	s	Purple	Insol	-1515.5
hydroxide	$Pu(OH)_3$	s	Black	Insol	-1157.0
iodide	PuI_3	s	Green	Sol	-579.8
oxalate	$Pu_2(C_2O_4)_3$	s		Insol	
oxide	Pu_2O_3	s	Black	Insol	-1722.6
sulfate	$Pu_2(SO_4)_3$	s	Green	Sol	
Plutonium(IV)					
fluoride	PuF_4	s	Brown	Insol	-1753.0
hydroxide	$Pu(OH)_4$	s	Yellow-brown	Insol	-1430.0
iodate	$Pu(IO_3)_4$	s		Insol	
nitrate	$Pu(NO_3)_4$	s	Green	Hydrol	
oxide	PuO_2	s	Yellow-brown	Insol	-998.3
phosphate	$Pu_3(PO_4)_4$	s		Insol	
sulfate	$Pu(SO_4)_2$	s	Pink	Hydrol	-2012.5
Plutonium(VI)					
fluoride	PuF_6	s	Red-brown		-1730.0

Table 18.2
(Continued)

Name	Formula	State	Color	Solubility	$\Delta G°$ (kJ/mole)
Plutonyl(VI)					
carbonate	PuO_2CO_3	s			
hydroxide	$PuO_2(OH)_2$	s	Brown		−1210.4
nitrate	$PuO_2(NO_3)_2$			Hydrol	
Americium	Am	s	Silvery-white		0.0
Americium(II) ion	Am^{+2}	s			−376.1
Americium(III) ion	Am^{+3}	aq	Pink-yellow		−599.1
Americium(IV) ion	Am^{+4}	aq			−346.4
Americinyl(V)	AmO_2^+	aq	Yellow		−741.0
Americinyl(VI)	AmO_2^{+2}	aq	Amber		−587.4
Americium(II)					
chloride	$AmCl_2$	s	Black	Decomp	
bromide	$AmBr_2$	s	Black	Decomp	
iodide	AmI_2	s	Black	Decomp	
Americium(III)					
bromide	$AmBr_3$	s	White-yellow	Sol	−786.6
carbonate	$Am_2(CO_3)_3$	s	Pink	Insol	−2996.2
chloride	$AmCl_3$	s	Pink-yellow	Sol	−910.9
fluoride	AmF_3	s	Pink	Insol	−1523.0
hydroxide	$Am(OH)_3$	s	Red	Insol	−1162.3
iodide	AmI_3	s	Yellow	Sol	−613.8
oxide	Am_2O_3	s	Tan	Insol	
sulfate	$Am_2(SO_4)_3$	s	Pink	Sol	
Americium(IV)					
fluoride	AmF_4	s	Tan	Insol	−1628.0
oxide	AmO_2	s	Black	Insol	−879.9
Americinyl(V)					
hydroxide	$AmO_2(OH)$	s	Yellow	Insol	−964.4
Americinyl(VI)					
hydroxide	$AmO_2(OH)_2$	s		Insol	−1015.5
Curium	Cm	s	Silvery-white		0.0
Curium(III) ion	Cm^{+3}	aq	Colorless		−595.9
Curium(IV) ion	Cm^{+4}	aq			−296.6
Curium(II)					
chloride	$CmCl_2$	s	Red	Decomp	
bromide	$CmBr_2$	s	Amber	Decomp	
iodide	CmI_2	s	Black	Decomp	
Curium(III)					
bromide	$CmBr_3$	s	Yellow-green	Sol	
carbonate	$Cm_2(CO_3)_3$	s		Insol	
chloride	$CmCl_3$	s	White	Sol	−902.5
fluoride	CmF_3	s	White	Sol	
hydroxide	$Cm(OH)_3$	s	White	Insol	−1153.1

Continued

Table 18.2
(Continued)

Name	Formula	State	Color	Solubility	$\Delta G°$ (kJ/mole)
Curium(III)—*cont'd*					
iodide	CmI_3	s	White	Sol	
oxide	Cm_2O_3	s	White	Insol	
Curium(IV)					
fluoride	CmF_4	s	Gray-green	Insol	
oxide	CmO_2	s	Black	Insol	−854.3
Berkelium	Bk	s	Silvery-white		0.0
Berkelium(II) ion	Bk^{+2}	aq			−297.1
Berkelium(III) ion	Bk^{+3}	aq	Green		−567.4
Berkelium(IV) ion	Bk^{+4}	aq	Yellow		−406.3
Berkelium(III)					
bromide	$BkBr_3$	s	Green	Sol	
chloride	$BkCl_3$	s	Green	Sol	
fluoride	BkF_3	s	Yellow-green	Insol	
hydroxide	$Bk(OH)_3$	s	Yellow-green	Insol	−1142.2
iodide	BkI_3	s	Yellow	Sol	
oxide	Bk_2O_3	s	Yellow-green	Insol	
Berkelium(IV)					
fluoride	BkF_4	s	Yellow-green	Insol	
oxide	BkO_2	s	Yellow-brown	Insol	−824.2
Californium	Cf	s	Silvery-white		0.0
Californium(II) ion	Cf^{+2}	aq			−380.2
Californium(III) ion	Cf^{+3}	aq	Green		−552.9
Californium(IV) ion	Cf^{+4}	aq	Yellow		−244.1
Californium(II)					
iodide	CfI_2	s	Violet		
Californium(III)					
bromide	$CfBr_3$	s	Green	Sol	
chloride	$CfCl_3$	s	Green	Sol	
fluoride	CfF_3	s	Green	Insol	
iodide	CfI_3	s	Orange	Sol	
oxide	Cf_2O_3	s	Green	Insol	
Californium(IV)					
fluoride	CfF_4	s	Green	Insol	
oxide	CfO_2	s	Black	Insol	
Einsteinium	Es	s	Silvery-white		0.0
Einsteinium(II) ion	Es^{+2}	aq			−424.6
Einsteinium(III) ion	Es^{+3}	aq	Colorless		−573.2
Einsteinium(III)					
bromide	$EsBr_3$	s	Tan		
chloride	$EsCl_3$	s	White-orange		
iodide	EsI_3	s	Amber-yellow		
oxide	Es_2O_3	s	White		

Table 18.2
(Continued)

Name	Formula	State	Color	Solubility	$\Delta G°$ (kJ/mole)
Fermium	Fm	s	Silvery-white		0.0
Fermium(II) ion	Fm^{+2}	aq			−482.5
Fermium(III) ion	Fm^{+3}	aq			−599.3
Mendelevium	Md	s	Silvery-white		0.0
Mendelevium(II) ion	Md^{+2}				−488.3
Mendelevium(III) ion	Md^{+3}				−503.7
Nobelium	No	s	Silvery-white		0.0
Nobelium(II) ion	No^{+2}	aq			−501.8
Nobelium(III) ion	No^{+3}	aq			−503.7
Lawrencium	Lr	s	Silvery-white		0.0
Lawrencium	Lr^{+3}	aq			−608.0

5. E–pH Diagrams

The E–pH diagram for Ac at $10^{-7.0}$ M is presented in Figure 18.1. This concentration is employed since $10^{-7.0}$ M is a reasonable value which takes into consideration the amount of the element available and safety with regard to its radioactivity. It is to be noticed that the diagram of Ac shows a strong resemblance to that of La , both the E° value and the pH of precipitation of Ac(OH)$_3$ being comparable if proper account of the differing concentrations is taken ($10^{-7.0}$ versus $10^{-1.0}$). As for La, the aqueous species Ac^{+3} is probably more accurately written as $Ac(HOH)_9^{+3}$.

Figure 18.2 represents the E–pH diagram for Th at $10^{-1.0}$ M. The cation is probably better represented as $Th(HOH)_9^{+4}$, and undoubtedly in the region of $Th(OH)_4/Th^{+4}$ there are hydrolysis products, both mononuclear and polynuclear. Comparison with the E–pH diagrams of Zr and Hf (Figures 13.2 and 13.3) and that for Ce (Figure 12.3) clearly indicate that Th more nearly belongs with the former. With adequate base, Th^{+4} precipitates as a hydrated form of ThO_2 which redissolves in acid. But when aged or dried or heated, HOH is lost, the crystal structure changes, and the compound becomes more resistant to dissolution.

The E–pH diagram for Pa is set out in Figure 18.3. The concentration value of $10^{-3.0}$ M is given as being a reasonable concentration taking into account availability and radiation safety. The resemblance to the Ta E–pH diagram (Figure 13.5) is not as close as similar comparisons in the two previous instances. This reflects the setting in of the tendency toward lanthanoid resemblance. The diagram reflects several aspects of Pa chemistry: Pa^{+3} is unstable in HOH, Pa^{+4} is stable in strong acid but tends to be readily oxidized to PaOOH^{+2}, and all solution species are subject to hydrolysis at low pH values.

The display in Figure 18.4 is the E–pH diagram for U at $10^{-1.0}$ M. The three cationic species are all hydrated with coordination numbers of 6, 7, 8, or higher. The diagram shows the instability of U^{+3}, the stability of U^{+4} at low pH, the ease of hydrolysis of U^{+4}, and the ready oxidation of U^{+4} to UO$_2{}^{+2}$. Species UO$_2{}^{+}$ does not appear on the diagram because of its

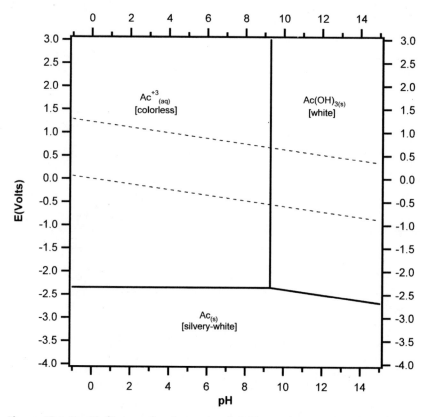

Figure 18.1 E–pH diagram for Ac species. Soluble species concentrations (except H^{+}) = $10^{-7.0}$ M. Soluble species and most solids are hydrated. No agents producing complexes or insoluble compounds are present other than HOH and OH^{-}.

Species ($\Delta G°$ in kJ/mol): Ac (0.0), Ac(OH)$_3$ (−1231.4), Ac^{+3} (−639.3), HOH (−237.2), H^{+} (0.0), and OH^{-} (−157.3).

Equations for the lines:

$$Ac^{+3}/Ac \qquad E = -2.21 + 0.020 \log [Ac^{+3}]$$

$$Ac(OH)_3/Ac \qquad E = -1.80 - 0.059 \, pH$$

$$Ac(OH)_3/Ac^{+3} \qquad 3 \, pH = 21.0 - \log [Ac^{+3}]$$

disproportionation into U^{+4} and UO_2^{+2}. At a pH around 2.5, it can be prepared since its rate of disproportionation is slow. Along the vertical lines, there are also numerous hydroxy species.

Np at $10^{-2.0}$ M is the subject of Figure 18.5. Its strong resemblance to U is clear when Figures 18.4 and 18.5 are compared. However, an important difference is that Np^{+3}, unlike U^{+3} is stable in HOH. The increased stability of the An^{+3} ion indicates the trend toward more lanthanoid behavior.

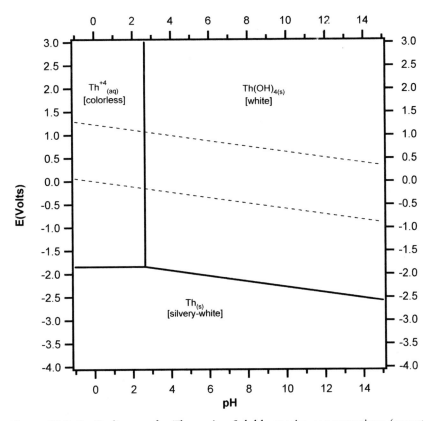

Figure 18.2 E–pH diagram for Th species. Soluble species concentrations (except H^+) $= 10^{-1.0}$ M. Soluble species and most solids are hydrated. No agents producing complexes or insoluble compounds are present other than HOH and OH^-.

Species ($\Delta G°$ in kJ/mol): Th (0.0), $Th(OH)_4$ (−1599.5), Th^{+4} (−705.4), HOH (−237.2), H^+ (0.0), and OH^- (−157.3)

Equations for the lines:

$$Th^{+4}/Th \qquad E = -1.83 + 0.015 \log [Th^{+4}]$$

$$Th(OH)_4/Th \qquad E = -1.69 - 0.015 \, pH$$

$$Th(OH)_4/Th^{+4} \qquad 4 \, pH = 9.6 - \log [Th^{+4}]$$

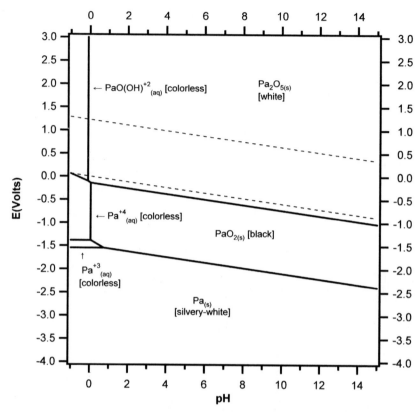

Figure 18.3 E–pH diagram for Pa species. Soluble species concentrations (except H^+) = $10^{-3.0}$ M. Soluble species and most solids are hydrated. No agents producing complexes or insoluble compounds are present other than HOH and OH^-.

Species ($\Delta G°$ in kJ/mol): Pa (0.0), Pa^{+3} (−430.5), Pa^{+4} (−564.0), PaO_2 (−1054.4), Pa_2O_5 (−2373.2), $PaO(OH)^{+2}$ (−1049.8), HOH (−237.2), H^+ (0.0), and OH^- (−157.3)

Equations for the lines:

$PaO(OH)^{+2}/Pa^{+4}$ $E = -0.12 - 0.177\ \text{pH}$

Pa^{+4}/Pa^{+3} $E = -1.38$

Pa^{+3}/Pa $E = -1.49 + 0.020\ \log [Pa^{+3}]$

Pa_2O_5/Pa^{+4} $E = -0.30 - 0.295\ \text{pH} - 0.059\ \log [Pa^{+4}]$

Pa_2O_5/PaO_2 $E = -0.14 - 0.059\ \text{pH}$

PaO_2/Pa^{+3} $E = -1.55 - 0.236\ \text{pH} - 0.059\ \log [Pa^{+3}]$

PaO_2/Pa $E = -1.50 - 0.059\ \text{pH}$

$Pa_2O_5/PaO(OH)^{+2}$ $4\ \text{pH} = -6.2 - 2\ \log [PaO(OH)^{+2}]$

PaO_2/Pa^{+4} $4\ \text{pH} = -2.8 - \log [Pa^{+4}]$

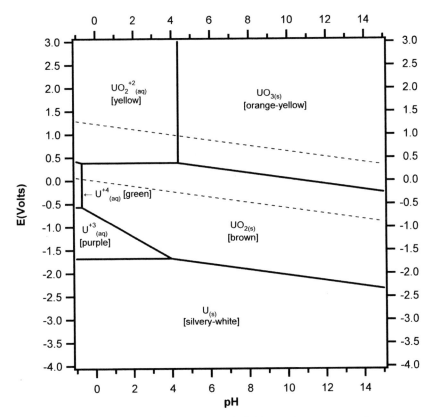

Figure 18.4 E–pH diagram for U species. Soluble species concentrations (except H^+) = $10^{-1.0}$ M. Soluble species and most solids are hydrated. No agents producing complexes or insoluble compounds are present other than HOH and OH^-.

Species ($\Delta G°$ in kJ/mol): U (0.0), U^{+3} (−480.7), U^{+4} (−535.6), UO_2 (−1031.8), UO_3 (−1146.0), UO_2^{+2} (−952.7), HOH (−237.2), H^+ (0.0), and OH^- (−157.3)

Equations for the lines:

$$UO_2^{+2}/U^{+4} \quad E = 0.30 - 0.118 \ pH$$

$$U^{+4}/U^{+3} \quad E = -0.57$$

$$U^{+3}/U \quad E = -1.66 + 0.020 \ \log [U^{+3}]$$

$$UO_2^{+2}/UO_2 \quad E = 0.41 + 0.030 \ \log [UO_2^{+2}]$$

$$UO_2/U^{+3} \quad E = -0.79 - 0.236 \ pH - 0.059 \ \log [U^{+3}]$$

$$UO_2/U \quad E = -1.44 - 0.059 \ pH$$

$$UO_3/UO_2 \quad E = 0.64 - 0.059 \ pH$$

$$UO_3/UO_2^{+2} \quad 2 \ pH = 7.7 - \log [UO_2^{+2}]$$

$$UO_2/U^{+4} \quad 4 \ pH = -3.8 - \log [U^{+4}]$$

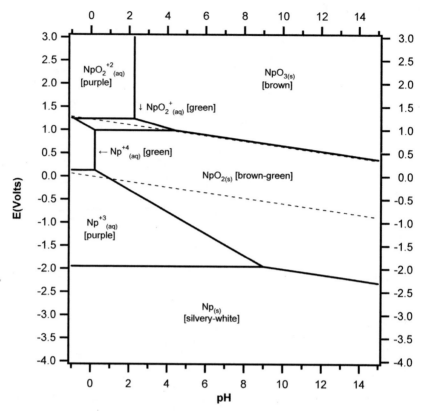

Figure 18.5 E–pH diagram for Np species. Soluble species concentrations (except H^+) = $10^{-2.0}$ M. Soluble species and most solids are hydrated. No agents producing complexes or insoluble compounds are present other than HOH and OH^-.

Species ($\Delta G°$ in kJ/mol): Np (0.0), Np^{+3} (−553.1), Np^{+4} (−540.6), NpO_2 (−1021.7), NpO_3 (−1018.0), NpO_2^{+2} (−795.4), NpO_2^+ (−915.0), HOH (−237.2), H^+ (0.0), and OH^- (−157.3)

Equations for the lines:

$$NpO_2^{+2}/NpO_2^+ \quad E = 1.24$$

$$NpO_2^+/Np^{+4} \quad E = 1.04 - 0.236 \text{ pH}$$

$$NpO_2^{+2}/Np^{+4} \quad E = 1.14 - 0.118 \text{ pH}$$

$$NpO_2^+/NpO_2 \quad E = 1.10 + 0.059 \log [NpO_2^+]$$

$$Np^{+4}/Np^{+3} \quad E = 0.13$$

$$NpO_3/NpO_2^+ \quad E = 1.39 - 0.118 \text{ pH} - 0.059 \log [NpO_2^+]$$

$$NpO_3/NpO_2 \quad E = 1.25 - 0.059 \text{ pH}$$

$$NpO_2/Np^{+3} \quad E = 0.06 - 0.236 \text{ pH} - 0.059 \log [Np^{+3}]$$

$$NpO_2/Np \quad E = -1.42 - 0.059 \text{ pH}$$

$$Np^{+3}/Np \quad E = -1.91 + 0.020 \log [Np^{+3}]$$

$$NpO_3/NpO_2^{+2} \quad 2 \text{ pH} = 2.6 - \log [NpO_2^{+2}]$$

$$NpO_2/Np^{+4} \quad 4 \text{ pH} = -1.2 - \log [Np^{+2}]$$

Further NpO_2^+ appears on the diagram and as the pH drops, it dispropor-tionates into Np^{+4} and NpO_2^{+2}. As in U, the vertical lines show hydroxy species of various sorts.

Figure 18.6 depicts the E–pH diagram for Pu at $10^{-1.0}$ M. Again, there are notable resemblances to both U and Np, but evidence of a trend to lanthanoid resemblance is present. These include further stabilization of the An^{+3} ion with an oxide Pu_2O_3 making its appearance, and a decreasing stability of the Pu^{+4} cation.

As one moves through the actinoids, the trend toward lanthanoid characteristics becomes more evident with Am as Figure 18.7 (at $10^{-4.0}$ M) illustrates. A much larger area in the figure is occupied by the An^{+3} ion, but resemblance to U, Np, and Pu remains in the upper regions of the E–pH diagram.

The last two E–pH diagrams, for Cm and Bk, are set out in Figures 18.8 and 18.9. The concentrations are low in accord with their availabilities, radioactivities, and needed safety aspects. The diagram for Cm is to be compared with that for Tb which is given in Figure 12.10, and that for Bk with Dy which is given in Figure 12.11. These comparisons indicate that the approach of the actinoid elements to lanthanoid behavior is getting closer. Upper regions in the diagrams of the actinoids series are still encumbered with tetravalent species.

For the remaining elements, Cf through Lr, lanthanoid characteristics are expected to become more evident. This indicates that their E–pH diagrams are expected to resemble those of Ho through Lu (Figures 12.12 through 12.17) ever more closely. Especially interesting is the case of No which shows at a pH $= 0.0$ the transition from No^{+2}/No at -2.60 v and the transition from No^{+3}/No^{+2} at 1.45 v. This indicates an amazing stability of the No^{+2} ion as compared with its counterpart Yb^{+2}. A glance at Figure 12.16 gives the corresponding transitions Yb^{+2}/Yb at -2.67 v and Yb^{+3}/Yb^{+2} at -1.09 v.

6. Complexes

The members of the actinoids have a somewhat greater tendency to form complexes than those of the lanthanoids. They also show a wider variety of complexes due to their more numerous oxidation states. The cations of the actinoids display coordination numbers which are often greater than 6. Although not many data are available, MO_2^{+2} and MO_2^+ ions attach 5 or 6 HOH molecules giving the central atom a coordination number of 7 or 8, and the M^{+4} and M^{+3} ions attach 9 or 10 HOH molecules. The coordination numbers with ligands other than HOH are often of these magnitudes, but sometimes may be somewhat smaller.

The tendencies to complex ion formation and hydrolysis ordinarily increase in the series $M^{+2} < AnO_2^+ < An^{+3} < AnO_2^{+2} < An^{+4}$ and with decreasing ion size from Ac to Lr. For complexation with some univalent

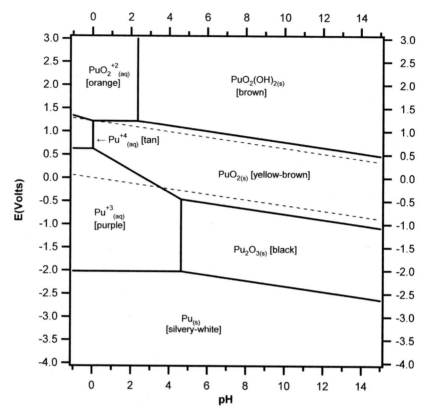

Figure 18.6 E–pH diagram for Pu species. Soluble species concentrations (except H^+) = $10^{-1.0}$ M. Soluble species and most solids are hydrated. No agents producing complexes or insoluble compounds are present other than HOH and OH^-.

Species ($\Delta G°$ in kJ/mol): Pu (0.0), Pu^{+3} (−579.1), Pu^{+4} (−518.8), PuO_2 (−998.3), Pu_2O_3 (−1722.6), PuO_2^{+2} (−757.3), $PuO_2(OH)_2$ (−1210.4), HOH (−237.2), H^+ (0.0), and OH^- (−157.3)

Equations for the lines:

PuO_2^{+2}/Pu^{+4}	$E = 1.22 - 0.118 \text{ pH}$
PuO_2^{+2}/PuO_2	$E = 1.25 + 0.030 \log [PuO_2^{+2}]$
PuO_2/Pu^{+3}	$E = 0.57 - 0.236 \text{ pH} - 0.059 \log [Pu^{+3}]$
PuO_2/Pu_2O_3	$E = -0.19 - 0.059 \text{ pH}$
Pu^{+4}/Pu^{+3}	$E = 0.62$
Pu^{+3}/Pu	$E = -2.00 + 0.020 \log [Pu^{+3}]$
Pu_2O_3/Pu	$E = -1.75 - 0.059 \text{ pH}$
$PuO_2(OH)_2/PuO_2$	$E = 1.36 - 0.059 \text{ pH}$
$PuO_2(OH)_2/PuO_2^{+2}$	$2 \text{ pH} = 3.7 - \log [PuO_2^{+2}]$
PuO_2/Pu^{+4}	$4 \text{ pH} = -0.9 - \log [Pu^{+4}]$
Pu_2O_3/Pu^{+3}	$6 \text{ pH} = 25.8 - 2 \log [Pu^{+3}]$

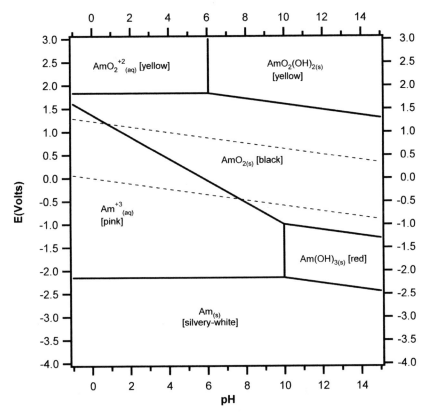

Figure 18.7 E–pH diagram for Am species. Soluble species concentrations (except H^+) = $10^{-4.0}$ M. Soluble species and most solids are hydrated. No agents producing complexes or insoluble compounds are present other than HOH and OH^-.

Species ($\Delta G°$ in kJ/mol): Am (0.0), Am^{+2} (−376.1), Am^{+3} (−599.1), Am^{+4} (−346.4), AmO_2 (−879.9), $Am(OH)_3$ (−1162.3), AmO_2^{+2} (−587.4), $AmO_2(OH)_2$ (−1015.5), HOH (−237.2), H^+ (0.0), and OH^- (−157.3)

Equations for the lines:

AmO_2^{+2}/AmO_2	$E = 1.52 + 0.030 \log [AmO_2^{+2}]$
AmO_2/Am^{+3}	$E = -2.91 - 0.236 \, pH - 0.059 \log [Am^{+3}]$
$AmO_2/Am(OH)_3$	$E = 0.47 - 0.059 \, pH$
Am^{+3}/Am	$E = -2.07 + 0.020 \log [Am^{+3}]$
$Am(OH)_3/Am$	$E = -1.56 - 0.059 \, pH$
$AmO_2(OH)_2/AmO_2$	$E = 1.76 - 0.059 \, pH$
$AmO_2(OH)_2/AmO_2^{+2}$	$2 \, pH = 8.1 - \log [AmO_2^{+2}]$
$Am(OH)_3/Am^{+3}$	$3 \, pH = 26.0 - \log [Am^{+3}]$

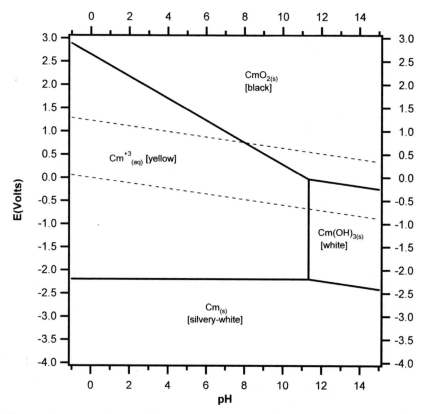

Figure 18.8 E–pH diagram for Cm species. Soluble species concentrations (except H^+) = $10^{-7.0}$ M. Soluble species and most solids are hydrated. No agents producing complexes or insoluble compounds are present other than HOH and OH^-.

Species ($\Delta G°$ in kJ/mol): Cm (0.0), Cm(OH)$_3$ (−1153.1), Cm^{+3} (−595.9), Cm^{+4} (−296.6), CmO_2 (−854.3), HOH (−237.2), H^+ (0.0), and OH^- (−157.3)

Equations for the lines:

CmO_2/Cm^{+3} $\qquad$ $E = 2.24 - 0.236\ pH - 0.059\ \log\ [Cm^{+3}]$

$CmO_2/Cm(OH)_3$ $\quad$ $E = 0.64 - 0.059\ pH$

Cm^{+3}/Cm $\qquad\qquad$ $E = -2.06 + 0.020\ \log\ [Cm^{+3}]$

$Cm(OH)_3/Cm$ $\qquad$ $E = -1.53 - 0.059\ pH$

$Cm(OH)_3/Cm^{+3}$ $\quad$ $3\ pH = 27.1 - \log\ [Cm^{+3}]$

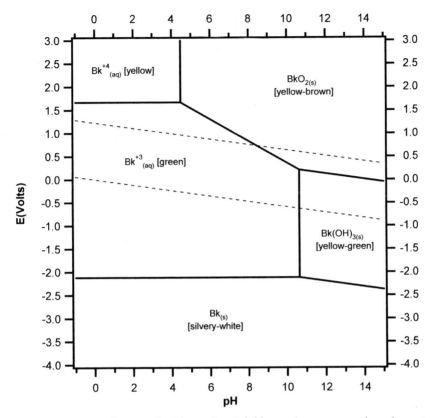

Figure 18.9 E–pH diagram for Bk species. Soluble species concentrations (except H^+) $= 10^{-8.0}$ M. Soluble species and most solids are hydrated. No agents producing complexes or insoluble compounds are present other than HOH and OH^-.

Species ($\Delta G°$ in kJ/mol): Bk (0.0), $Bk(OH)_3$ (−1142.2), Bk^{+3} (−567.4), Bk^{+4} (−406.3), BkO_2 (−824.2), HOH (−237.2), H^+ (0.0), and OH^- (−157.3)

Equations for the lines:

$$Bk^{+4}/Bk^{+3} \qquad E = 1.67$$

$$Bk^{+3}/Bk \qquad E = 1.96 + 0.020 \log [Bk^{+3}]$$

$$BkO_2/Bk^{+3} \qquad E = 2.25 - 0.236 \, pH - 0.059 \log [Bk^{+3}]$$

$$BkO_2/Bk(OH)_3 \qquad E = 0.84 - 0.059 \, pH$$

$$Bk(OH)_3/Bk \qquad E = -1.49 - 0.059 \, pH$$

$$BkO_2/Bk^{+4} \qquad 4 \, pH = 9.9 - \log [Bk^{+4}]$$

$$Bk(OH)_3/Bk^{+3} \qquad 3 \, pH = 24.0 - \log [Bk^{+3}]$$

anions, the general sequence is $F^- > NO_3^- > Cl^- > ClO_4^-$ and that for some divalent anions is $CO_3^{-2} > C_2O_4^{-2} > SO_4^{-2}$. Increased stability can be attained by use of chelating agents such as α-hydroxycarboxalate$^-$, acetylacetonate$^-$, cupferrate$^-$, 8-hydroxyquinolinate$^-$, nitrilotriaiacetate^{-3}, and ethylenediaminetetraacetate^{-4}. Of considerable interest is the strong complexing ability of these cations with the O in $R_3P=O$, $R(RO)(HO)P=O$, $R(HO)_2P=O$, and similar compounds, where R is a large organic hydrocarbon group. Such compounds may be dissolved in organic solvents, then put into contact with immiscible aqueous phases of An cations, which results in the partition into the organic phase of the actinoid complexes. Hydrogen-ion control permits the separation of the various An cations because they exhibit slightly different complexation stabilities.

Among the known $\log \beta_n$ values for the An species are the following.

- Th^{+4}: SCN^-(1.1, 1.8), NO_3^-(0.7), OH^-(10.8, 21.1, 30.3, 40.1), SO_4^{-2} (3.2, 5.5), F^-(8.4, 15.1,19.8, 23.2), Cl^- (1.4), 8-hydroxyquino linate$^-$ (10.5, 20.4, 29.9, 38.8), formate$^-$ (3.1, 5.2, 6.7), oxalate^{-2}(8.2, 16.8, 22.8), acetate$^-$ (3.9, 6.9, 9.0, 10.3, 11.0), hydroxyacetate$^-$ (4.0, 7.4, 10.0, 12.0), malonate^{-2}(7.4, 12.7), phthalate^{-2} (5.9, 10.1), edta^{-4} (23.2), acetylacetonate$^-$ (8.8, 16.2, 22.5, 26.7)
- U^{+4}: SCN^- (1.5, 2.0, 2.2), NO_3^-(0.2), OH^- (13.3, 25.4, 36.2, 45.7, 54.0), SO_4^{-2} (3.4, 5.8), F^- (9.0, 15.7, 21.2), Cl^- (0.3), edta^{-4} (25.8), acetylacetonate$^-$ (8.6, 17.0, 23.4, 29.5)
- UO_2^{+2}: CO_3^{-2} (—, 14.6, 18.3), SCN^-(1.0, 0.7, 0.2), OH^- (8.2, 9.6, $\log \beta_{22} = 22.4$), SO_4^{-2} (3.0, 4.0, 3.7), F^- (4.3, 8.0, 10.6, 12.0), Cl^- (0.2), formate$^-$ (1.9, 3.0, 3.5), oxalate^{-2} (6.0, 10.6, 11.0), acetate$^-$ (2.4, 4.4, 6.4), lactate$^-$ (2.8, 4.4, 5.8), citrate^{-3}(7.4, $\log \beta_{22} = 18.9$), salicylate^{-2} (12.1, 20.8), nta^{-3} (9.6), edta^{-4} (19.7), acetylacetonate$^-$ (7.7, 14.1)
- Np^{+4}: NO_3^-(1.7, 0.1), OH^-(12.5), SO_4^{-2}(3.5, 5.4), F^-(8.3, 14.5, 20.3, 25.1), Cl^-(0.2), nta^{-3} (17.3, 32.1), edta^{-4}(24.6), acetylacetonate$^-$ (8.6, 17.2, 23.9. 30.2)
- NpO_2^+: OH^-(5.1), phthalate^{-2}(2.2), iminodiacetate^{-2}(6.3), nta^{-3} (6.8), edta^{-4}(7.3), tropolonate$^-$(5.5, 9.8)
- NpO_2^{+2}: OH^-(8.9, $\log \beta_{22} = 21.6$), SO_4^{-2}(2.2, 3.8), F^-(4.1, 7.0), acetate$^-$(2.3, 4.2, 6.0)
- Pu^{+3}: OH^-(6.7), SO_4^{-2}(1.2), tropolonate$^-$(7.2)
- Pu^{+4}: OH^-(13.5, 25.7, 36.7, 46.5, 55.0), SO_4^{-2}(3.7), F^-(6.8), nta^{-3} (17.3, 32.1), edta^{-4}(24.6), acetylacetonate$^-$(8.6, 17.2, 23.9. 30.2)
- PuO_2^+: OH^-(4.3), iminodiacetate^{-2}(6.2) nta^{-3}(6.9)
- PuO_2^{+2}: $C_2O_4^{-2}$ ($\log \beta_2 = 9.4$), CO_3^{-2} ($\log \beta_2 = 15.1$), OH^-(8.4, $\log \beta_{22} = 19.6$), F^-(5.1, 10.1, 15.0, 18.1), acetate$^-$(2.1, 3.5, 5.0), phthalate^{-2} (4.1), edta^{-4}(7.3), tropolonate$^-$(5.5, 9.8).

Table 18.3
Trans-lawrencium Elements

Atomic Number	Name	Symbol	Longest lived isotope	Half life	Periodic Table Group	Congener of
104	Rutherfordium	Rf	267	1.3 h	4	Ti–Hf–Zr
105	Dubnium	Db	268	1.2 days	5	V–Nb–Ta
106	Seaborgium	Sg	271	1.9 min	6	Cr–Mo–W
107	Bohrium	Bh	272	9.6 s	7	Mn–Tc–Re
108	Hassium	Hs	277	17 min	8	Fe–Ru–Os
109	Meitnerium	Mt	276	0.7 s	9	Co–Rh–Os
110	Darmstadtium	Ds	281	11 s	10	Ni–Pd–Pt
111	Roentgenium	Rg	280	3.6 s	11	Cu–Ag–Au
112	Copernicium	Cp	285	34 s	12	Zn–Cd–Hg
113	—	—	284	0.5 s	13	Ga–In–Tl
114	—	—	289	2.6 s	14	Ge–Sn–Pb
115	—	—	288	87 m	15	As–Sb–Bi
116	—	—	293	61 m	16	Se–Te–Po
118	—	—	294	0.9 ms	18	Kr–Xe–Rn

7. Trans-lawrencium Elements

Beyond Lr, a number of further elements have been identified. These are displayed in Table 18.3. The data on the longest lived isotope illustrates one of the difficulties in chemical investigations of these elements. To the extent that the chemistries of the elements Rf, Db, Sg, Bh, and Hs have been investigated, they fit into the congener categories indicated above. However, predictions by extrapolation from the congeners can differ from the actual properties, this being due to the relativistic effect. This effect involves the velocities of the s and p electrons, which approach the speed of light. The consequence is that electrons in s levels tend to be stabilized, those in p levels tend to be slightly stabilized, and those in d and f levels are destabilized. These considerations are important for all elements treated in this chapter.

References

1 Clark, W. M. and Cohen, B. (1923). *Studies on oxidation-reduction. II. An analysis of the theoretical relations between reduction potentials and pH. Public Health Rep.* **38**, 666–83.

2 Clark, W. M. (1928).*The determination of hydrogen ions; an elementary treatise on electrode, indicator and supplementary methods.* The Williams & Wilkins Company, Baltimore.

3 Pourbaix, M. (1966). *Atlas of Electrochemical Equilibria in Aqueous Solutions* (Translated by J. A. Franklin). Pergamon, New York.

4 Santoma, L. (1973). *Application of computers to the construction of Eh–pH diagrams of mineralogical stability. Boletin Geologico y Minero.* **84**(2), 156–61; Williams, B.G. and Patrick, W.H., Jr. (1977). *A computer method for the construction of Eh–pH diagrams., J. Chem. Ed.* **54**(2), 107; Linkson, P. B., Phillips, B. D., and Rowles, C. D. (1979). *Computer methods for the generation of Eh–pH diagrams. Miner. Sci. and Eng.* **11**(2), 65–79; Osseo-Asare, K., Xue, T., and Ciminelli, V. S. T. (1984). *Solution chemistry of cyanide leaching systems. Precious Met.: Min., Extr., Process., Proc. Int. Symp.* 173–97; Mao, M. and Peters, E. (1984). *Computer method for calculating and plotting high-temperature Eh–pH diagrams. Jisuanji Yu Yingyong Huaxue.* **1**(3), 161–72; Drewes, D. R. (1985). *Computer code for producing Eh–pH plots of equilibrium chemical systems. J. of Chem. Inform. and Comp. Sci.* **25**(2), 73–7; Huang, H.-H. and Young, C. A. (1996). *Mass-balanced calculations of EH–pH diagrams using STABCAL. Proc. Electrochem. Soc., Electrochemistry in Mineral and Metal Processing,* 227–238; Birk, J. P.

and Tayer, L. L. (1997). *Computer-generated Eh–pH diagrams for teaching descriptive inorganic chemistry.* Abstracts, 213th ACS National Meeting, San Francisco, CHED-381; Glasby, G. P. and Schulz, H. D. (1999). *Eh, pH diagrams for Mn, Fe, Co, Ni, Cu and As under seawater conditions: application of two new types of Eh, pH diagrams to the study of specific problems in marine geochemistry.* Aquatic Geochem. **5**(3), 227–248; *New method for computer drawing Eh–pH diagrams.* Feng, Qi-ming, Ma, Yun-zhu, Wang, Yu-hua., and Lu, Yi.-ping. (2000). *Zhongnan Gongye Daxue Xuebao,* **31**(4), 297–299; Huang, H.-H.; Twidwell, L. G.; Young, C. A. (2005). *Speciation for aqueous systems—an equilibrium calculation approach.,* Computational Analysis in Hydrometallurgy, Proc. Int. Symp. on Computational Analysis in Hydrometallurgy, 295–310.

5 Speight, J. (2004). *Lange's Handbook of Chemistry.* McGraw-Hill Professional, New York.

6 Lide, D. R. (2009). *CRC Handbook of Chemistry and Physics.* CRC Press/Taylor and Francis, Boca Raton, FL.

7 Kotrly, S. and Sucha, L. (1985). *Handbook of Chemical Equilibria in Analytical Chemistry.* Ellis Horwood Ltd, New York.

8 Smith, R. M. (2009). *NIST Critically Selected Stability Constants of Metal Complexes: Version 8.0.* National Institute of Standards and Technology, Gaithersburg, MD.

Index